LEAK-BEFORE-BREAK
IN WATER REACTOR PIPING AND VESSELS

This volume contains papers presented at the OECD–NEA Committee on the Safety of Nuclear Installations, CANDU Owners Group, Specialists Meeting, Leak-Before-Break in Water Reactor Piping and Vessels, held on 24–27 October 1989 in Toronto, Canada

LEAK-BEFORE-BREAK IN WATER REACTOR PIPING AND VESSELS

Edited by

C. E. COLEMAN

AECL Research,
Chalk River Nuclear Laboratories,
Chalk River, Ontario, Canada

Reprinted from
The International Journal of
Pressure Vessels and Piping
Vol. 43 Nos 1–3

ELSEVIER APPLIED SCIENCE
LONDON and NEW YORK

ELSEVIER SCIENCE PUBLISHERS LTD
Crown House, Linton Road, Barking, Essex IG11 8JU, England

Sole Distributor in the USA and Canada
ELSEVIER SCIENCE PUBLISHING CO., INC.
655 Avenue of the Americas, New York, NY 10010, USA

British Library Cataloguing in Publication Data

Leak-before-break in water reactor piping and vessels.
1. Nuclear reactors. Pressure vessels & pressure piping systems
I. Coleman, C. E. II. The international journal of pressure vessels and piping
621.4832

ISBN 1–85166–541–2

Library of Congress Cataloging-in-Publication Data

Leak-before-break in water reactor piping and vessels/edited by C. E. Coleman.
p. cm.
"Reprinted from the International journal of pressure vessels and piping, vol. 43, nos. 1–3."
Includes bibliographical references and index.
ISBN 1–85166–541–2
1. Nuclear reactors—Pipe lines—Testing—Congresses. 2. Nuclear pressure vessels—Testing—Congresses. 3. Pressure vessels—Cracking—Congresses. 4. Pipe lines—Cracking—Congresses. I. Coleman, C. E. II. International journal of pressure vessels and piping.
TK9202.L34 1991
621.48'35—dc20 90–42003
CIP

Printed in Great Britain by Galliard (Printers) Ltd, Great Yarmouth

Foreword

by L. A. Simpson, Chairman, LBB–89

Leak-Before-Break (LBB) is now widely applied in the nuclear industries of the Member countries of the Organization for Economic Cooperation and Development (OECD) as a means of assessing the susceptibility of pressurized components to failure by unstable crack propagation. While the detailed nature of the applications differ widely, depending on the country and the component, the basic concept does not. A crack is postulated to exist in a structure and assumed to grow stably by some mechanism to a size at which it penetrates the wall of the component resulting in leakage of the pressurizing fluid. When the crack, through continued growth, reaches a size at which leakage is deemed to be detectable (with margin), an assessment is done to show the margin against failure. The margin can be expressed either as a factor of safety on the stress or crack length for instability or on the time available to effect a safe shutdown of the reactor before the crack grows to a critical size.

There have been several meetings on the topic of LBB prior to this one. A list of four is given below with the sponsoring organizations indicated and the proceedings identified.

Other LBB Conference Proceedings

CSNI REPORT NO. 82	USNRC/CSNI	Sept. 1983	Monterey, CA, USA
NUREG/CP-0077	USNRC/IPIRG	Nov. 1985	Columbus, OH, USA
NUREG/C-0092	USNRC/IPIRG	May 1987	Tokyo, Japan
To be published	USNRC/IPIRG	May 1989	Taipei, Taiwan

A key difference between those meetings and this one is that the proceedings of previous meetings were published in media which do not normally receive wide distribution. It was then felt that, with the current widespread interest in applications of LBB within the nuclear industries of the OECD countries, and a burgeoning of interest in other industries, the time was appropriate for a meeting that reviewed the state of the art and published the results in a form that would be readily available. The matter was raised by Canada in the Nuclear Energy Agency of the OECD (OECD/NEA) and the Agency's Committee on the Safety of Nuclear Installations (CSNI) accepted a recommendation from its Principal Working Group No. 3, Reactor Component Integrity (PWG3) to sponsor a Specialist Meeting. This sponsorship ensured the submission of papers and the participation of the leading experts in LBB applications from the OECD Member countries. The Canadian Organizing Committee was established in early 1988 and was co-sponsorship of the meeting by the CANDU Owners Group provided the necessary organizational support.

Papers were invited through the OECD/NEA-CSNI on topics that would lead to an appreciation of how LBB is applied in the various member countries and, in particular, the national regulatory views on the application of LBB. Over 50 abstracts were submitted from which the final programme of 29 papers and 5 poster presentations was selected by the following Programme Committee of CSNI-PWG3.

L. A. Simpson	AECL, Canada Chairman
C. E. Coleman	AECL, Canada
B. Mukherjee	AECL, Canada
C. Z. Serpan	USNRC, USA
R. W. Nichols	Nichols Consultancies
P. Milella	ENEA/DISP, Italy
S. Shibata	JAERI, Japan
N. R. McDonald	OECD-NEA

The papers selected were those having the most direct relevance to a LBB issue and which could be accommodated in the five topical sessions of the meeting timetable with adequate time for presentation and discussion. The Canadian Organizing Committee arranged for the proceedings to be published by Elsevier Applied Science both as a special issue of the International Journal of Pressure Vessels and Piping and as a hard cover book with worldwide distribution.

The final, sixth session of the meeting was a directed discussion under the control of Professor E. Smith of the University of Manchester, UK. Following a summary of the issues as Professor Smith saw them, the delegates were invited to give their views on these issues. A summary of the discussion appears at the end of these proceedings and has constituted a report of the meeting to the NEA-CSNI and PWG3.

The credit for the obvious success of the meeting attended by 100 delegates from ten countries is due to many people. The generosity of a number of organizations enabled the fixed costs of the meeting as well as two lunches and a conference banquet to be covered. We are very grateful to these organizations for their support.

CANDU Owners Group (Working Party 31)	Fixed costs and overhead expenses
CANDU Owners Group (Operations)	Banquet
Teledyne Wah Chang Albany	Wednesday luncheon
MTS (Canada Ltd.)	Thursday luncheon
Phillips Electronics Ltd	Pre-Banquet reception
New Brunswick Electric Power Comm.	Welcoming reception
Canatom	Operating costs
Donlee Manufacturing Industries	Operating costs
Intertechnology Inc.	Operating costs
Canadian Society for Research on the Strength and Fracture of Materials	Operating costs

Support from the Canadian Nuclear Society in the form of assistance of their members is also gratefully acknowledged.

I am grateful to the members of PWG3 of the OECD/NEA for their enthusiastic support of this meeting and in particular to the Secretary, Neil R. McDonald, for tireless efforts in handling the international aspects of the organization of such an event. His tolerance of my ability to track him down by telephone, wherever in the world he happened to be, was much appreciated.

Special thanks are due to my Deputy Chairman and Editor of these proceedings, Dr C. E. Coleman for his tireless effort in finding volunteers to review the papers to the stringent standards of the International Journal of Pressure Vessels and Piping and his persistence in ensuring that most of the papers were submitted and reviewed by the time of the meeting. The very high quality of all of the presentations was clearly a direct result of the quality of the reviews carried out before the meeting.

Finally much credit must go to the members of the Canadian Organizing Committee for their efforts, especially during the two month period prior to the meeting. We were scattered over distances up to 1800 km at four separate locations but managed to meet seven times in the course of two years and function between meetings via the wonders of electronic mail. The committee members are listed below.

Local Organizing Committee

L. A. Simpson	Chairman
C. E. Coleman	Deputy Chairman
E. G. Price	Local Arrangements
B. Mukherjee	Program Committee Chairman
G. D. Moan	Treasurer
G. J. Field	Sponsors/Publicity
N. R. McDonald	OECD/NEA

Sponsoring Organizations

OECD/NEA PRINCIPAL WORKING GROUP 3

The Committee is composed of all interested member countries of the Nuclear Energy Agency (NEA). It was established in 1981 to address topics in the area of reactor component integrity.

The Group investigates selected high-priority topics related to the structural integrity and safety assessment of the primary coolant circuit boundary and pressurized components in nuclear reactors.

With the assistance of groups of experts, meetings of specialists, preparation of reports and exchange of information, attention is given to:

- flaw characterization by non-destructive examination;
- properties and behaviour of structural materials;
- data on loading and service conditions affecting integrity; and
- fracture analysis methodologies and their application to safety assessments.

Particular emphasis is given to practical reactor safety issues, the assessment and use of operational experience with materials and component integrity, and the degradation of materials properties and component integrity with service life.

CANDU OWNERS GROUP

The CANDU Owners Group (COG) was formed in 1984 by the Canadian CANDU-owning utilities (Ontario Hydro, Hydro Quebec, and New Brunswick Electric Power Commission) and Atomic Energy of Canada Limited (AECL). The purpose of COG is to provide a framework that will promote closer cooperation among utilities worldwide that own and operate CANDU stations in matters relating to plant operation and maintenance, and to foster cooperative research and development programmes leading to improved plant performance.

The primary funding for the meeting was supplied by COG Working Party 31 which sponsors a programme of research on delayed hydride cracking and fracture of zirconium alloys at the research laboratories of AECL and the member utilities.

Contents

List of Contributors

Y. ASADA
Tokyo University, 3-1 Hongo 7-Chome, Bunkyo-ku, Tokyo 113, Japan.

H. BABA
Power Reactor and Nuclear Fuel Development Corp., Oarai Engineering Center, 4002, Narita-cho, Oarai-machi, Higashi Ibaraki-gun, Ibaraki 311-13, Japan.

V. K. BAJAJ
CANDU Operations, Atomic Energy of Canada Ltd, Sheridan Park, Mississauga, Ontario, Canada L5K 1B2.

B. F. BEAUDOIN
Robert L. Cloud and Associates, Inc., 125 University Avenue, Berkeley, California 94710, USA.

S. BELICZEY
Gesellschaft für Reaktorsicherheit (GRS)mbH, Schwertnergasse 1, D-5000 Köln 1, Germany

S. BHANDARI
Framatome, Tour Fiat, La Defense, 92084 Paris, Cedex 16, France.

J. G. BLAUEL
Fraunhofer-Institut für Werkstoffmechanik, Wöhlerstrasse 11, D-7800 Freiburg, Germany

J. M. BOAG
Ontario Hydro Research Division (OHRD), 800 Kipling Avenue, Toronto, Ontario, Canada M8Z 5S4.

M. F. BOLAÑOS
Consejo de Seguridad Nuclear (CSN), 11 Justo Dorado, 28040 Madrid, Spain.

B. BRICKSTAD
Royal Institute of Technology, S100 44, Stockholm, Sweden.

W. BROCKS
Bundesanstalt für Materialforschung und-prüfung (BAM), Unter den Eichen 87, D-1000 Berlin 45, Germany.

A. M. CLAYTON
UKAEA, Structural Integrity Centre, NRL-Risley, Warrington, Cheshire WA3 6AT, UK.

B. COCHET
Framatome, Tour Fiat, La Défense, 92040 Paris, Cedex 16, France.

C. E. COLEMAN
Reactor Materials Division, Atomic Energy of Canada Ltd, Chalk River Nuclear Laboratories, Chalk River, Ontario. Canada K0J IJ0.

M. A. DURAND
Ontario Hydro, 700 University Avenue, Toronto, Ontario, Canada M5G 1X6.

A. ESTEBAN
Consejo de Seguridad Nuclear (CSN), 11 Justo Dorado, 28040 Madrid, Spain.

C. FAIDY
EDF-SEPTIN, 12–14 Avenue Dutrievoz, 69628 Villeurbanne Cedex, France.

J. M. FIGUERAS
Consejo de Seguridad Nuclear (CSN), 11 Justo Dorado, 28040 Madrid, Spain.

M. T. FLAMAN
Ontario Hydro Research Division (OHRD), 800 Kipling Avenue, Toronto, Ontario, Canada M8Z 5S4.

B. FLESCH
Electricité de France, SPT, 13–27 Esplanade Charles-de-Gaulle, La Défense, 92060 Paris, Cedex 57, France.

PH. GILLES
Framatome, Tour Fiat, La Defense, 92084 Paris, Cedex 16, France.

N. GOTOH
Nuclear Plant Maintenance Service Department, Hitachi Ltd, Saiwai-cho 3-1-1, Hitachi-shi, Ibaraki-ken 317, Japan.

D. B. GRAHAM
Ontario Hydro, 700 University Avenue, Toronto, Ontario, Canada M5G 1X6.

D. GUERRIERI
Battelle, 505 King Avenue, Columbus, Ohio 43201, USA.

T. C. HARDIN
Robert L. Cloud and Associates, Inc., 125 University Avenue, Berkeley, California 94710, USA.

H. HATA
Mitsubishi Atomic Power Industries Inc., 4-1 Shibakouen 2-Chome, Minato-ku, Tokyo 105, Japan.

L. HODULAK
Fraunhofer-Institut für Werkstoffmechanik, Wohlerstrasse 11, D-7800 Freiburg, Germany.

M. HUTERER
Ontario Hydro, 700 University Avenue, Toronto, Ontario, Canada M5G 1X6.

M. IIDA
Ishikawajima-Harima Heavy Industries Co. Ltd, 3-1-15, Toyosu, Koto-ku, Tokyo 135, Japan.

T. ISOZAKI
Department of Reactor Safety Research, Japan Atomic Energy Research Institute, Tokai-mura, Ibaraki-ken 319-11, Japan.

P. JAMET
CEA-DEMT, Centre d'Etudes Nucléaires de Saclay, 91191 Gif-sur-Yvette, France.

D. JONES
Battelle, 505 King Avenue, Columbus, Ohio 43201, USA.

K. KASHIMA
Central Research Institute of Electric Power Industries, Komae-shi, Tokyo 201, Japan.

M. H. KOIKE
Power Reactor and Nuclear Fuel Development Corp., Oarai Engineering Center, 4002, Narita-cho, Oarai-machi, Higashi Ibaraki-gun, Ibaraki 311-13, Japan.

M. KOJIMA
Ishikawajima-Harima Heavy Industries Co. Ltd, 3-1-15, Toyosu, Koto-ku, Tokyo 135, Japan.

H. KRAFKA
Bundesanstalt für Materialforschung und -prüfung (BAM), Unter den Eichen 87, D-1000 Berlin 45, Germany.

G. KÜNECKE
Bundesanstalt für Materialforschung und -prüfung (BAM), Unter den Eichen 87, D-1000 Berlin 45, Germany.

S. LEE
Office of Nuclear Reactor Regulation, US Nuclear Regulatory Commission, Washington, District of Columbia 20555, USA.

O. E. LEPIK
Fracture Mechanics Unit, Nondestructive and Fracture Evaluation Section, Metallurgical Research Department, Ontario Hydro, 800 Kipling Avenue, Toronto, Ontario, Canada M8Z 5S4.

C. MARICCHIOLO
ENEA/DISP, Via V. Brancati 48, 00144 Rome, Italy.

P. P. MILELLA
ENEA/DISP, Via V. Brancati 48, 00144 Rome, Italy.

B. E. MILLS
Ontario Hydro Research Division (OHRD), 800 Kipling Avenue, Toronto, Ontario, Canada M8Z 5S4.

S. MIYAZONO
Department of Reactor Safety Research, Japan Atomic Energy Research Institute, Tokai-mura, Ibaraki-ken 319-11, Japan.

G. D. MOAN

CANDU Operations, Atomic Energy of Canada Ltd, Sheridan Park, Mississauga, Ontario, Canada L5K 1B2.

B. MUKHERJEE

Fracture Mechanics Unit, Nondestructive and Fracture Evaluation Section, Metallurgical Research Department, Ontario Hydro, 800 Kipling Avenue, Toronto, Ontario, Canada M8Z 5S4.

G. NAGEL

Preussen-Elektra AG, Treskowstrasse 5, D-3000 Hannover, Germany.

J. S. NATHWANI

Nuclear Safety Department, Design and Development Division—Generation, Ontario Hydro, 700 University Avenue, Toronto, Canada M5G 1X6.

F. NILSSON

Royal Institute of Technology, S100 44, Stockholm, Sweden.

A. OCKEWITZ

Fraunhofer-Institut für Werkstoffmechanik, Wohlerstrasse 11, D-7800 Freiburg, Germany.

R. OLSON

Battelle, 505 King Avenue, Columbus, Ohio 43201, USA.

A. PINI

ENEA/DISP, V. Via Brancati 48, 00144 Rome, Italy.

C. O. POIDEVIN

Ontario Hydro, 700 University Avenue, Toronto, Ontario, Canada M5G 1X6.

E. G. PRICE

CANDU Operations, Atomic Energy of Canada Ltd, Sheridan Park, Mississauga, Ontario, Canada L5K 1B2.

D. F. QUIÑONES

Robert L. Cloud and Associates, Inc., 125 University Avenue, Berkeley, California 94710, USA.

D. K. RODGERS

Reactor Materials Division, Atomic Energy of Canada Ltd, Chalk River Nuclear Laboratories, Chalk River, Ontario, Canada K0J 1J0.

M. RODRIGUEZ
Consejo de Seguridad Nuclear (CSN), 11 Justo Dorado, 28040 Madrid, Spain.

S. SAGAT
Reactor Materials Division, Atomic Energy of Canada Ltd, Chalk River Nuclear Laboratories, Chalk River, Ontario, Canada K0J 1J0.

W. SCHMITT
Fraunhofer-Institut für Werkstoffmechanik, Wohlerstrasse 11, D-7800 Freiburg, Germany.

P. SCOTT
Battelle, 505 King Avenue, Columbus, Ohio 43201, USA.

H. SCHULZ
Gesellschaft für Reaktorsicherheit (GRS)mbH, Schwertnergasse 1, D-5000 Köln 1, Germany.

J. K. SHARPLES
UKAEA, Structural Integrity Centre, NRL-Risley, Warrington WA3 6AT, Cheshire, UK.

G. K. SHEK
Ontario Hydro, 700 University Avenue, Toronto, Ontario, Canada M5G 1X6.

K. SHIBATA
Department of Reactor Safety Research, Japan Atomic Energy Research Institute, Tokai-mura, Ibaraki-ken 319-11, Japan.

L. A. SIMPSON
Atomic Energy of Canada Ltd, Whiteshell Nuclear Research Establishment, Pinawa, Manitoba, Canada R0E 1L0.

H. SINOKAWA
Department of Reactor Safety Research, Japan Atomic Energy Research Institute, Tokai-mura, Ibaraki-ken 319-11, Japan.

L. SKÅNBERG
Royal Institute of Technology, S100 44, Stockholm, Sweden.

E. SMITH
14 The Meade, Wilmslow, Cheshire SK9 2JF, UK.

J. D. STEBBING
Nuclear Safety Department, Design and Development Division—Generation, Ontario Hydro, 700 University Avenue, Toronto, Canada M5G 1X6.

W. STOPPLER
Staatliche Materialprüfungsanstalt (MPA), University of Stuttgart, Pfaffenwaldring 32, D-7000 Stuttgart 80, Germany.

D. STURM
Staatliche Materialprüfungsanstalt (MPA), University of Stuttgart, Pfaffenwaldring 32, D-7000 Stuttgart 80, Germany.

I. TAKAHASHI
Ishikawajima-Harima Heavy Industries Co. Ltd, 3-1-15, Toyosu, Koto-ku, Tokyo 135, Japan.

T. TAKAHASHI
Power Reactor and Nuclear Fuel Development Corp., Oarai Engineering Center, 4002, Narita-cho, Oarai-machi, Higashi Ibaraki-gun, Ibaraki 311-13, Japan.

K. TAKUMI
Nuclear Power Engineering Test Center, Shuwa Kamiyacho Building, 3–13 Toranomon 4-Chome, Minato-ku, Tokyo 115, Japan.

PH. TAUPIN
Framatome, Tour Fiat, La Defense, 92084 Paris, Cedex 16, France.

T. UMEMOTO
Ishikawajima-Harima Heavy Industries Co. Ltd, Isogoku, Yokohama-shi 235, Japan.

M. L. VANDERGLAS
Research Division, Ontario Hydro, 800 Kipling Avenue, Toronto, Canada M8Z 5S4

J. R. WALKER
Materials and Mechanics Branch, Atomic Energy of Canada Ltd, Whiteshell Nuclear Research Establishment, Pinawa, Manitoba, Canada R0E 1L0. Present address: *Fuel Engineering Branch, AECL Research, Chalk River Laboratories, Chalk River, Ontario, Canada K0J 1J0.*

K. WICHMAN
Office of Nuclear Reactor Regulation, US Nuclear Regulatory Commission, Washington, District of Columbia 20555, USA.

G. M. WILKOWSKI
Battelle, 505 King Avenue, Columbus, Ohio 43201, USA.

K. WOBST
Bundesanstalt für Materialforschung und -prüfung (BAM), Unter den Eichen 87, D-1000 Berlin 45, Germany.

H. W. WONG
CANDU Operations, Atomic Energy of Canada Ltd, Sheridan Park, Mississauga, Ontario, Canada L5K 1B2.

Y. YAMAMOTO
Toshiba Corp., 8 Shinsugitacho, Isogo-ku, Yokohama 230, Japan.

K. YOSHIDA
Ishikawajima-Harima Heavy Industries Co. Ltd, Yokohama Nuclear and Chemical Components Works, 1 Shir-nakahara-cho, Isogo-ku, Yokohama 235, Japan.

LEAK-BEFORE-BREAK–89, TORONTO, OCTOBER 1989

Front Row from Left to Right: Awadalla, Price, Mukherjee, Field, Duarte, McDonald, Simpson, Coleman, Moan, Koike; Second Row: Swamy, Shalaby, Meranda, Vijay, Endoh, Lin, Darlaston, McClanahan, Bajaj; Third Row: Monette, Schwenk, Smith, Cueto-Felgueroso, Bartholome, Huterer, Gilles, Misra, Poidevin, Muzumdar, Bonechi; Fourth Row: Shek, Hosbons, Kittmer, Stoppler, Schmitt, Xanthopoulos, Brickstad, Holman, Clendening, Stebbing; Fifth Row: Beaudoin, Yamamoto, Kozluk, Ishiwata, Hata, Kishida, Kashima, Flesch, Lee, Esteban, Maricchiolo, Dumas, Richards; Sixth Row: Hardin, Vanderglas, Mills, Furuya, Harrop, Sharples, Stewart, Rodriguez, Cochet, Bolanos, Hardie, Richinson; Seventh Row: Gabbani, Wong, Graham, Isozaki, Beliczey, Horowitz, Scarth, Phillips, Leger, Walker, Faidy; Eighth Row: Durand, Van Maaren, Fischer, Gerard, Wardle, Venter, Moles, Lepik, Cheadle, Rodgers; Ninth Row: Manning, Sagat, Brocks.

Int. J. Pres. Ves. & Piping **43** (1990) 1–21

Leak-Before-Break in the Pressure Tubes of CANDU Reactors

G. D. Moan,[a] C. E. Coleman,[b] E. G. Price,[a]
D. K. Rodgers[b] & S. Sagat[b]

[a] CANDU Operations, Atomic Energy of Canada Ltd, Sheridan Park, Mississauga, Ontario, Canada L5K 1B2
[b] Reactor Materials Division, Atomic Energy of Canada Ltd, Chalk River Nuclear Laboratories, Chalk River, Ontario Canada K0J 1J0

ABSTRACT

In the CANDU nuclear reactor, pressure tubes made from Zr–2·5 Nb contain the fuel and the heat-transport heavy water. At each end, the tube is connected to a stainless steel end fitting by a rolled joint. A thin Zircaloy tube surrounds the pressure tube and the annular space between the tubes contains an inert gas. If a crack develops in a pressure tube at a rolled joint, unstable fracture is avoided by the application of Leak-Before-Break (LBB). In some early CANDU reactors, a few pressure tubes leaked through delayed hydride cracks. High residual stresses, produced when tubes were incorrectly rolled into end fittings, contributed to the cracking. The cracks were easily detected by their leakage into the gas annulus, the reactor was safely shut down and the leaking tubes were identified and replaced, thus providing a clear demonstration of LBB. To ensure the margins against unstable fracture continue to be adequate, we must demonstrate the validity of leak-before-break. The method depends on showing that the time available *to detect a crack is much greater than the time* required *to detect the crack. The time available is estimated from measurements of crack velocities, crack lengths at instability and crack lengths at penetration of the tube wall. A leak is detected by the dewpoint and the rate of dewpoint increase of the gas in the annulus. Action time for operators is shown to have a usable margin before tube rupture, supporting continued use of LBB as an operating criterion.*

INTRODUCTION

Within the normal operating limits of a pressure-retaining component, cracking may be initiated when a combination of the three main elements—stress, environment and material sensitivity—reaches a critical value. Prevention of crack initiation can be achieved by design, manufacturing procedures, or operational practice. Some structures remain subject to crack initiation because of an initial lack of recognition of potential mechanisms. The characteristics of individual cracking mechanisms may differ in the mode of cracking and in the crack growth rates, but the modes of stable cracking of most concern are those that propagate quickly without deformation.

Fatigue cracks usually propagate slowly, typically 10^{-8} m/cycle. Safety approaches based on periodic inspections at intervals of a few cycles have been developed to detect propagating or incipient defects. For modes of cracking that do not depend on cycling and propagate at up to 10^{-6} m/s, such as stress corrosion cracking, corrosion fatigue or hydrogen assisted cracking, the probability of detecting a propagating crack during a periodic examination is small. Thus a methodology must be developed for these modes of cracking that accommodates the cracking mechanism. In this paper we describe how the design of the CANDU reactor facilitates a Leak-Before-Break (LBB) methodology based on the characteristics of cracking in the pressure tube material and on leak detection.

THE CANDU REACTOR

A CANDU reactor consists of a large tank (the calandria) containing D_2O moderator at 70°C and is penetrated by about 400 horizontal fuel channels each 6 m long. Each channel consists of a pressure tube, containing the UO_2 fuel and heat transport D_2O at a pressure of 10 MPa and at a temperature ranging from 250°C at the inlet end to 300°C at the outlet end. The pressure tube is surrounded, and insulated from the cold moderator, by a calandria tube (Fig. 1). The space between the pressure tube and the calandria tube is filled with gas and is described as the gas annulus. The pressure tubes are made from cold-worked Zr–2·5 Nb, with a wall thickness of 4 mm and inside diameter of 103 mm, and the calandria tubes are made from annealed Zircaloy-2, with a wall thickness of 1·4 mm and inside diameter of 129 mm.

The annulus gas system has been developed into a system sensitive to the presence of moisture resulting from a penetration of the primary heat

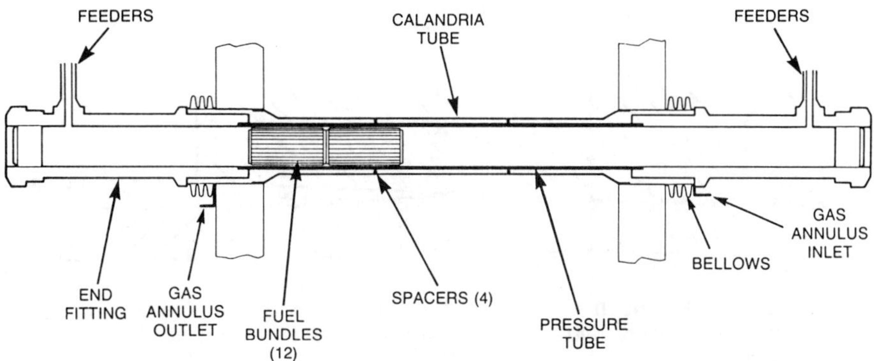

Fig. 1. Simplified description of a fuel channel. Each fuel channel consists of a zirconium alloy pressure tube, sealed at each end with end fittings that have side port connections to the heat transport system. The gap, or annulus, between the pressure tube and the surrounding Zircaloy-2 calandria tube is filled with an insulating gas and contains four close-coiled helical spring spacers that provide physical separation between the two tubes and partial support for the pressure tube. The annulus is sealed at each end by bellows that accommodate the relative axial movement between the fuel channel and the reactor shields.

transport pressure boundary that passes through the reactor core (and also leakage from the surrounding calandria and lattice tubes, although this type of leakage has not been significant).

The pressure tubes are rolled into end fittings at each end of the channel. The end fittings contain connections to the heat transport system and closures to enable fuelling to be done on power. The rolled joint fabrication produces residual stresses in the pressure tube due to the wall thinning and tube expansion. These stresses have been a source of crack initiation; in two instances manufacturing flaws provided the initiation sites.

LEAK-BEFORE-BREAK EXPERIENCE IN CANDU REACTORS

In early power reactors incorrect rolling produced excessive tensile residual stresses in the pressure tube (Fig. 2). The hoop tensile stresses had high values over a length of 10 mm on the inside surface. Revised installation procedures have eliminated the high stresses in current reactors. At some of the rolled joints in these early reactors, the total tensile hoop stresses were sufficiently large that a stable crack propagation process called delayed hydride cracking (DHC) initiated on the inside surface of the pressure tube. The mechanism and its characteristics are discussed elsewhere;[1] at present, it is sufficient to indicate that DHC in a Zr–Nb pressure tube requires (1) the presence of a large hoop tensile stress and (2) the presence of hydrides.

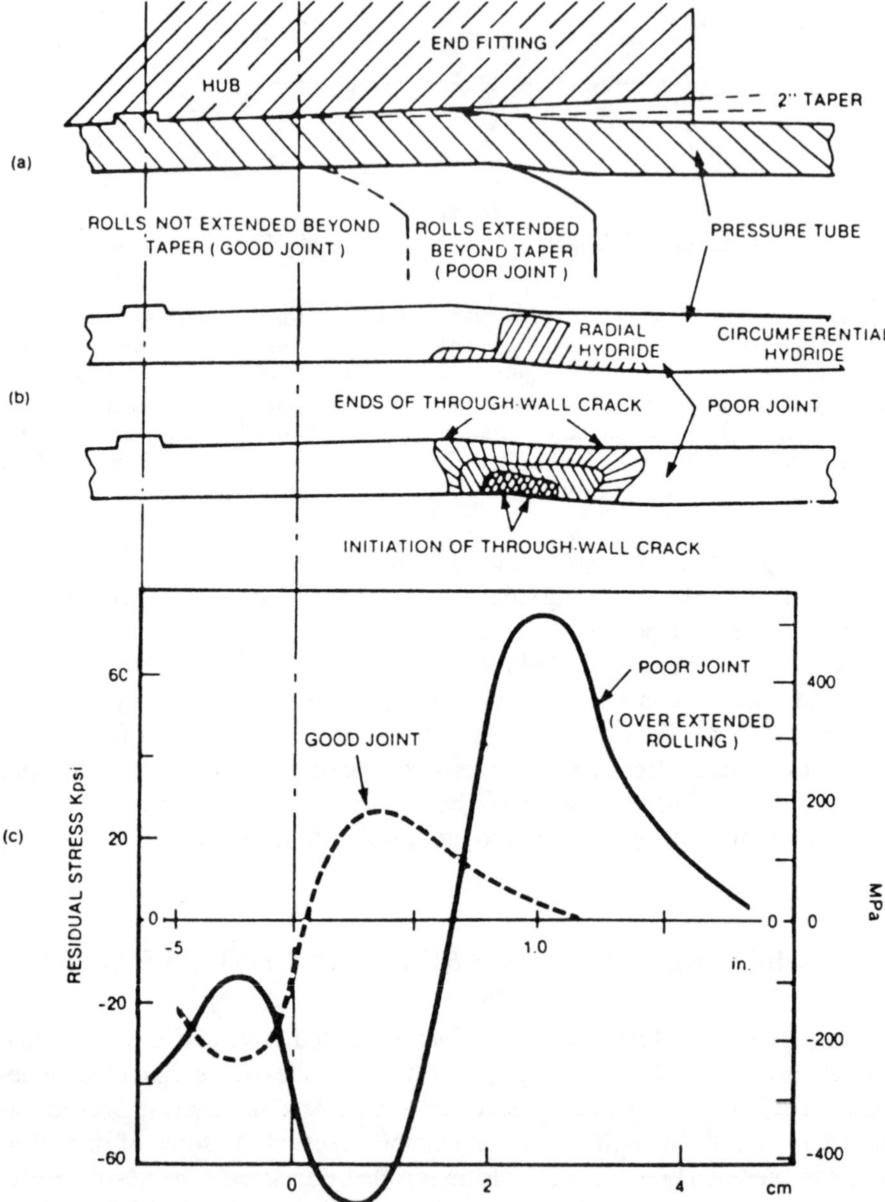

Fig. 2. The effect of roll extension on the residual hoop stresses in the pressure tube in rolled joints.

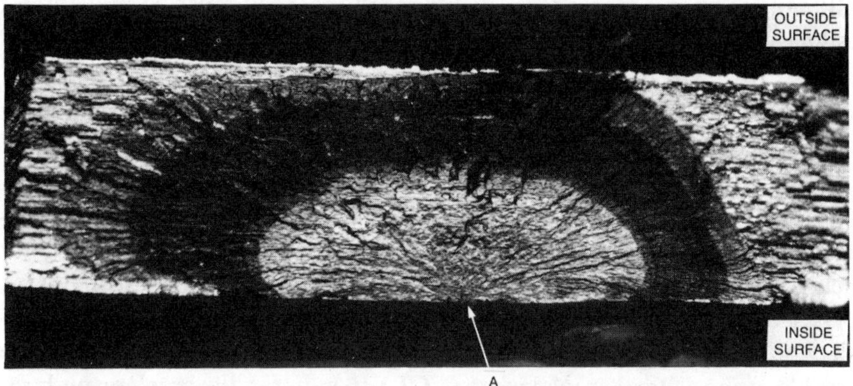

Fig. 3. Composite macrograph showing the crack face in a leaking Zr–2·5Nb pressure tube removed from Pickering. The crack nucleated at the inside surface at (a) and grew by delayed hydride cracking in the radial and axial directions. The different lines represent positions at which the crack had stopped. The wall thickness was 4 mm.

The principal features of a through-wall DHC crack are shown in Fig. 3. Following nucleation, the cracks grew in the radial-axial plane under the hoop tensile stress when hydrides were present when the reactor was cold. When the reactor was hot, the hydrides were dissolved and the crack growth stopped, but the crack surfaces were oxidized by the heat transport D_2O. During later shutdowns the hydrides reprecipitated and the cracking recommenced. When the reactor was again returned to power the hydrides redissolved, the cracking stopped and the crack oxidation continued, with the newest part of the crack having the oxide with the smallest thickness. Eventually the crack growth in the radial direction led to the penetration of the tube wall so that the heavy water of the pressurized heat transport (PHT) water leaked into the gas annulus.

Since crack growth continues in the axial direction it is essential that the operators detect the presence of the leakage in the annulus and take action before the crack reaches its critical crack length and grows in an unstable manner.

Heavy water leaked into the annuli of 24 channels in early CANDU reactors. The leakage was detected, the reactors were shut down, the leaking tubes identified and replaced and the reactors returned to power. The experience demonstrated that LBB was an effective defence against unstable cracking in CANDU pressure tubes, but to ensure that there are adequate margins we are studying the elements that affect the time available for operators to take action following leakage. There are, however, situations where through-wall thermal gradients may prevent delayed hydride cracking in the radial direction while axial crack growth continues. Since

LBB cannot be guaranteed for such conditions, they must be prevented by design and installation practice.

DELAYED HYDRIDE CRACKING

For Zr–Nb there is a threshold stress intensity factor, K_{1H}, below which DHC does not occur. K_{1H} has values between 4·5 and 7·0 MPa$\sqrt{m}$ and is insensitive to temperature, hydrogen concentration and neutron fluence. For K_1 values greater than K_{1H} the crack growth rate, V, is essentially independent of K_1 over a wide range, until the crack growth becomes unstable at the critical crack length (CCL). In general, the crack growth rate,

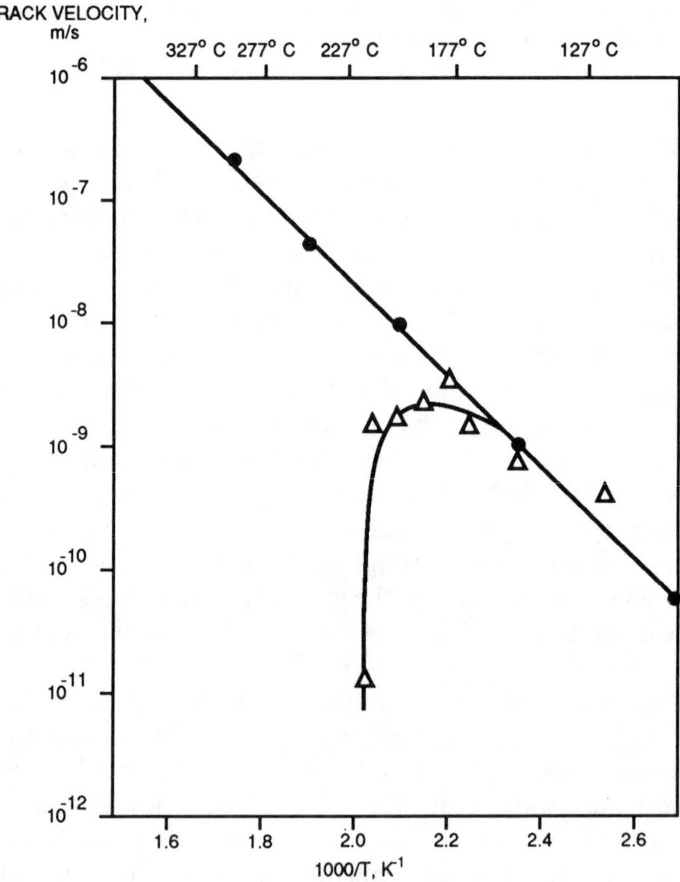

Fig. 4. Graph showing the temperature dependence of the crack growth rate in Zr–2·5Nb. The two lines show the effect of approaching the test temperature by heating (△), or by cooling (●).

V, is temperature dependent, being described by the Arrhenius relationship (provided hydrides are present at the temperature T). Crack growth rates measured when the test temperature was approached by heating have a temperature dependence that is significantly different from that measured when the test temperature was approached by cooling. The results in Fig. 4 show the Arrhenius relationship, and the difference in crack growth behaviour when the test temperature is approached by heating or cooling.

The crack growth rate, V, is also affected by the neutron fluence to which the material has been subjected and by the temperature of irradiation. Values of V have been determined from tests on small specimens cut from pressure tubes removed from power reactors (Fig. 5). The tubes have been

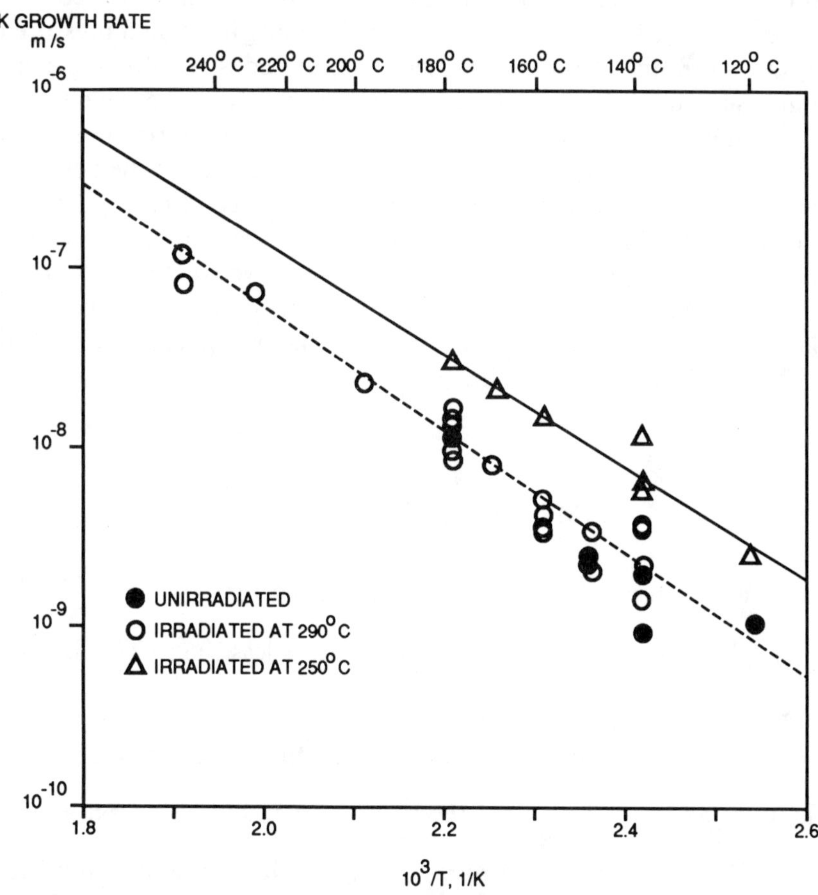

Fig. 5. Comparison of crack growth rate data for Zr–Nb pressure tube material, (○) irradiated to $2 \times 10^{25}\,\mathrm{n/m^2}$ at 290°C, (△) irradiated to $2 \times 10^{25}\,\mathrm{n/m^2}$ at 250°C, (●) unirradiated. In all cases the crack growth rates were measured in the radial direction on small specimens, containing up to 0·41 at % hydrogen/deuterium, at $K_1 = 17\,\mathrm{MPa}\sqrt{\mathrm{m}}$, with the test temperature being approached by cooling.

subjected to neutron fluences up to $8{\cdot}2 \times 10^{25}$ n/m^2 and had hydrogen isotope concentrations up to 0·41 atomic %. Cantilever beam specimens were prepared from different sections close to the inlets and outlets of tubes and crack growth rates were measured at a nominal K_1 value of 17 MPa$\sqrt{m}$. The maximum test temperature was limited by the requirement that hydrides be present; this temperature depended on the concentration of hydrogen isotopes. The test temperature was approached by cooling to maximize the cracking.

Annealing of irradiation damage and other temperature-dependent microstructural changes that could affect DHC may vary along the pressure tube because the temperature during reactor operation is about 40°C higher at the outlet end than at the inlet end. X-ray diffraction studies confirm that there are indeed changes in microstructure along irradiated pressure tubes. In the unirradiated tubes the microstructure consists of elongated alpha grains surrounded by a continuous network of beta phase, which is thought to be an important component in the diffusivity of hydrogen in zirconium alloys. Prolonged exposure to reactor conditions causes the beta phase to decompose and become discontinuous. The degree of change in the beta-phase is higher and the dislocation density is lower at the outlet end than at the inlet end. These studies suggest that the observed factor of two difference in crack velocities between the inlet and outlet ends could be caused by different amounts of irradiation hardening and beta phase decomposition at each end of the pressure tube.

TIME AVAILABLE FOR ACTION

To assure LBB in pressure tubes it is required that:

(1) the crack length at wall penetration is less than CCL for unstable propagation;
(2) the leak is detected;
(3) action is taken before the crack length exceeds the CCL.

The important questions that the reactor operators need answering are:

(a) How much time is *available* to detect the leak and to take action?
(b) How much time is *required* to detect the leak?

The time available can be estimated using the simple model shown in Fig. 6. The tube wall thickness is W, the crack length at leakage is L and the CCL is C, assuming $C > L$. In the cold-worked Zr–2·5Nb pressure tubes, cracks tend

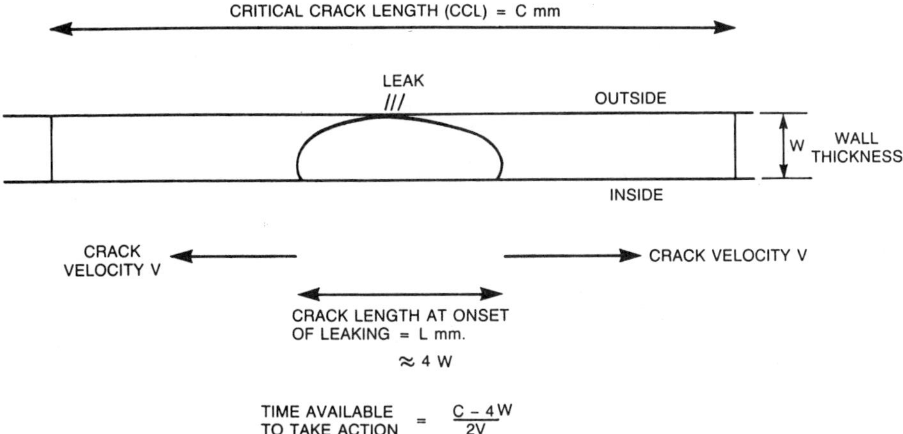

Fig. 6. Schematic diagram to show crack dimensions at onset of leakage.

to grow about twice as fast axially as radially. After penetration, if the crack continues to grow in both directions axially, the amount of crack growth that can occur in each direction before CCL is reached is $0{\cdot}5\,(C-L)$. With a crack velocity V the time t available between first penetration to the attainment of the CCL is

$$t = \frac{C-L}{2V} \tag{1}$$

Thus to estimate t, we need to know C and V and whether the penetration length L is always $4W$.

The critical crack length C has been determined using slit burst tests on tube sections, and fracture tests on small specimens; data have been obtained at different temperatures for unirradiated material and for pressure tubes removed from power reactors after up to 12 years of operation. For the inlet end of the pressure tube at 250°C the 95% lower confidence value for C is 50 mm.

The crack length L at penetration is also required for tubes that have leaked during reactor service. Because the cracks grew only during shutdowns and were oxidized by the PHT during operation, the history of the crack growth is recorded in the oxide. Two hundred measurements of crack shapes in pressure tubes at different stages of their growth in reactors show that most of the cracks initiated at a point and the average crack length at wall penetration L was $3{\cdot}6W$, with the upper bound of $4W$ (Fig. 7).

Using the above approach, the time available for operator action is 18 h, based on $L=4W$, and the crack velocity for irradiated material of $2{\cdot}7 \times 10^{-7}$ m/s (extrapolated from tests at temperatures below 200°C).

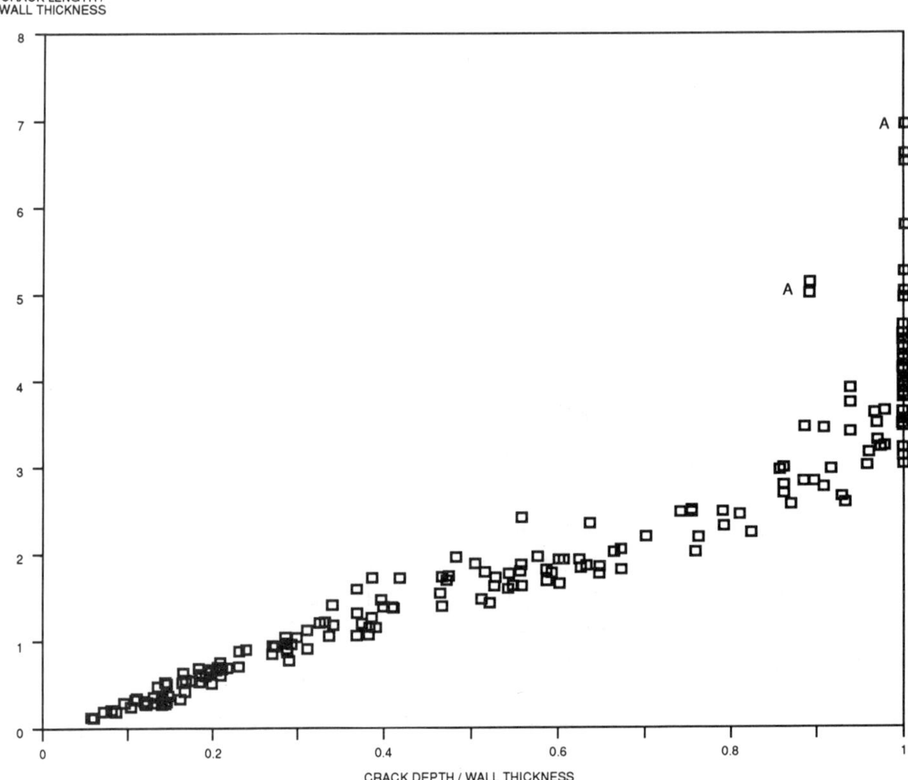

Fig. 7. Graph showing the relationship between crack length and crack depth for cracks found in pressure tubes removed from reactors because of leakage. Since they were oxidized before wall penetration, each crack provided several data points. The crack dimensions are normalized relative to the wall thickness.

In a rolled joint, the situation is more complicated. Some factors shorten the available action time while others can lengthen it. The former factors include:

(1) In a leaking crack, a temperature gradient is established in the metal by latent heat loss where the water flashes to steam across the crack face. Thus the temperature of the metal at the crack edge on the outside of the pressure tube is less than that at the inside.

Consequences of the temperature gradient may be that more axial cracking continues close to the outside surface of the tube with little growth at the inside surface. This asymmetry of crack growth allows the crack length to increase without the leak rate increasing. Figure 8 shows a crack that grew in such a manner, with a final crack length at removal of $7W$.

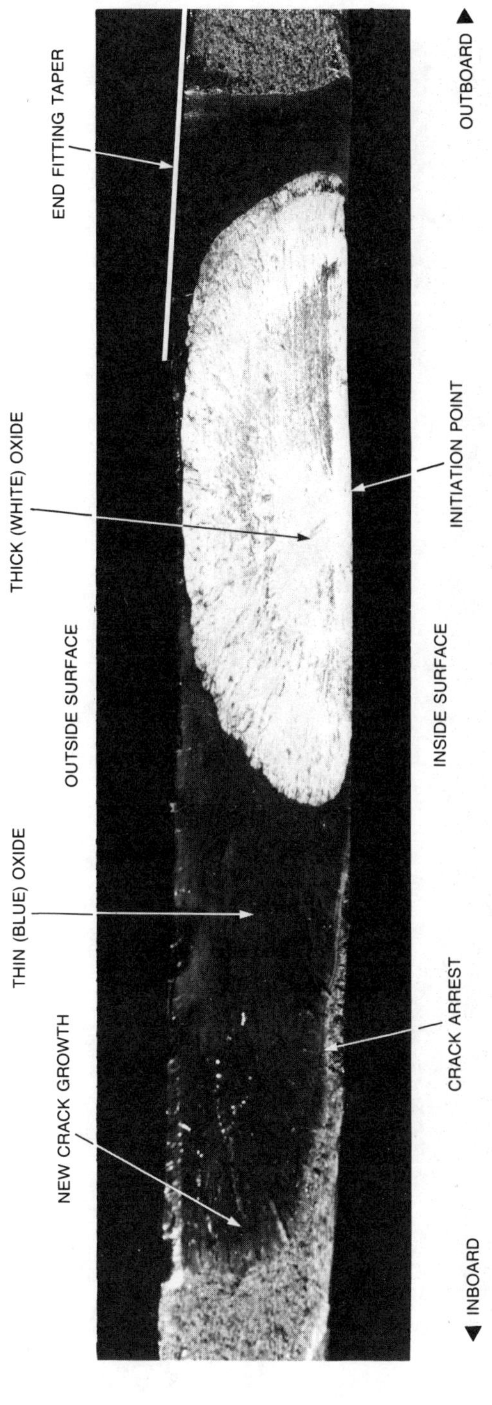

Fig. 8. Crack in a leaking pressure tube showing more axial growth at the outside surface. In this case the leak rate is controlled by the crack length at the inside surface. Wall thickness = 4 mm.

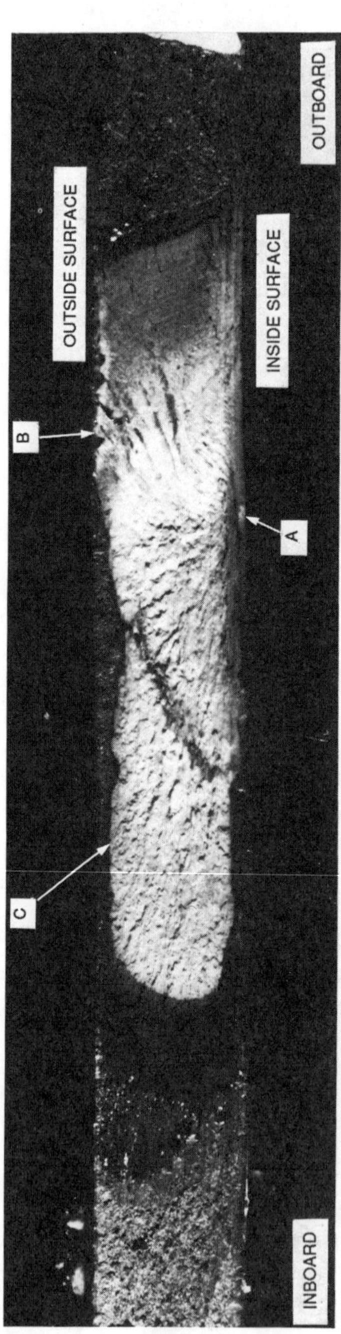

Fig. 9. Detail of crack face in a pressure tube that leaked in a reactor. The crack nucleated at A and grew axially and radially. An initial wall penetration and leakage is suspected to have occurred at B, but that leak path was clogged during reactor shutdown. Axial crack growth continued at later times and led to a second wall penetration at C. Wall thickness = 4 mm.

(2) During reactor operation the crack surfaces are oxidized and the leak path becomes restricted. Reactor shutdown may cause a leaking crack to become clogged with debris so that the leakage is stopped. Axial crack growth may continue so that the leakage rate increases again when the reactor is returned to power. In some cases the axial crack growth might lead to penetration at a new location. In one tube removed from a reactor in 1985 because of leakage detection, examination of the crack showed that the crack length at removal was $7W$. Penetration may have occurred much earlier, at $L = 4W$; the leak path becoming clogged and the leakage stopping. The macrograph in Fig. 9 shows the crack face, with the penetrations indicated; the data for this crack are shown as point A in Fig. 7.

Other features of the cracking at rolled joints render eqn (1) conservative, leading to underestimates of the time available:

(1) After a certain amount of crack growth, the outboard end of the crack is influenced by the compressive stress at the rolled joint and the crack growth outwards slows down and may eventually stop. An example is seen in Fig. 9, where the amount of crack growth in the later stages is much smaller at the outboard end than at the inboard end of the crack. Once the crack is stopped at the rolled joint, it becomes single-ended and eqn (1) requires modification to:

$$t = \frac{2C - L - CR}{2V} \tag{2}$$

where CR = length of crack when stopped at the rolled joint, as shown in Fig. 10.

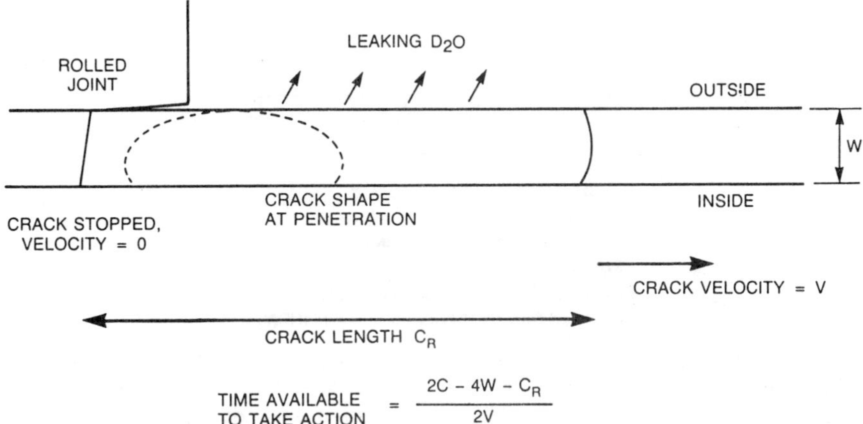

Fig. 10. Schematic diagram to show crack dimensions when crack growth in outboard direction is stopped by the end fitting.

(2) Delayed hydride crack growth rates measured on irradiated pressure tubes removed from power reactors are lower when coolant is leaking through the crack than the results of small specimen tests would predict (Fig. 11).

(3) At a rolled joint the pressure tube picks up more hydrogen than in the body of the tube. The hydrogen concentration thus decreases from the value found under the end fitting to lower values in the main body

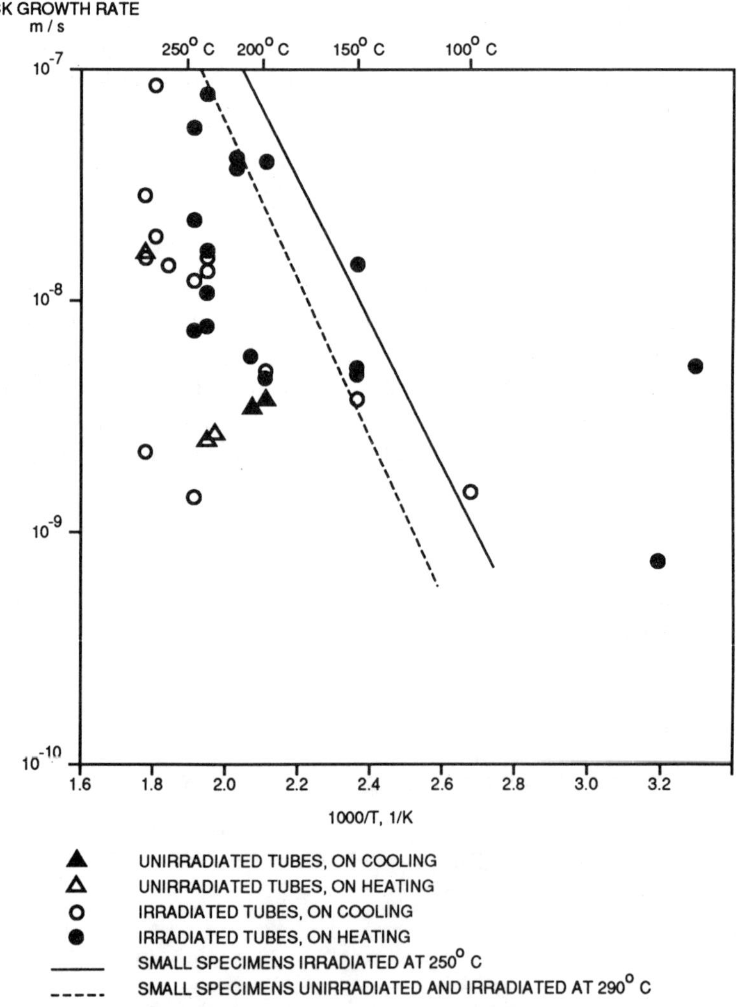

Fig. 11. Graph showing the difference in crack growth rates measured using small specimens and using tube samples. The data for the tube samples were obtained in tests on leaking cracks in pressure tubes, heated by pressurized hot water, during heating and cooling cycles when the crack lengths were 25–50 mm. The lines summarize the data from the small specimen tests, and the data points are the results from the tube tests,

of the tube. A crack growing inboard in this hydrogen concentration gradient will stop when hydrides (which are required for DHC) are no longer present at the operating temperature. This has been demonstrated in a laboratory experiment using a rolled joint from a power reactor.

In summary, more detailed calculations of the time available for operators to take action will be different from the 18 h derived earlier. Using crack velocities for irradiated material, a crack length at detection of $7W$ (rather than the $4W$ used earlier) and a value of the crack length CR (when outward growth stops at the rolled joint) of 22 mm, eqn (2) based on Fig. 10 leads to an available time of 25 h. This value does not include the benefit that can be obtained by using the crack growth rates measured on leaking pressure tubes instead of those measured on small specimens. Their inclusion would lead to times greater than 85 h.

LEAK RATES

To determine if the times calculated above are acceptable they must be compared with the time required to detect a leak. This time depends on the characteristics of the leak.

Information on the leak rate behaviour of axial cracks has been obtained from experiments using rolled joints removed from power reactors containing through-wall cracks in the high stress region of the pressure tubes.[2] These cracks, stressed by hot pressurized water, grew by delayed hydride cracking. The fluid leak rates through the cracks were measured under different conditions.

The results can be summarized as follows:

(1) At low pressure (less than operating pressure) the leak rates decrease with time probably because of clogging by oxide and debris.

(2) At higher pressures (similar to operating pressure), the leak rate appears to increase slowly, probably because the crack is extending sufficiently to compensate for the clogging.

(3) Crack growth rates measured on irradiated pressure tubes removed from the reactor are lower when the coolant is leaking than if the coolant is prevented from leaking (Fig. 11); under these conditions the temperatures used for the comparison are the coolant temperatures.

(4) Tests show that for a crack length greater than some threshold value, the leak rate rises rapidly with small increases in crack length. However, this threshold value is variable from tube to tube. Thus it is

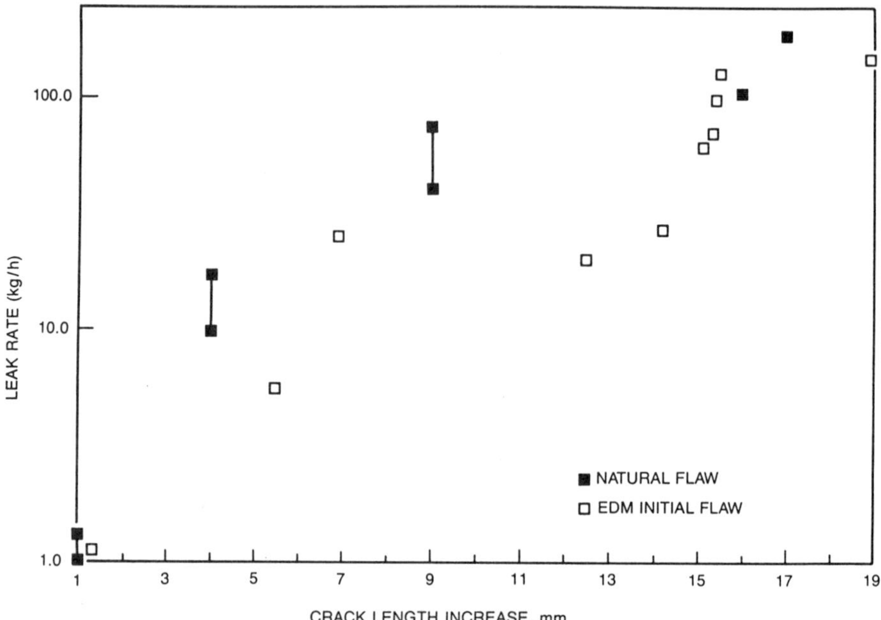

Fig. 12. Graph showing the effect of crack length increase on the water leak rate measured from cracks in pressure tubes in laboratory tests (at 220–280°C) after tubes were removed from the reactor.

necessary in any analysis to use the maximum threshold value which is considered to be 7*W*. For such cracks the leak rate is always a strong function of crack length (Fig. 12).

In summary, the leak rate experiments show that cracks maintained at operating pressure will develop a leak rate much larger than 1 kg/h at some length well below the CCL, but that the leak rate will diminish if the pressure is reduced and that short cracks may clog when maintained at reduced pressure.

MOISTURE DETECTIBILITY

The annulus gas system (AGS) is a closed loop, with header and intermediate tubing connections, that ensure as uniform a flow as possible through each annulus.[3] In recent reactors up to 11 annuli may be connected in series between sets of headers. In early reactors a parallel rather than a series connection limited the number of annuli to two in a series between headers.

Compressors recirculate CO_2 (or N_2) through the system, which is operated at a pressure of 70 kPa(g) and a dewpoint range of −30°C to 0°C

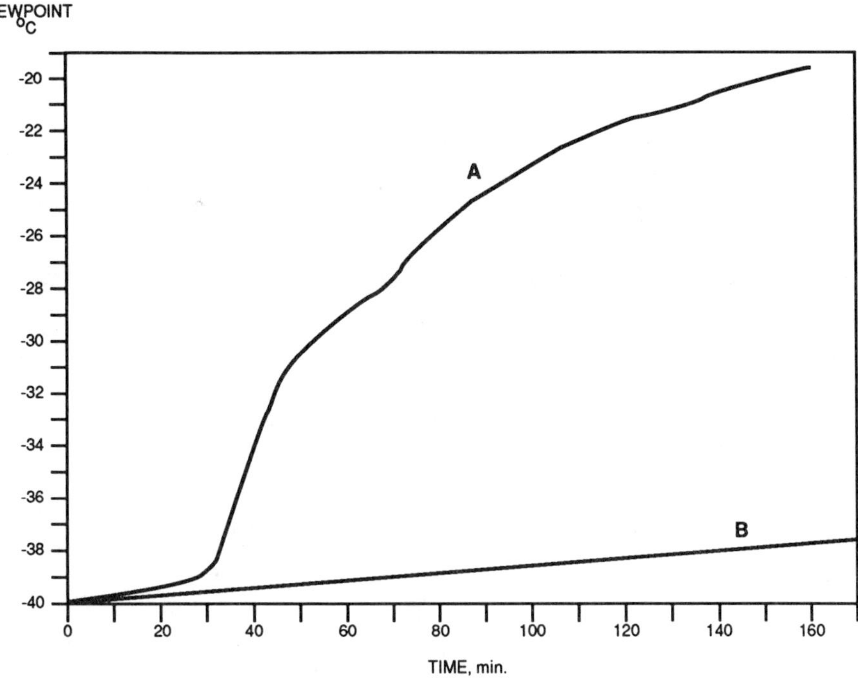

Fig. 13. Graph showing the variation of dewpoint with time. (A) Following the addition of D_2O at a rate of 10 g/h to the annulus in Pickering A, at an initial dewpoint of $-40°C$ (based on data in Ref. 4). (B) Arising from the formation of D_2O in the annulus by the oxidation of deuterium.

in Bruce A and $-40°C$ to $-18°C$ in Pickering A. The dewpoint (DP) shows a gradual increase due to the production of moisture from the reactions

$$CO_2 + D_2 = D_2O + CO$$
$$O_2 + 2D_2 = 2D_2O$$

The deuterium in the annulus, which can be present in concentrations up to 1000 ppm by volume, has diffused through the stainless steel end fittings from the PHT system. The deuterium and the D_2O are removed from the AGS by periodic purging to a low dewpoint. Following the purge the increase of dewpoint is characteristic of the above reactions (shown by line B in Fig. 13) and there is an interval of several days between purges.

The AGS is instrumented outside containment with dewpoint indicators, 'beetles', sight glasses and cold finger moisture traps, as shown in Figure 14. The dewpoint indicators are aluminum oxide hygrometers calibrated for D_2O in CO_2 or N_2; 'beetles' make use of the limited conductivity of water between electrodes to detect the presence of liquid water; sight glasses are used to indicate the string of channels in which the leak is located. The cold

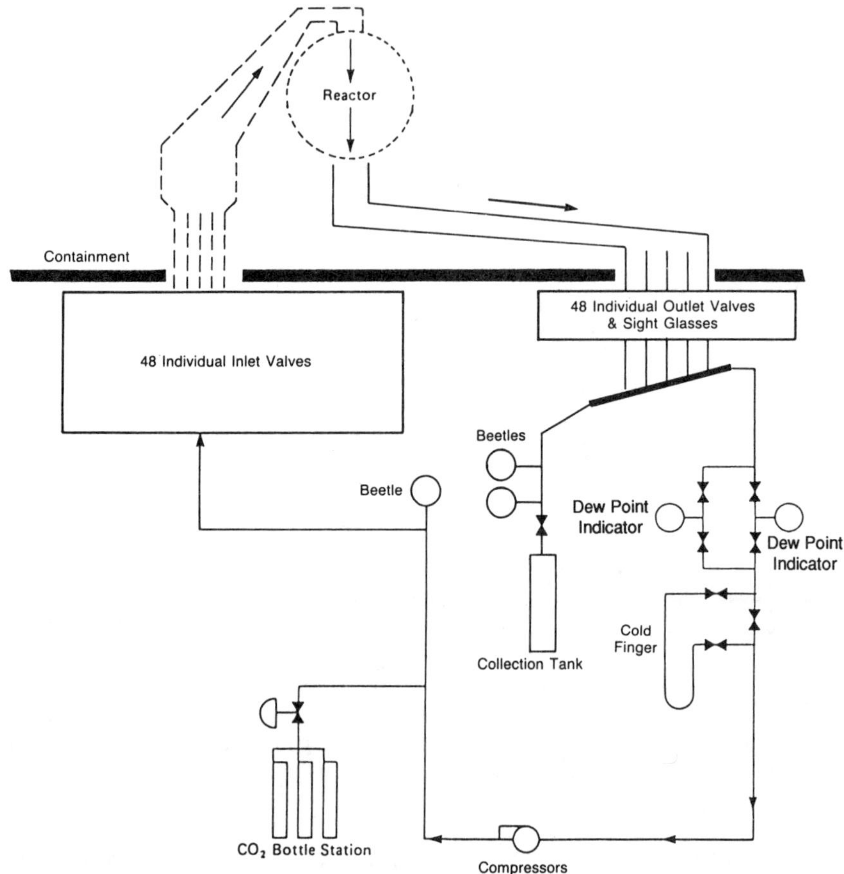

Fig. 14. Schematic diagram of gas flow in current annulus gas systems.

finger traps are used to obtain moisture samples to establish the source of the heavy water (water from the moderator contains more tritium than the primary heat transport water).

Testing and analysis of the AGS shows that the system dewpoint is sensitive to the presence of small quantities of moisture, of the order of a few g/h. The response of the system to a small leak depends on the size of the leak, on its location, on the dewpoint of the system when the leak occurred and on the flow rate of the CO_2 gas.

When a leak occurs its presence is revealed by the following:

(1) The dewpoint increases.
(2) The rate of dewpoint increase is much faster than that expected from deuterium ingress and water formation.
(3) When the dewpoint exceeds a specified value or the rate of rise of

dewpoint, d(DP)/dt, exceeds a specified value the dewpoint alarms are triggered.

(4) At later times, the 'beetle' alarms will confirm the presence of liquid water.

The response times of the AGS at Pickering A and Bruce A have been measured using calibrated water additions or using CH_4 tracer techniques.[4] The experiments at Pickering A have included calibrated D_2O additions of 10 g/h at an initial dewpoint of −40°C. Note that these addition rates are about 1% of the leak rates measured in Fig. 12. The dewpoint increased quickly as shown by curve A in Fig. 13. Curve B is the dewpoint change expected in the same time interval from the oxidation of the background D_2O. Figure 15 shows that rate of change of dewpoint, d(DP)/dt, went through a maximum value greater than 30°C/h in less than 1 h after the start of the D_2O addition. The operator has to take action if the rate of change of dewpoint, d(DP)/dt exceeds 7°C/h.

Beetle alarms require the presence of liquid water and respond more slowly than dewpoint monitors. In a second experiment carried out at Pickering A the response of the beetle alarms was examined. Additions of D_2O at a rate of 2·3 kg/h at an initial dewpoint of −40°C caused the beetle

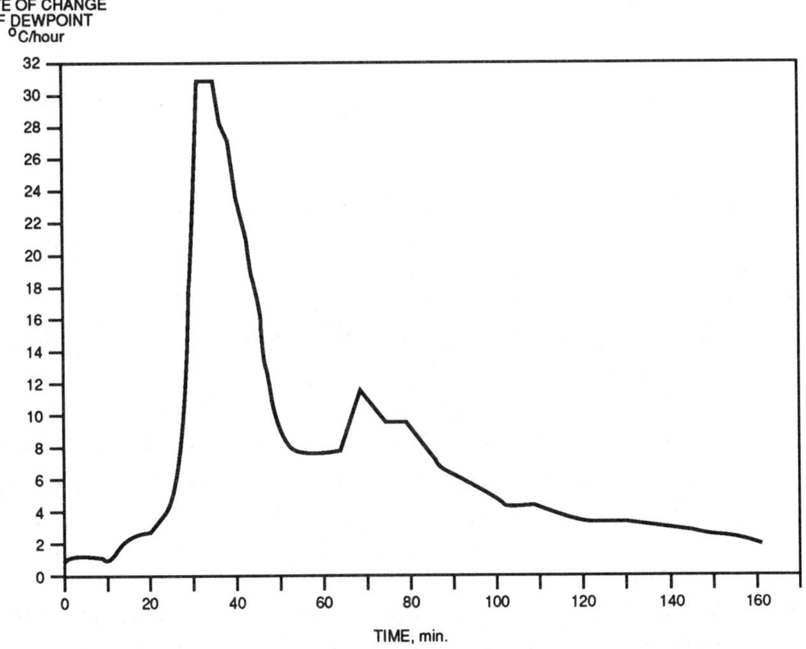

Fig. 15. Graph showing the rate of change of dewpoint, d(DP)/dt, with time following the addition of D_2O to the annulus at a rate of 10 g/h at an initial dewpoint of −40°C (based on curve A in Fig. 13).

alarm to indicate a leak in less than 1 h. Before that beetle alarm occurred the dewpoint monitors had given alarms at shorter times.

The results of the experiments carried out on the AGS response to small leaks show that small leaks of a few grams of water per hour can be detected in a few hours from the onset of leakage. The time required for this detection is much less than the time available for a crack to grow to its critical length.

OTHER METHODS OF ANALYSIS

In the preceding sections the approach to LBB analysis has been deterministic. That is, the worst combination of parameters has been used to determine the minimum operator action time. This approach has given satisfactory predictions for station use. However, we believe that the circumstances leading to an adverse combination of controlling parameters has a very low probability of occurrence. There is no obvious reason why a relationship between properties such as crack growth rate, CCL and crack length at penetration should exist. Thus the analysis of action time may be amenable to a probabilistic approach. Such an approach is described in the paper by Walker.[5]

CONCLUSIONS

(1) A methodology based on crack growth behaviour and leak detection capability is available to operators of CANDU stations to ensure LBB for pressure tube cracking.

(2) The effectiveness of the AGS has been shown a number of times. Shutdown procedures based on LBB are incorporated into CANDU station operating procedures.

(3) The time available to the operators of CANDU stations to take action following the onset of a pressure tube leak is the time needed for the crack to grow in a stable manner by delayed hydride cracking from its length at penetration to the critical crack length. The time depends on the critical crack length, the crack length at wall penetration and the crack growth rate.

(4) The annulus gas system is sensitive to the presence of small leaks, of a few grams per hour. At operating conditions, small leaks cause the dewpoint to change rapidly. Large leakage rates, >1 kg/h, occur when the crack lengths are well below the critical crack length.

ACKNOWLEDGEMENTS

Some of the data contained in this paper were obtained in experiments funded by CANDU Owners Group, whose support is recognized and appreciated.

REFERENCES

1. Coleman, C. E. & Ambler, J. F. R. Delayed hydride cracking in Zr–2·5%Nb alloy. *Reviews of Coatings and Corrosion*, 3 (1979) 105–57.
2. Coleman, C. E. & Simpson, L. A. Evaluation of a leaking crack in an irradiated CANDU pressure tube. AECL Report, AECL 9733, 1988.
3. Kenchington, J. M., Ellis, P. J. & Meranda, D. G. An overview of the development of leak detection monitoring for Ontario Hydro Nuclear Stations. *Proc. of the Eighth Annual Conference*, Canadian Nuclear Society, St John, NB, 1987.
4. Kenchington, J. M., Ellis, P. J. & Meranda, D. G. Annulus Gas System response tests for Pickering NGS A, Unit 1 commissioning. *Proc. of the Ninth Annual Conference*, Canadian Nuclear Society, Winnipeg, 1988.
5. Walker, J. R. A probabilistic approach to Leak-Before-Break in CANDU pressure tubes. *Int. J. Pres. Ves. & Piping*, **43** (1990) 229–39.

Int. J. Pres. Ves. & Piping **43** (1990) 23–37

The Role of Leak-Before-Break in Assessments of Flaws Detected in CANDU Pressure Tubes

H. W. Wong, V. K. Bajaj, G. D. Moan,

CANDU Operations, Atomic Energy of Canada Ltd, Sheridan Park, Mississauga, Ontario, Canada L5K 1B2

M. Huterer & C. O. Poidevin

Ontario Hydro, 700 University Avenue, Toronto Ontario, Canada M5G 1X6

ABSTRACT

This paper reviews the role of the Leak-Before-Break (LBB) concept in the Fitness for Service Guidelines being developed for cold worked (cw) Zr–2·5Nb pressure tubes in a CANDU reactor. The guidelines complement the rules of Section XI of the ASME Code and the requirements of Canadian Standards Association (CSA) CAN 3-N285.4-M83.

The evaluation procedures in the guidelines consist of a flaw growth analysis to determine the maximum size of the flaw at the end of the evaluation period. It must then be demonstrated that the flaw is stable with adequate margins of safety for the various loading conditions.

For the delayed hydride cracking failure mode LBB is used as defense in depth against unstable rupture. First the flaw must be shown to be non-susceptible to propagation by delayed hydride cracking during normal operating conditions. In addition, it must then be demonstrated that, if the flaw were to penetrate the tube wall, the leaking coolant would be detected and the reactor shutdown before the postulated crack became unstable.

The Guidelines contain criteria for performing both deterministic and probabilistic LBB analyses.

1 INTRODUCTION

The CANDU reactor utilizes horizontal fuel channels which are installed in a low pressure calandria tank filled with cool heavy water. A simplified

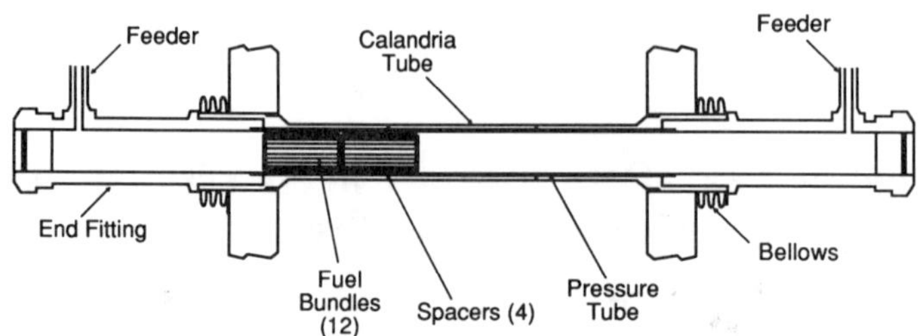

Fig. 1. Simplified illustration of a CANDU fuel channel.

illustration of a typical CANDU fuel channel is shown in Fig. 1. Each channel consists of a pressure tube containing natural uranium fuel and heat transport heavy water at a pressure of about 10 MPa and temperature ranging from 250 to 310°C. The pressure tubes are fabricated from cold worked (cw) Zr–2·5Nb alloy and each tube is 6 mm long with an internal diameter of 103 mm and a wall thickness of 4·0 mm.

CAN3-N285.4[1] of the Canadian Standards Association (CSA) requires periodic inspection of pressure tubes in operating CANDU nuclear reactors. These inspections are intended to ensure that unacceptable degradation in component quality is not occurring and that the probability of failure remains acceptably low for the life of the reactor. If flaws exceeding the acceptance standards of CSA N285.4 are revealed, an evaluation is required to determine if the pressure tube is acceptable for continued service. In this paper the authors describe evaluation procedures and acceptance criteria which are being developed as part of the Fitness for Service Guidelines for pressure tubes by the CANDU Owners Group (COG).

Part of the fitness procedures will be centered around the demonstration that tubes will operate in a condition that if they do fail, they will leak at a rate sufficiently large that the leak will be detected and the reactor shutdown before unstable crack propagation occurs. This condition is known as Leak-Before-Break (LBB) and the proposed role of the LBB concept in the flaw evaluation methodology is discussed in this paper. Although other jurisdictions in the nuclear industry outside Canada are also developing procedures for the application of LBB, these are for application to stainless and ferritic steels in light water reactor pressure vessel reactors and are not directly applicable to cw Zr–2·5Nb pressure tubes because of several reasons:

(a) The consequences of failures in steel vessels and piping in light water reactors are different from those of a single pressure tube failure in a CANDU reactor.
(b) The annulus gas monitoring system in CANDU fuel channels provides a very sensitive leak detection capability which allows for a rapid response.

2 FLAW EVALUATION METHODOLOGY

To develop flaw evaluation procedures and acceptance criteria it is necessary to define appropriate flaw-related failure mechanisms, including those that control crack growth rate and stability.

For cw Zr–2·5Nb pressure tubes the two primary mechanisms of stable crack growth are fatigue and delayed hydride cracking (DHC). Because the fatigue properties of irradiated Zr–2·5Nb indicate that failure by fatigue is less likely than failure by DHC, and because DHC has been associated with all pressure tube failures to date[2–4] DHC is considered to be the dominant mechanism by which flaws can grow to a size which exceeds the critical crack length.

The conditions necessary for DHC are discussed in Refs 5 and 6. For a pre-existing flaw, which is the case being addressed in this paper, the key conditions necessary for DHC to occur are summarized below:

(a) hydrides must be present at the flaw tip;
(b) the stress intensity factor K_I at the flaw tip must exceed the threshold value, K_{IH}, below which DHC does not occur;
(c) the material must be maintained for a sufficient time at suitable temperature and stress conditions.

The dependence of DHC crack growth velocity on the stress intensity factor is shown in Fig. 2. At low K_I values (less than K_{IH}) there is no crack growth. Over a wide K_I range, the crack growth is stable and almost independent of the crack driving force. When the crack driving force reaches a critical value, crack growth is unstable.

In the Fitness for Service Guidelines, the DHC evaluation procedures are based on the requirement that for DHC to occur, K_I must exceed K_{IH} in the presence of hydrides. Therefore, the analytical method uses linear elastic fracture mechanics, with the crack growth rate being described by Fig. 2.

2.1 Flaw evaluation procedures and acceptance criteria

The evaluation procedures consist of an analysis of subcritical flaw growth to determine the maximum size of the flaw at the end of the evaluation period. It must then be demonstrated that this flaw size is stable with adequate margins of safety for the various loading conditions. The acceptance criteria contained in the Guidelines are intended to prevent failure by brittle fracture, plastic collapse of the ligament, and DHC.

The concept of service level categories is used to adjust the margin of safety based upon the likelihood of an event. The categories are: Service Level A—normal operating conditions; Service Level B—upset conditions;

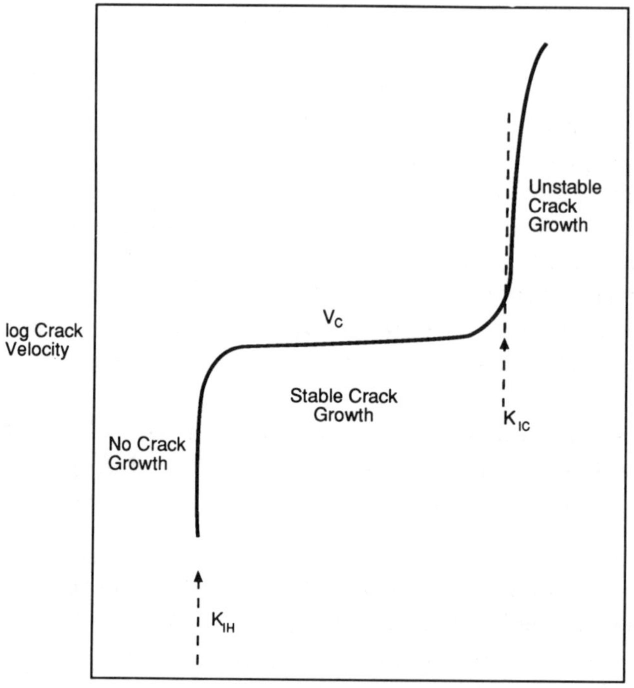

Fig. 2. Schematic diagram of the dependence of delayed hydride crack growth rate on the stress intensity factor.

Service Level C—emergency conditions; and Service Level D—faulted conditions.

The proposed flaw evaluation sequence is illustrated in Fig. 3 and summarized below:

(1) Characterise the flaw shape and size.
(2) Identify the design basis and operating transients for Levels A, B, C and D service conditions.
(3) Perform a subcritical flaw growth analysis based on flaw growth due to fatigue and DHC.
(4) Evaluate the flaw based on the calculated maximum flaw size (a_f).

A pressure tube containing a flaw is assessed as being acceptable for continued operation if the following criteria are satisfied.

2.1.1 Acceptance criteria for Level A and B conditions

(a) Demonstrate that brittle fracture will not occur. The following conservative method is recommended:

$$K_I < K_i/\sqrt{10}$$

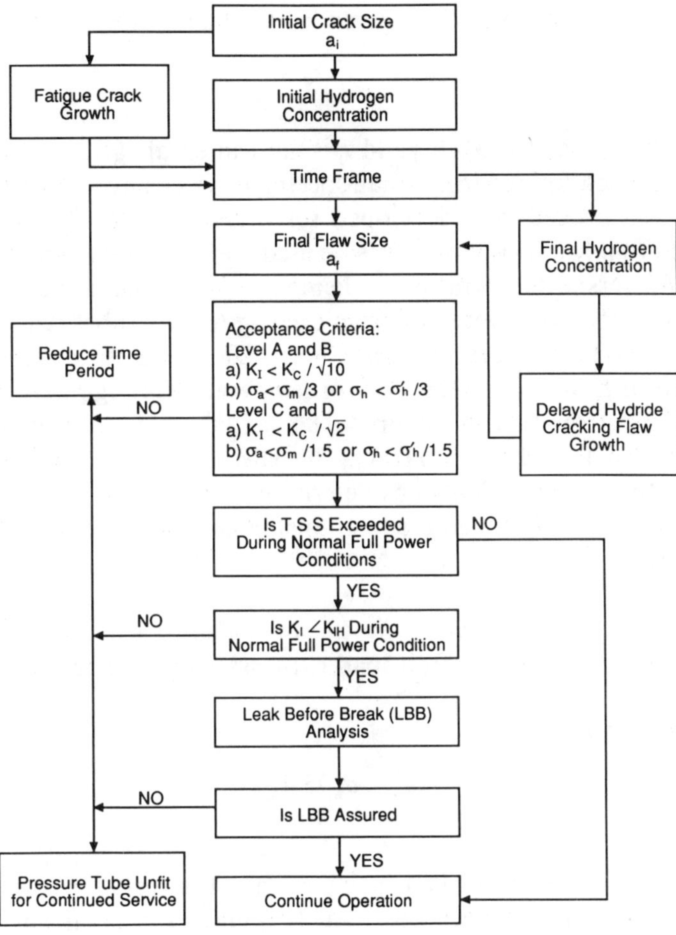

Fig. 3. Proposed flaw evaluation sequence.

where:

K_I = the maximum applied stress intensity factor for the loading condition being evaluated and the flaw size a_f;

a_f = the maximum depth to which a detected flaw is calculated to grow during the evaluation period;

K_i = the critical intensity factor for fracture initiation based on temperature, neutron fluence and predicted hydrogen concentration.

The margin of safety recommended for the Level A and B conditions are consistent with that contained in ASME IWB-3610[7] for reactor pressure vessels.

The fracture toughness, K_i, is given by the plane–stress relation:

$$K_i = \sqrt{J_i E}$$

where E is the Young's modulus and J_i is the J-integral at fracture initiation. The value J_i is determined from small specimen tests using standard fracture toughness test procedures[8] developed for pressure tubes. The Guidelines recommend that lower bound values be used and a recommended reference curve of K_i versus temperature is being prepared. Data being used to produce this curve were obtained from eleven pressure tubes removed from various CANDU reactors. To date, over 200 small specimen fracture toughness tests have been performed for temperatures ranging from 30°C to 300°C and neutron fluence of up to $8{\cdot}8 \times 10^{25}$ n/m^2. To determine the reference curve for use in analysis, the effects of neutron fluence and temperature on K_i are evaluated by determining the relationship between K_i and neutron fluence and then the relationship between K_i and temperature for appropriate fluence levels.

(b) Demonstrate that failure by plastic collapse will not occur. The following method based on limit load expressions developed for piping is recommended.[9]

Circumferential flaws:

$$\sigma_a < \sigma_m / 3{\cdot}0$$

where:

σ_a = the maximum applied primary axial stress for the loading condition being evaluated. Note that diametral expansion and wall thinning due to irradiation enhanced creep and growth must be taken into account in conjunction with corrosion and wear allowances when determining the stresses.

σ_m = membrane stress corresponding to plastic collapse for the crack depth and angle being evaluated. The value σ_m is to be calculated as follows:

$$\sigma_m = \sigma_f(1 - (a/t)(\theta/\pi) - 2\gamma/\pi)$$

$$\gamma = \mathrm{Arcsin}\left(\frac{a \sin\theta}{2t}\right)$$

The definition of a, t, θ and σ_f are illustrated in Fig. 4.

Longitudinal flaws:

$$\sigma_h < \sigma_h^1 / 3{\cdot}0$$

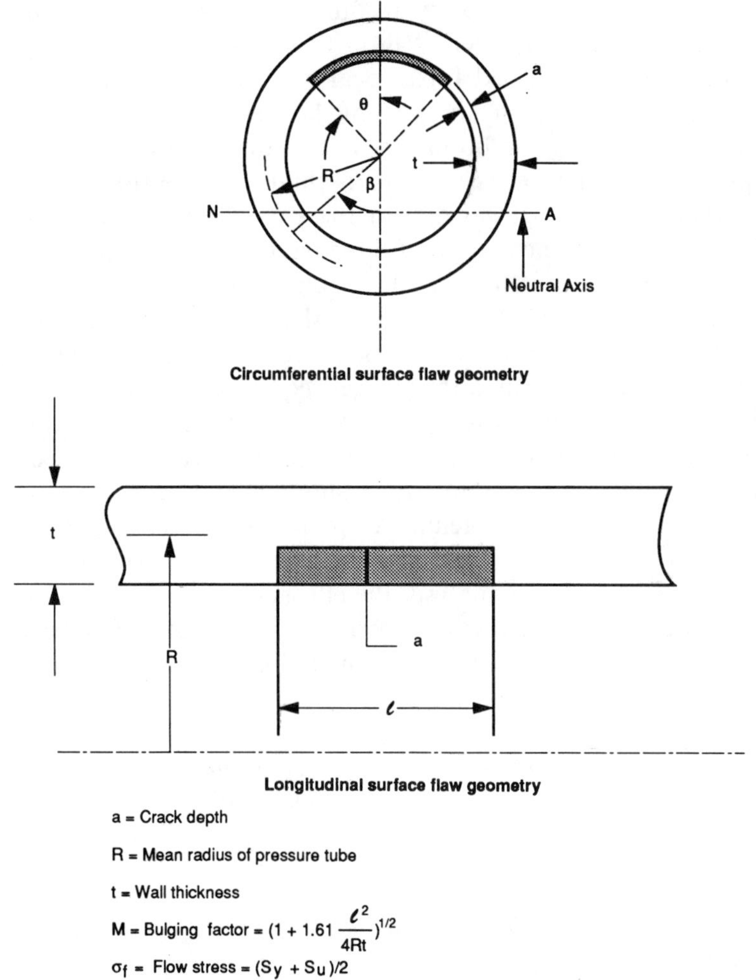

Fig. 4. Illustration of the symbols used in the plastic collapse stress calculation.

where:

σ_h = the maximum applied primary hoop stress for the loading condition being evaluated. Note that diametral expansion and wall thinning due to irradiation enhanced creep and growth must be taken into account in conjunction with corrosion and wear allowances.

σ_h^1 = the hoop stress at plastic collapse for the crack depth and length being evaluated. The value σ_h^1 is to be calculated as follows:

$$\sigma_h^1 = \sigma_f[(1 - a/t)/(1 - a/tM)]$$

The definition of a, t, M and σ_f are illustrated in Fig. 4.

The minimum margin of safety recommended is consistent with that contained in ASME Section XI IWB-3610[7] which states that the primary stress limits must be satisfied for a local area reduction of the pressure retaining membrane equal to the area of the characterized flaw.

For normal operating conditions, the minimum factor of safety provided by the primary membrane stress limits is three because the code requires $P_m < S_m$ where S_m is the lower of $\frac{2}{3}$ of the yield strength (S_y) or $\frac{1}{3}$ of the ultimate strength (S_u). The margin of three recommended in the Guidelines is, therefore, consistent with the code requirements. The same margin of safety has been conservatively applied to both the Level A and B conditions.

In determining the collapse load, the flow stress (σ_f), which is defined as the average of the yield and ultimate stresses, is required. The yield and ultimate strength of cw Zr–2·5Nb pressure tube material increases due to neutron irradiation. The guidelines state that this increase in strength may be used in determining σ_f but lower bound values should be used. A recommended reference curve of σ_f versus temperature is being prepared using data obtained from surveillance tubes removed from reactor. To determine the reference curve(s) for use in analysis, the effects of neutron irradiation and temperature on σ_f are evaluated by determining the relationship between σ_f and neutron fluence and then the relationship between σ_f and temperature for appropriate fluence levels.

2.1.2 Acceptance criteria for Level C and D conditions

(a) Demonstrate that brittle fracture will not occur.

$$K_I < K_i/\sqrt{2}$$

(b) Demonstrate that failure by plastic collapse will not occur.

Circumferential flaws:

$$\sigma_a < \sigma_m/1{\cdot}5$$

Longitudinal flaws:

$$\sigma_h < \sigma_h^1/1{\cdot}5$$

These safety margins are consistent with the requirements of ASME Section XI IWB-3610.[7]

2.1.3 Acceptance criteria if TSS is exceeded at normal full power conditions

If the calculated equivalent hydrogen concentration at a flaw location at the end of the evaluation period will exceed the Terminal Solid Solubility (TSS) limit for normal full power conditions, the following two additional criteria must be satisfied:

(a) Demonstrate that DHC will not occur:

$$K_I < K_{IH}$$

where:

K_I = the applied stress intensity factor for normal full power conditions and the final flaw size, a_f;
K_{IH} = the threshold stress intensity factor for DHC.

The Guidelines reccomend that a lower bound value of K_{IH} should be used. K_{IH} has values between 4·5 and 7·0 MPa$\sqrt{m}$ and is insensitive to temperature, hydrogen concentration and neutron fluence.[10] There is currently an extensive research program being funded by COG to provide additional data on K_{IH} and these results will also be evaluated for use in the Guidelines.

(b) Demonstrate that even if the flaw did propagate through the wall because of DHC, LBB is assured. Either deterministic or probabilistic LBB analyses may be completed.

To assure LBB in CANDU pressure tubes it is required that:

—the crack length at wall penetration be less than the Critical Crack Length (CCL) for unstable propagation;
—the leak be detected and the reactor put into a cold, depressurized condition before the crack length exceeds the CCL.

Therefore, the objective of an LBB assessment of CANDU pressure tubes is to determine:

(1) how much time is *available* to detect the leak and to take action;
(2) How much time is *required* to detect the leak.

The following is a description of the acceptance criteria for the deterministic and probabilistic LBB approaches.

2.1.3.1 Deterministic LBB approach. The proposed acceptance criterion for the deterministic LBB approach is:

$$t > T/SF$$

where:

t = the calculated time for an assumed through-wall crack to grow from its length at penetration to the critical crack length;
T = the response time of the leak detection system which includes the time required to confirm the presence of a leak at normal operating

conditions and the time required to reach a cold depressurized condition;

SF = safety factor.

The methodology for calculating the time t is described in Refs 2, 10 and 11 and summarized below

$$t = \frac{CCL - L_p}{2V}$$

where

L_p = the initial length of the crack at leakage;

CCL = the critical crack length which is defined as the minimum length of an axial through-wall crack that would be unstable at the temperature and pressure being evaluated;

V = the delayed hydride crack growth rate in the axial direction at the temperature being evaluated.

Each of these parameters in this equation is discussed below.

Crack length at penetration (L_p): Evaluation of up to two hundred measurements of crack shapes in pressure tubes at different stages of their growth in reactor by DHC has indicated that the average crack length at wall penetration was 3·6 times the thickness of the pressure tube.[10] However, the relationship between leak rate and crack size and the leak detection capability must also be considered when determining the value of L_p for use in the LBB assessment. In a leaking crack, a temperature gradient is established in the metal by latent heat loss where the water flashes to steam across the crack face. Consequences of this temperature gradient may be that more axial cracking continues close to the outside surface of the tube with little growth at the inside surface. This asymmetry of crack growth allows the effective crack length to increase without the leak rate increasing. Also during reactor operation the crack surfaces are oxidized and the leak path could become restricted.

In the guidelines it is recommended that L_p be conservatively taken to be at least seven times the wall thickness of the tube. This value was recommended based on the following reasons:

—It is the upper bound of data of DHC crack growth in pressure tubes[10] and addresses conditions such as asymmetric crack growth and clogging of the leak path due to oxidation.

—Correlations between leak rate and crack length data and leak rate with crack length increase data[10] indicate that, for cracks of initial throughwall length of seven times the wall thickness, the leak rate is always more than sufficient to ensure leak detection by the annulus gas system.

Critical Crack Length (CCL): The critical crack length has been determined using slit burst tests on tube sections and fracture tests on small specimens. To date over 70 burst tests have been performed and as discussed in Section 2.1.1, over 200 small specimen fracture tests have been performed to estimate the CCL. These results are being used to evaluate the relationship between CCL and neutron fluence, the relationship between CCL and temperature, and to quantify the conservatisms associated with the CCL estimated from small specimens.

In the Guidelines it is currently recommended that the lower bound CCL obtained from the burst tests be used in the analysis for the temperature, neutron fluence and loading conditions under consideration.

Delayed Hydride Cracking Growth Rate (V): Over 80 crack growth rate tests have been carried out using small specimens prepared from Zr–2·5Nb pressure tubes removed from reactor after irradiation to fluences up to 8×10^{25} n/m^2.[10] Pressure tube sections have also been used to measure the crack growth rates.[11]

These results are being evaluated, and a reference curve showing the temperature dependence of V is being prepared.

Safety factor: The proposed acceptance criteria for addressing the delayed hydride cracking failure mode uses LBB as defense in depth against unstable rupture. If TSS is exceeded at full power steady state conditions, the flaw must first be shown to be non-susceptible to propagation by DHC (i.e. $K_I < K_{IH}$). Then, it must be demonstrated that, if the flaw were to penetrate the tube wall because of DHC, the leaking coolant would be detected and the reactor shutdown before the postulated through-wall crack became unstable (i.e. LBB). Therefore, LBB is not used as the sole or primary defense against unstable rupture. Also the deterministic LBB assessment incorporates a conservative approach by using the worst combination of parameters as summarized below:

—Initial penetration length of at least 7 times the wall thickness which addresses the most pessimistic conditions with respect to crack growth, leak rate, and assurance of leak detection.
—Lower bound critical crack lengths.
—Upper bound crack growth rates.

Therefore, it is currently recommended that no additional factor of safety need be employed for the deterministic LBB approach.

As part of the continuing development and verification of the Guidelines, sensitivity analyses using the database on initial penetration lengths, CCLs and crack growth rates will be completed to quantify the inherent margins associated with the conservative deterministic approach.

2.1.3.2 Probabilistic LBB approach. The proposed acceptance criterion for the probabilistic LBB approach is to demonstrate that the probability of ($t < T$) is less than 1×10^{-3} which corresponds to an event frequently of less than 10^{-3} occurrences/reactor-year.

In the CANDU design, failure of a pressure tube is not assumed to be an incredible event as the design incorporates a number of safety systems to accommodate such an event. In the safety report, which is an essential part of CANDU licensing, an event frequency of $<2 \times 10^{-3}$ occurrences/reactor-year is used for the scenario in which both the pressure tube and calandria tube are assumed to fail. The proposed LBB acceptance criterion of $\leq 10^{-3}$ occurrences/reactor-year was conservatively selected so as not to contravene the safety analyses. Reference 12 contains a description of the probabilistic LBB methodology and computer program developed for CANDU pressure tubes.

3 LEAK DETECTION CAPABILITY

The previous sections have discussed the recommended approach for determining 'how much time is available to detect a pressure tube leak and to take action before a crack becomes unstable'. This philosophy is applicable to CANDU pressure tubes because the annulus gas system has been developed into a leak detection system which is very sensitive to the presence of moisture resulting from a break of the primary heat transport pressure boundary that passes through the reactor core (and also leakage from the surrounding calandria and lattice tubes, although this type of leakage has not been significant).

The space between the pressure tube and the calandria tube is filled with gas (Fig. 1) and is thus called the gas annulus. These gas annuli are part of a closed loop, having header and intermediate tubing connections, which ensure a uniform flow through each annulus.

Compressors recirculate CO_2 (or N_2) through the system which is operated at a pressure of 70 kPa(g) and a dewpoint range of -30°C to -18°C. The system is instrumented outside containment with dewpoint indicators, 'beetles' (moisture detectors), sight glasses, and cold finger moisture traps (see Fig. 5). Testing and analysis of the system show the system dewpoint is sensitive to ingresses of moisture as small as one gram per hour. The annulus gas system response to a leak is a function of leak rate, leak location, initial system dewpoint and CO_2 flow rate. Beetle alarms indicate liquid water collection, and have a slower response than the dewpoint indicators. Cold finger traps are used to obtain samples of moisture to establish the source of the heavy water (water from the

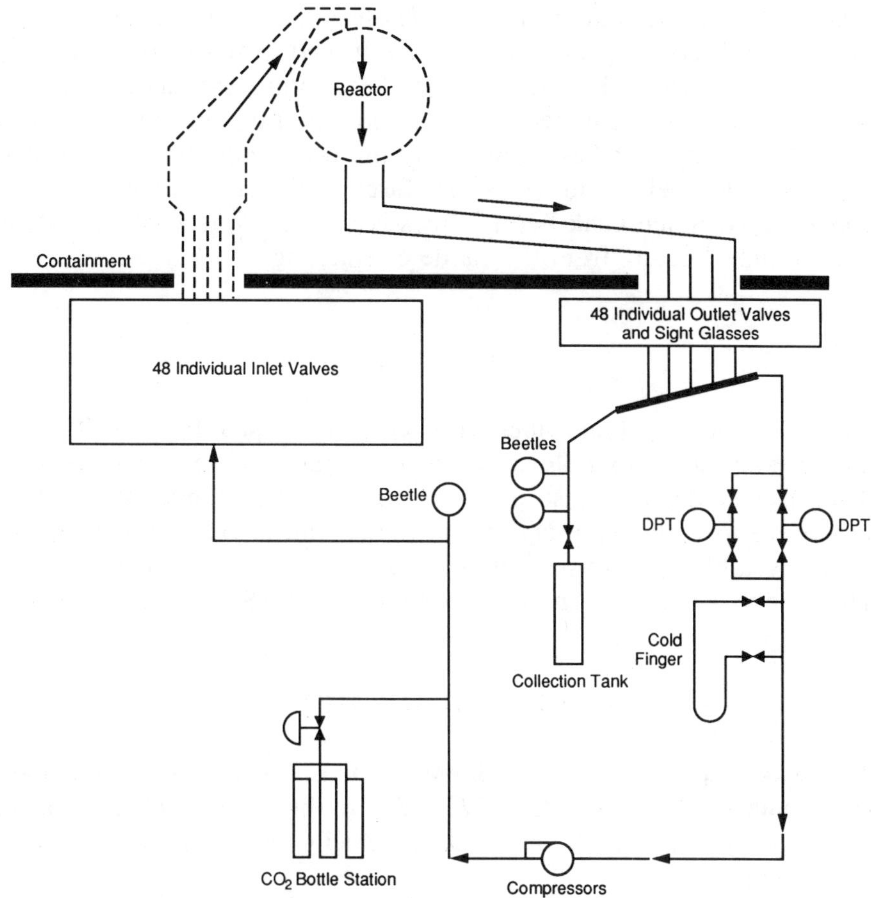

Fig. 5. Schematic diagram of gas flow in current Annulus Gas Systems of CANDU reactors.

moderator contains more tritium than the primary heat transport water).

In LBB assessments of CANDU pressure tubes the response of the annulus gas system and the operating procedures of the reactor unit being evaluated need to be identified to determine the time (T) required for detecting a leak and putting the reactor in a cold depressurized condition. This value can vary from reactor to reactor and is not addressed in this paper.

4 EXPERIENCE OF LEAK-BEFORE-BREAK

There have been 24 instances of pressure tubes that have leaked from rolled joint cracks with the reactor being taken to a cold shutdown condition without the crack becoming unstable.[2] The leaking pressure tubes were then removed, and replaced, and the reactor returned to full power. The leaks were detected when the dewpoint in the annulus increased and were confirmed by the water indicators. Examination of the pressure tubes after removal from reactor has shown that the crack initiation and growth occurred by delayed hydride cracking. The experience with these 24 pressure tubes has shown that Leak-Before-Break is a valid defense against unstable pressure tube fracture from this mode of cracking.

5 CONCLUSIONS

Fitness for Service Guidelines are being developed for CANDU cw Zr–2·5Nb pressure tubes. For flaws detected during in-service inspections. LBB will be used as a defense in depth against unstable rupture when TSS is exceeded at normal full power conditions. In addition to demonstrating that the flaw is stable and non-susceptible to DHC, it must also be demonstrated that if the flaw were to penetrate the tube wall, LBB would be assured.

6 ACKNOWLEDGMENTS

The assistance provided by L. A. Simpson, D. A. Scarth, C. E. Coleman and other members of Working Parties 22 and 31 of the CANDU Owners Group Fuel Channel Research and Development Program is gratefully acknowledged.

REFERENCES

1. Canadian Standards Association, CAN3-N285.4-M83. Periodic inspection of CANDU nuclear power plant components, December 1983.
2. Price, E. G., Moan, G. D. & Coleman, C. E., Leak-Before-Break experience in CANDU reactors. Presented at ANS-ASME Topical Meeting, Myrtle Beach, April 1988. Atomic Energy of Canada Limited Report AECL-9609.
3. Price, E. G. & Cheadle, B. A., Fast fracture of a zirconium alloy pressure tube. Presented at the International Conference and Exposition on Fatigue, Corrosion Cracking, Fracture Mechanics and Failure Analysis, Salt Lake City, UT, USA, December 1985. Atomic Energy of Canada Limited Report AECL-7542.
4. Cheadle, B. A. & Price, E. G., Operating performance of CANDU pressure tubes. Presented at IAEA Technical Committee Meeting on the Exchange in Operational Safety Experience of Heavy Water Reactors, Vienna, Austria, February 1989. Atomic Energy of Canada Limited Report AECL-9939.

5. Coleman, C. E., Cheadle, B. A., Ambler, J. F. R., Lichtenberger, P. C. & Eadie, R. L., Minimizing Hydride Cracking in Zirconium Alloys. *Can. Met. Quart.*, **24** (1985) 245–50.
6. Cheadle, B. A., Coleman, C. E. & Ambler, J. F. R., Prevention of delayed hydride cracking in zirconium alloys. *ASTM STP 939*. American Society for Testing and Materials, Philadelphia, PA, 1985, pp. 224–40.
7. ASME Boiler and Pressure Vessel Code, Section III, 1986 Edition.
8. Simpson, L. A., Chow, C. K. & Davies, P. H., Standard test method for fracture toughness of CANDU pressure tubes, COG-89-110-1, September 1989.
9. Novetech Corporation, Evaluation of flaws in ferritic piping, EPRI Report No. NP-6045, October 1988.
10. Moan, G. D., Coleman, C. E., Price, E. G., Rodgers, D. K. & Sagat, S., Leak-Before-Break in the pressure tubes of CANDU reactors. *Int. J. Pres. Ves. & Piping*, **43** (1990) 1–21.
11. Coleman, C. E. & Simpson, L. A., Evaluation of a leaking crack in an irradiated CANDU pressure tube. Presented at IAEA Specialists' Meeting on Fracture Mechanics Verification by Large Scale Testing, Stuttgart, FRG, May 1988. Atomic Energy of Canada Limited Report AECL-9733.
12. Walker, J., A probabilistic approach to Leak-Before-Break in CANDU pressure tubes. *Int. J. Pres. Ves. & Piping*, **43** (1990) 229–39.

Int. J. Pres. Ves. & Piping **43** (1990) 39–56

Leak-Before-Break Experiments on Heat-Treated Zr–2·5 wt% Nb Pressure Tubes

M. H. Koike, T. Takahashi & H. Baba

Power Reactor and Nuclear Fuel Development Corp., Oarai Engineering Center, 4002, Narita-cho, Oarai-machi, Higashi Ibaraki-gun, Ibaraki 311-13, Japan

ABSTRACT

Pressure tubes of Advanced Thermal Reactor (boiling-light-water-cooled, heavy-water-moderated, pressure tube-type reactor) in Japan are made of heat-treated Zr–2·5wt% Nb alloy and both ends are mechanically joined with stainless steel extension tubes. Sharp artificial cracks were introduced in the rolled joint region of pressure tube specimens. The cracks were propagated, and penetrated the tube wall by fatigue and DHC in a high-temperature, high-pressure water loop. From the results, it was shown that the LBB phenomena were valid for the rolled joint region of the pressure tube under the reactor operating conditions and that the critical crack length was more than 50 mm. Calculations were performed for the subsequent leak rate, using critical flow data.

1 INTRODUCTION

The Power Reactor and Nuclear Fuel Development Corporation (PNC) in Japan has developed the ATR Fugen, a 165 MWe prototype boiling-light-water-cooled, heavy-water-moderated, pressure tube-type reactor of Japan, which has operated satisfactorily since the start of commercial operation in March 1979. It achieved the total electrical generation of $9{\cdot}8 \times 10^9$ KW h in July 1989. A 606 MWe ATR Demonstration Plant has been designed on the basis of the experience of the Fugen and is scheduled to begin commercial operation at the end of the 1990s. The ATR is a unique reactor designed mainly to use plutonium–uranium mixed oxide (Mox) fuels.[1]

A vertical section of the reactor and primary cooling system of the Fugen is shown in Fig. 1. The Fugen has 224 pressure tubes made of heat-treated (to

Int. J. Pres. Ves. & Piping 0308-0161/90/$03·50

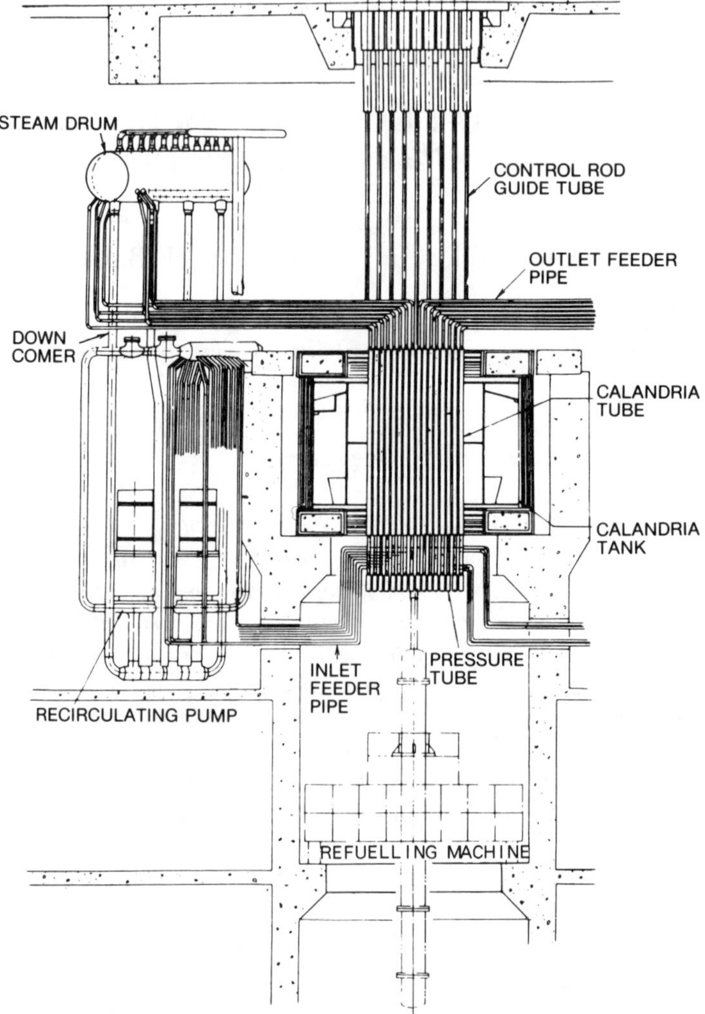

Fig. 1. Vertical section of reactor and primary cooling system of the Fugen.

be referred to as HT hereafter) Zr–2·5wt%Nb alloys. One fuel assembly, consisting of a bundle of 28 fuel rods, is held in each pressure tube. The pressure tube is a zirconium alloy tube of approximately 5 m length with an i.d. of 117·8 mm and a wall thickness of 4·3 mm, which is mechanically joined with stainless steel pressure tube extensions by rolling at the top and bottom, as shown in Fig. 2(a). There has been no trouble with pressure tubes of the Fugen since the start of operation.

As the pressure tubes are located at the reactor core with fuel assemblies, and constitute the pressure boundary of the high-temperature, high-pressure reactor cooling water, material that has a low absorption rate

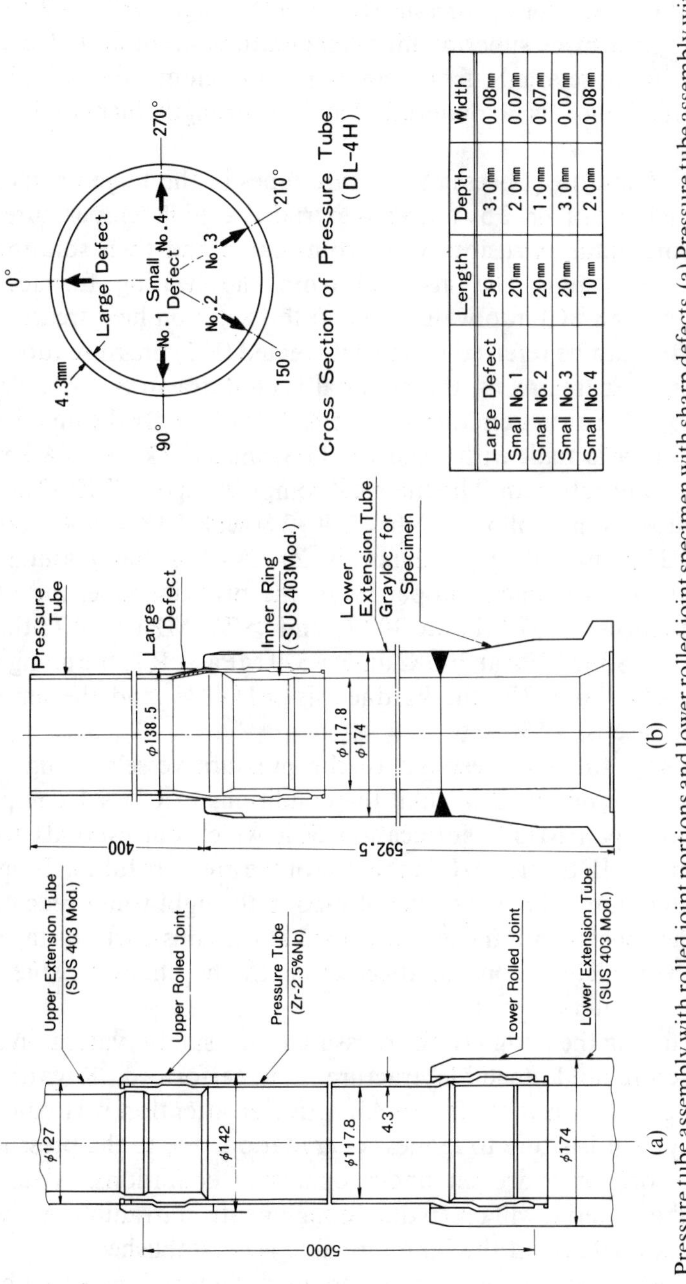

Fig. 2. Pressure tube assembly with rolled joint portions and lower rolled joint specimen with sharp defects. (a) Pressure tube assembly with rolled joint; (b) lower rolled joint specimen with sharp defects.

of thermal neutrons and possesses high strength at high temperature under high irradiation is required for the pressure tubes. A zirconium alloy satisfies these conditions, and for the pressure tubes of the Fugen, HT Zr–2·5wt%Nb alloy, which possesses superior high-temperature strength and corrosion resistance, was selected from among zirconium alloys. The HT Zr–2·5wt%Nb alloy is a material that has strength increased by heat treatment.

The manufacturing process of pressure tubes of the Fugen reactor is as follows. The billet of Zr–2·5wt%Nb is extruded at high temperature by the use of a horizontal extrusion press, cold-drawn and then solution-heat-treated by quenching the material from the heating furnace (at a temperature of 887°C) into water. Due to this solution heat treatment, the pressure tubes can be referred to as heat-treated (HT) pressure tubes. After the solution heat treatment, the material is cold-drawn 5–15%. Then the material is aged in a vacuum furnace at 500°C for 24 h. By the final forming, the pressure tube is made with an i.d. of 117·8 mm, a thickness of 4·3 mm and a length of about 5500 mm. The chemical composition of HT Zr–2·5wt%Nb pressure tube is as follows: Nb = 2·40–2·80wt%, O_2 = 900–1300 ppm, H_2 = max 25 ppm and the balance is Zr. As for the manufacturing specification of mechanical properties of the pressure tube, the ultimate tensile strength is $\geq$537 MPa at 300°C and $\geq$763 MPa at RT; the 0·2% proof stress is $\geq$392 MPa at 300°C and $\geq$529 MPa at RT; the elongation is $\geq$10% at 300°C and RT; the hardness is $\geq$HV235, and the amount of primary α-phase is $>$5%.

The pressure tubes correspond to the pressure vessels of light water reactors, and are designed to fulfil the conditions of Class 1 components regulated in Japan MITI Notification 501, which corresponds to those regulated in ASME Section III. In the case of the pressure tubes, the material used is a zirconium alloy which is not used in the light water reactors, and therefore design evaluation was performed at the construction stage of the Fugen with particular consideration given to the characteristics of the pressure tube material.

In undertaking the design of the pressure tubes, safety evaluation against ductile fracture and unstable fracture was performed. Regarding the evaluation against unstable fracture, particular attention was paid to the characteristics of fracture toughness degradation, due to the pressure tube material absorbing hydrogen under operating conditions. Also, as the pressure tube material will creep due to high neutron irradiation; the creep strain was evaluated and the limiting value was established.[2]

The pressure tubes for the Fugen were manufactured on a trial basis by Chase Brass Co. (Waterbury, CT, USA) who possessed manufacturing technology of zirconium alloy pressure tubes at the time of construction of the Fugen, and the manufacturing specifications of the pressure tubes were concluded using the specifications of Atomic Energy of Canada Limited, as reference. In advance of the manufacture of pressure tubes for use in the actual reactor, approximately 120 pressure tubes were manufactured on a trial basis to investigate tensile strength, fracture toughness and metallographic structure, and the manufacturing specifications were established as well as the characteristics of the material being confirmed.

More recently, the manufacturing technology of pressure tubes has been developed in Japan and domestically developed pressure tubes have been manufactured. To reduce the residual stress of the pressure tubes near the rolled joint, the structure of the rolled joint and the manufacturing conditions of the rolled joint for the Demonstration Plant have been slightly changed from those of the Fugen.

It is very important from the standpoint of the reactor design to show that the pressure tube with a rolled joint has naturally Leak-Before-Break (LBB) characteristics under the reactor operating conditions (temperature $\simeq 280°C$, inner pressure $\simeq 7$ MPa), because the pressure tubes are designed to be Class 1 vessels, like the pressure vessels of light water reactors.

Therefore, in the present study LBB experiments were conducted using mock-up rolled joint specimens under high-temperature, high-pressure water, and critical crack length (CCL) of the pressure tube with rolled joint was also estimated.

Pressure tubes might have very small initial cracks before use (postulated defect) due to the fabrication process. The maximum allowable size of the initial crack for the Fugen pressure tube is 3·3 mm in length and 0·15 mm in depth which was determined by the flaw inspection precision at the fabrication stage. The minimum detectable size of the crack by in-service-inspection (ISI) for the Fugen pressure tubes was determined at the construction stage of the Fugen to be 5·0 mm in length and 0·4 mm in depth. According to ASME Code Section III the conservative evaluation on the initial crack (5·0 mm in length, 0·4 mm in depth) propagation by fatigue cycles due to start-up and shutdown of the reactor was performed at the construction stage of the Fugen, and the propagated crack size after 30 years operation was estimated to be 6·3 mm in length and 1·05 mm in depth.

The crack initiation or propagation in the pressure tube might be induced by delayed hydride cracking (DHC), residual stress near the rolled joint portion of the pressure tube as well as fatigue and thermal cycles.

The fracture toughness of the pressure tube is degraded when it absorbs hydrogen under operating conditions. The design value of hydrogen concentration absorbed in the pressure tube after 30 years reactor operation is about 200 ppm for the Fugen, which was conservatively determined by reference to the Gentilly-1 reactor design. Actually, HT Zr–2·5wt%Nb pressure tubing is estimated to absorb less than 100 ppm hydrogen after 30 years operation by the use of Fugen surveillance test (PIE) data.[3]

For the straight portion of the pressure tube, fracture toughness values of both as-received tube (<25 ppm hydrogen) and after hydrogen was absorbed (about 200 ppm hydrogen) were obtained at about 280°C by internal pressuring tests and bend tests.[4] The value obtained is ≥ 80 MPa $\sqrt{m}$. The critical crack length (CCL) of the straight portion pressure tube is estimated to be ≥ 70 mm at an operating hoop stress of 100 MPa. Here, the stress intensity factor K_I is calculated by the use of Newman's equation[5] for a part-through-wall flow in cylindrical geometry as

$$K_I = F\sigma\sqrt{\pi C} \tag{1}$$

where

$$\sigma = \frac{PR}{t} \tag{2}$$

$$F = \sqrt{1 + 0{\cdot}52\lambda_t + 1{\cdot}29\lambda_t^2 - 0{\cdot}074\lambda_t^3} \tag{3}$$

$$\lambda_t = C/\sqrt{Rt} \tag{4}$$

$$0 \leq \lambda_t \leq 10 \tag{5}$$

where σ is the nominal stress, C is the half crack length, P is the internal pressure, t is the wall thickness and R is the internal radius of cylindrical pressure vessel. F is the boundary-correction factor expressed in eqn (3), which accounts for the effects of shell curvature[6,7] on stress intensity.

The shape of the rolled joint is a sandwich-type, as shown in Fig. 2(a). The pressure tube near the sandwiched rolled joint has a residual stress (circumferential stress: 50–200 MPa for the Demonstration Plant) due to rolled joint fabrication, which is more severe for crack initiation or propagation than that of the straight pressure tube portion. As the shape of the rolled joint is complicated and the residual stress distribution is complex, the fracture toughness value of the pressure tube near the rolled joint cannot be estimated with high precision. Therefore, the mock-up rolled joint specimens with axial sharp artificial cracks in the region of high residual stress were made as shown in Fig. 2(b), and LBB experiments were performed under high-temperature, high-pressure circulated water. Some sharp cracks were propagated to the penetration of the wall by pressure cycles (fatigue LBB experiments); other sharp cracks were propagated to the penetration of the wall by constant pressure (DHC LBB experiments). From these experiments LBB phenomena for the rolled joint portion can be observed for fatigue crack and DHC crack under mock-up conditions, and the CCL also can be obtained.

For the leak detection system of the Fugen, dried CO_2 gas flows in the annulus between pressure tube and calandria tube and the dew point of CO_2 gas is measured with a high sensitivity of 10^{-3} atm cc/s. In the LBB experiment, the leak detection was performed by the same CO_2 gas system. A TV observation system was also used.

2 EXPERIMENTAL

For the LBB experiments, specimens simulating the actual rolled joint portions were manufactured as shown in Fig. 2(b); a lower rolled joint specimen is an example.

Both ends of the specimen must be sealed for the experiments under high-temperature, high-pressure water. One end was sealed by the use of a Grayloc flange and the other was sealed by the use of copper seal packing. In the manufacturing process of the rolled joint, residual stress is produced in the pressure tube near the sandwiched rolled joint portion. In the residual stress region, a large, sharp artificial defect was made in the inner surface of the specimen by spark machining. The length of the large defect was 50 or 70 mm, the depth of it was 3 mm and the width (sharpness) of it was about 0·08 mm. Also, some small defects were made; a typical example of which is shown in Fig. 2(b). The length of the large defect (50 mm or 70 mm) was chosen with reference to the CCL of the straight pressure tube, as described in Section 1. It is estimated that the CCL of the pressure tube near the sandwiched rolled joint portion is slightly smaller than that of the straight portion pressure tube.

Rolled joint specimens consist of Fugen type lower rolled joint specimens (FL), Demonstration Plant type lower joint specimen (DL) and Demonstration Plant type upper rolled joint specimen (DU).

Table 1 lists the specimen number, type of rolled joint, the size of the large defect and hydrogen concentration. In the present study, LBB experiments were performed on nine rolled joint specimens. Six specimens, indicated by H in the specimen number, are high hydrogen concentration specimens (approximately 200 ppm, Fugen design value) in which hydrogen was intentionally absorbed before the experiments. Hydrogen was absorbed into the specimen by holding it in hydrogen gas at 350°C for about 15 h.

Figure 3 shows a schematic diagram of a high-temperature, high-pressure LBB experimental loop in which the rolled joint specimen is placed. Water chemistry, such as pH, conductivity (CON) and dissolved oxygen content (DO), is automatically controlled in this loop so that it is the same as that of

TABLE 1
Experimental Results for LBB, Hydrogen Concentration and Residual Stress

Specimen no.	*Large artificial defect (length × depth)*	*LBB or unstable fracture*	*Internal pressure at leaking (MPa)*	*Hydrogen concentration at large crack (ppm)*	*Maximum circumferential residual stress (MPa)*
FL-1	50 mm × 3 mm	LBB (Fatigue)	11·8	14 ~ 15	69
FL-2H	50 mm × 3 mm	Unstable fracture	10·3	59 ~ 233	380
DL-1	70 mm × 3 mm	LBB (Fatigue)	11·9	10 ~ 21	88
DL-2H	70 mm × 3 mm	LBB (DHC)	7·4	133 ~ 203	36
DL-3H	50 mm × 3 mm	LBB	8·8	43 ~ 254	74
DL-4H	50 mm × 3 mm	LBB	8·1	57–133[a]	70
DU-1H	50 mm × 3 mm	LBB (DHC)	7·8	262 ~ 269	88
DU-2	70 mm × 3 mm	LBB (Fatigue)	10·0	37 ~ 38	105
DU-3H	50 mm × 3 mm	LBB	8·3	211 ~ 254	108

[a] Maximum analysed hydrogen concentration of the pressure tube DL-4H is 188 ppm.

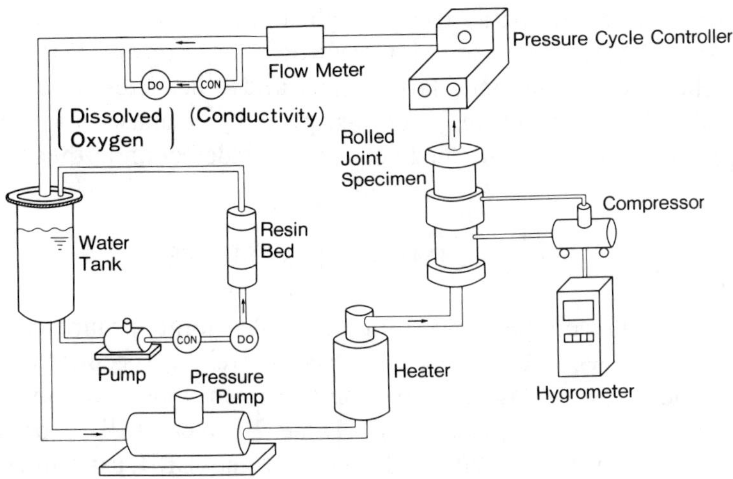

Fig. 3. Schematic diagram of high-temperature, high-pressure loop.

ATR (pH $\simeq$ 7, CON $\simeq$ 0·3 μS/cm, DO $\simeq$ 300 ppb). The flow rate in the loop is approximately 40 kg/h. Before the experiments, several strain gauges were pasted on the outer surface of the specimen near the crack, to measure the strain increase due to crack propagation during experiments. Also, acoustic emission probes were fixed on the outer surface of the specimen near the crack, to measure the acoustic emission events due to crack propagation during the experiments. To observe and examine the LBB phenomenon in which steam spouts from penetrated crack a TV camera was used, and the measurement of dew point of CO_2 gas which is circulated past the outer surface of the specimen was performed using a hygrometer as shown in Fig. 3.

The temperature and the pressure of the water in the loop were raised up to approximately 280°C and 7 MPa, respectively, which are the Fugen operating conditions (maximum design pressure: 8 MPa). For some of the specimens, pressure cycles of the water were applied automatically by the pressure cycle controller to propagate the crack in the specimens (Fatigue LBB experiments). For other specimens with absorbed hydrogen, the pressure was held constant (7 MPa) to propagate the crack of the specimens by the DHC mechanism (DHC LBB experiments). After the steam leakage was detected and observed from the penetrated wall, the leak rate was measured for some specimens, and for many specimens, the internal pressure was held constant (approximately 7 MPa) for some hours to maintain steam leakage.

After the LBB experiments, residual stress measurements of the pressure tube near the rolled joint were performed by cutting the specimen, on which many strain gauges were pasted, into small pieces. Fracture surface observations were conducted using a light microscope, scanning electron microscope (SEM) and transmission electron microscope (TEM). Metallographic photographs near the large defect and other sites were taken to examine the microstructure, such as primary α-phase and hydride. Hydrogen analysis was performed near the large defect and other portions.

3 EXPERIMENTAL RESULTS

LBB experiments were performed on nine rolled joint specimens listed in Table 1. LBB phenomena were observed for eight specimens, and an unstable fracture was observed for one specimen which was pressurised above reactor operating pressure. In Table 2 typical LBB experimental results are shown for Specimen DL-1 (Fatigue LBB experiment) and Specimen DU-1H (DHC LBB experiment). For Specimen DL-1, internal pressure cycles were applied to propagate the sharp artificial defect, from 7·4 MPa ↔ 7·8 MPa to 7·4 MPa ↔ 11·8 MPa at 280°C (as shown in Table 2).

TABLE 2

Typical LBB Experimental Results for Cycles and Holding Time (Test Temperature: 280°C, Water Chemistry: 300 ppb DO, 0·3 μS/cm)

Specimen no.	*Operational history of the specimen*	*Total cycles up to LBB*	*Total time at 280°C up to LBB*
DL-1	LBB 11.8 MPa 11.8 MPa 11.3 10.8 10.3 9.3 9.8 8.8 7.4 MPa 7.8 8.3 7.4 MPa 7.4 MPa 7.4 MPa 0 MPa 719 cyc-les 360 C 359 C 240 cycles 240 C 180 C 180 C 144 C 85 C 48 cyc-les 5 hrs 18 hrs 1 hrs 5 hrs 17 hrs 8 hrs 3 hrs	2 507 cycles	240 h
DU-1H	LBB 7.3MPa 6.4 MPa 5.9 MPa 7.3 MPa 5.7 MPa 6.5MPa 4.5 MPa 0 MPa 5 hrs 160 hrs 24 hrs 5 hrs 2 hrs 6 hrs 10 hrs 138 hrs →→ : pressure is changed spontaneously	0 cycles	160 h

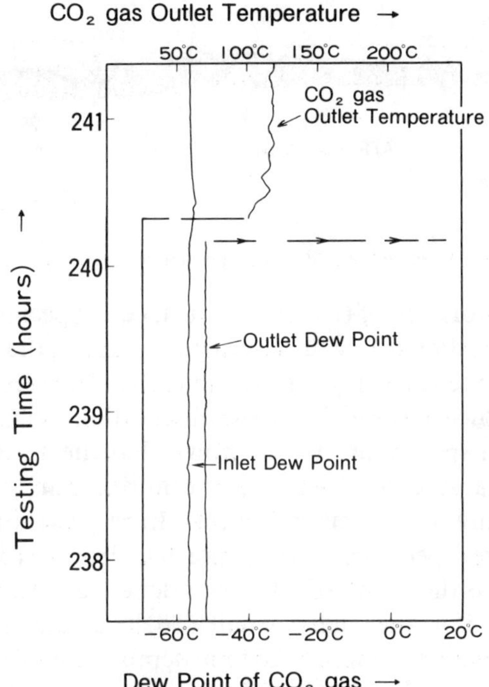

Fig. 4. Dew point change at leaking (LBB) for DL-1 specimen.

At the internal pressure of 11·8 MPa, steam leakage from the penetrated wall was detected and observed. For Specimen DU-1H, the internal pressure was maintained constant (7·3 MPa) at 280°C to propagate the defect by the DHC mechanism, and, after 160 h at 280°C, leaking was detected and observed.

Figure 4 illustrates the dew point change of CO_2 gas, which is circulated past the outer surface of the specimen, as a function of time. When the crack propagated and penetrated the tube wall by pressure cycles or by the constant pressure, water spouted as steam from the penetrated crack into the CO_2 gas flow, so that the dew point of CO_2 gas increased rapidly because it contained moisture, which is shown in Fig. 4 for Specimen DL-1 as an example. The leak detection capability of CO_2 system is sensitive, 1×10^{-3} atm cc/s of leak water, which is the same as that of the Fugen. Steam leak was also observed by the TV camera. Strain measurements and acoustic emission measurements were also conducted on the outer surface of the specimen near the crack during experiments. Circumferential strain at the center of the crack was observed to increase as the crack propagated because the crack opens circumferentially; however, axial strain on the center of the crack did not change as the crack propagated. The number of AE events above the background was counted as a function of time, and AE events increased when the crack penetrated the tube wall.

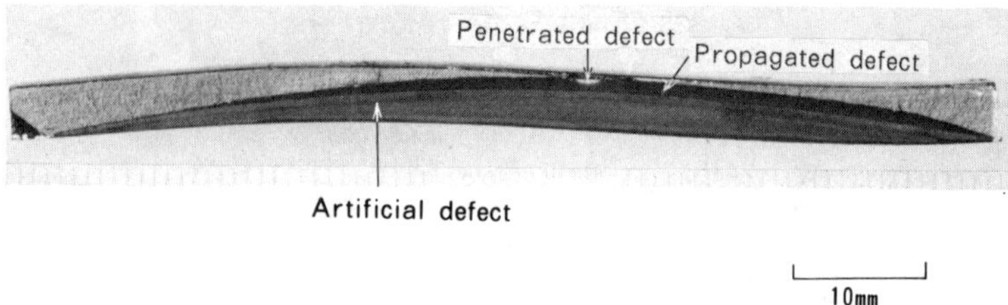

Fig. 5. Fracture surface of large defect for DL-1 specimen.

In Fig. 5, a photograph of the fracture surface of Specimen DL-1 after the LBB experiment is shown, in which the artificial crack, the propagated crack by fatigue and penetrated region are indicated. Fracture surface by SEM observation is shown in Fig. 6. From the observations on the fracture surface and shape for all specimens, it was shown that the defect length did not increase much; large defects were at the most 2 mm, and there was no propagation in any of the smaller defects. In respect of the sharp smaller defects for the three specimens without much hydrogen (FL-1, DL-1, DU-2), the depth as well as the length of all defects (length 5–10 mm, depth 1–2 mm) was unchanged. For the sharp smaller defects for hydrogen-enriched specimens, some defects (length 5–20 mm, depth 1–2 mm) did not propagate at all and other defects (length 10–20 mm, depth 0·5–3 mm) were observed to propagate through the wall; maximum depth increment 0·9 mm).

After the LBB experiments, residual stress in the pressure tube near the rolled joint was measured by the strain gauge method, which is shown in Fig. 7 for Specimen DL-1. Circumferential residual stress on the inner surface of the pressure tube was less than 88 MPa. In Table 1, the maximum circumferential residual stress is listed for all specimens. Also listed is the hydrogen concentration at the large crack for all specimens. The range of the hydrogen concentration for hydrogen-enriched specimens is large, which is considered to be due to the hydrogen absorption process. Metallographic

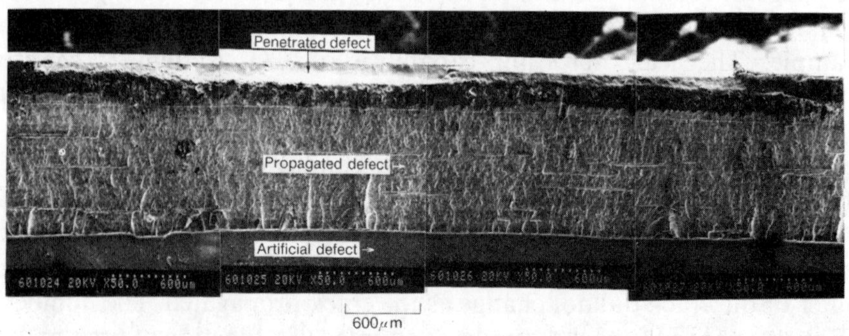

Fig. 6. SEM topography of large defect for DL-1 specimen.

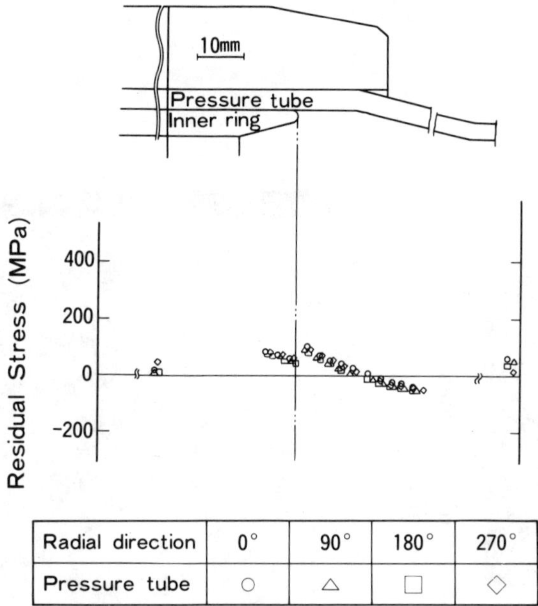

Fig. 7. Residual stress on the inner surface of the pressure tube near the rolled joint for DL-1 specimen.

photographs of the pressure tubes are shown in Fig. 8, in which the microstructure of the martensite phase (primary α-phase is indicated as white circles and hydride is indicated as black lines) are shown. For the hydrogen-enriched material, like Specimen DL-4H, many hydrides were observed; some of which were precipitated in the radial direction. In the remaining specimens, some hydrides were precipitated in the radial direction near the defect and inner wall portion of the tube.

LBB phenomena were observed on eight rolled joint specimens in the present experiments. Specimen FL-2H was broken by unstable fracture, in which the through-wall crack propagated rapidly in the axial direction and LBB phenomenon was not established because of the high hydrogen concentration in the material and the high applied internal pressure of 10·3 MPa. Specimen FL-2H was broken at an internal pressure of 10·3 MPa, which is higher than the reactor operating pressure (normal pressure 7·3 MPa, maximum design pressure 8·0 MPa). From the LBB data of Specimens DL-2H, DL-4H and DL-3H, the pressure tube of lower rolled joint with approximately 200 ppm hydrogen and 50 mm or 70 mm length crack is considered to have LBB phenomena under reactor operating conditions.

After the leak was detected and observed, all of the specimens with a 50 mm length defect could be held at an internal pressure above 7·3 MPa,

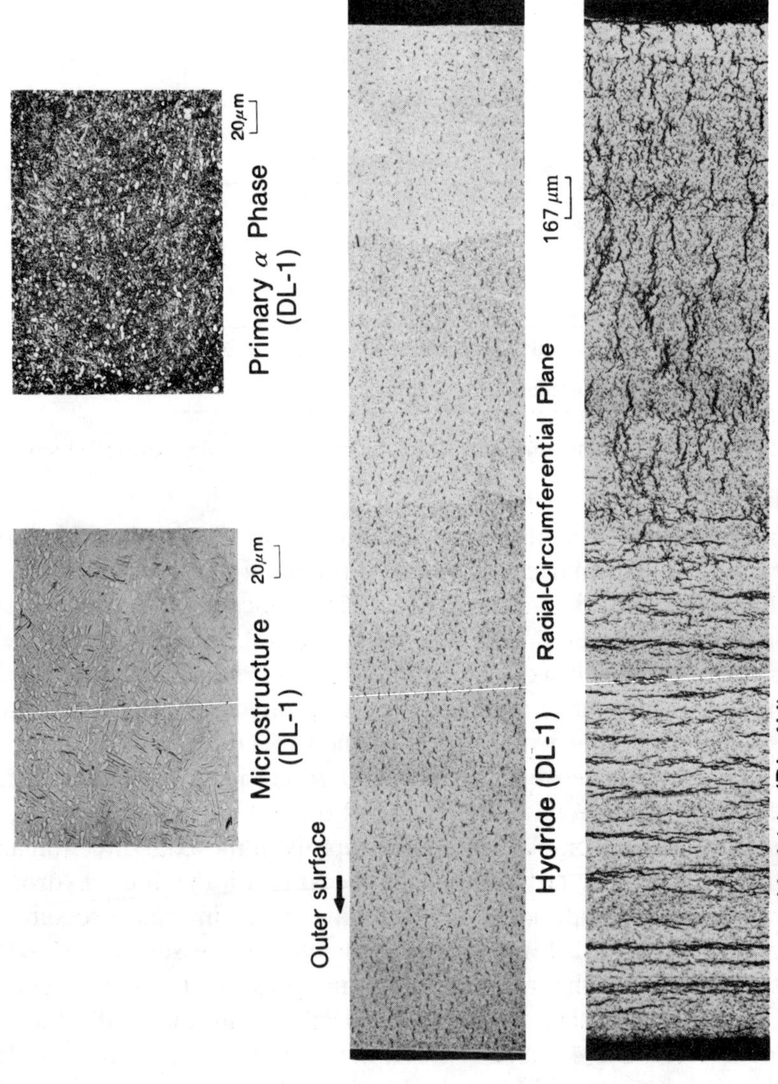

Fig. 8. Metallographic photographs of HT Zr–2·5wt%Nb pressure tube.

which is reactor operating pressure, and steam leakage was continued for 1–75 h. In Table 2 typical examples for Specimens DL-1 and DU-1H are shown. For Specimen DL-1, after the leakage the internal pressure above 7·3 MPa was held for 11 h. For Specimen DU-1H, after the leakage an internal pressure of 7·3 MPa was held for 6 h during which steam leakage continued.

4 DISCUSSION

The relationship between crack length and the pressure is shown in Fig. 9 when the crack penetrated the tube wall, and is also shown in the figure whether the LBB characteristic was valid or not (white circles indicate the LBB data and the solid circle indicates the unstable fracture). After the leakage, the inner pressure was maintained at 280°C and above 7 MPa (7·3 MPa is reactor operating pressure) for most of the specimens and leaking was continued. For Specimen DL-3H leaking (280°C, >7 MPa) continued for 75 h; for Specimen DL-4H for 8 h; for Specimen DU-1H (DHC) for 6 h and for Specimen DU-3H for 18 h. Also, the leaking was detected by CO_2 gas dew point measurement with high sensitivity of 10^{-3} atm cc/s.

As the shape of the rolled joint is complicated and the residual stress distribution is complex, the fracture toughness value of the pressure tube near the rolled joint cannot be estimated with high precision (calculations or

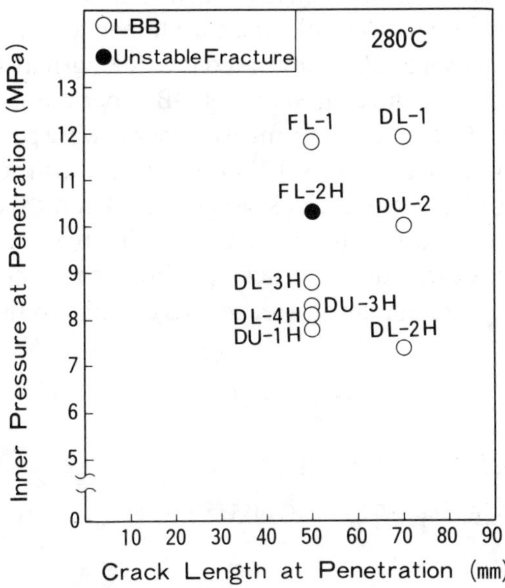

Fig. 9. Relation between crack length at penetration and pressure at penetration for nine specimens in the present study.

small size experiments are invalid for the rolled joint estimation). However, from the present LBB simulation experiment, the CCL can be obtained. The CCL is defined as the boundary crack length between LBB and unstable fracture under reactor operating conditions. For the pressure tube near the Demonstration Plant rolled joint with absorbed hydrogen up to approximately 200 ppm, CCL for fatigue and DHC was obtained to be more than 50 mm under reactor operating conditions (280°C, 7 MPa). On the other hand, the CCL for the straight portion pressure tube was already estimated to be more than 70 mm, as described previously. So that, CCL for the pressure tube near the rolled joint is considered to be comparable to that of the straight portion pressure tube.

The initial postulated crack of 5·0 mm in length and 0·4 mm in depth for the pressure tube is estimated to propagate by fatigue cycles to the crack of 6·3 mm in length and 1·05 mm in depth after 30 years reactor operation. The fatigue, propagated crack length of 6·3 mm after 30 years operation is much smaller than the CCL (> 50 mm) obtained here. With DHC, in general, the natural crack length, when it penetrates the pressure tube wall, is considered to be about four or five times the wall thickness (16–20 mm), which was shown by the fracture observations of Pickering DHC pressure tubes.[8] Besides, the minimum detectable length of the crack by ISI for ATR is less than 5 mm.

The leak detection sensitivity of dried CO_2 gas system (dew point measurement) is 10^{-3} atm cc/s, which is very high. In this experiment, leakage was detected by the CO_2 gas system about 30 s after the leakage, and generally for several hours the leak was maintained at the pressure of about 7·4 MPa described before. The penetrated crack length at the outer surface of the pressure tubes was measured after LBB experiments and obtained to be 5–50 mm in length. The leak rate measurements were performed for some specimens and leak rate of $> 30 \times 10^3$ atm cc/h (> 8 atm cc/s) was obtained for a penetrated outer surface crack length of 14 mm (Specimen DU-1H).

The leak rate estimation under actual ATR reactor conditions was conducted with the calculation of crack opening area A and critical flow rate Gc as follows. The crack opening area A was calculated using the equation by Tada[9] as

$$A = \frac{\sigma}{E}(2\pi Rt) \times G(\lambda) \tag{6}$$

$$\begin{aligned} G(\lambda) &= \lambda^2 + 0{\cdot}625\lambda^4 \qquad (0 \le \lambda \le 1) \\ &= 0{\cdot}14 + 0{\cdot}36\lambda^2 + 0{\cdot}72\lambda^3 + 0{\cdot}405\lambda^4 \qquad (1 \le \lambda \le 5) \end{aligned} \tag{7}$$

$$\lambda = C/\sqrt{Rt}$$

$$\sigma = PR/t$$

where E is Young's modulus. The critical flow rate Gc for two-phase flow was summarized by Akagawa[10] on experimental results and theories. As the critical pressure Pc is about 3·4 MPa in this case, the critical flow leak rate $Gc \times A$ was calculated for steam quality of 15% (ATR condition) as follows.

(i) $2C = 5$ mm: $Gc \times A = 1$ atm cc/s
(ii) $2C = 14$ mm: $Gc \times A = 8$ atm cc/s
(iii) $2C = 50$ mm: $Gc \times A = 240$ atm cc/s

The agreement between the calculated leak rate and the present experimental one for $2C = 14$ mm was good. The sensitivity for leak detection by dried CO_2 gas system is very high for the actual ATR reactor; therefore it is considered that there is no problem with the leak detection sensitivity.

5 CONCLUSIONS

According to the LBB experimental findings and analyses, the following conclusions can be drawn for the HT Zr–2·5wt%Nb pressure tube near the ATR Demonstration Plant type rolled joints with absorbed hydrogen approximately 200 ppm:

(1) The LBB phenomena for fatigue and DHC crack propagations were valid for the rolled jointed pressure tube under reactor operating conditions (temperature $\simeq 280°C$; inner pressure $\simeq 7$ MPa).
(2) The critical crack length for fatigue and DHC crack propagations was obtained as > 50 mm for the rolled jointed pressure tube under reactor operating conditions.

REFERENCES

1. Fugen HWR, *Nuclear Engineering International*, **24** (289) (1979) 33–49.
2. Koike, M. & Asada, T., Irradiation creep and growth of pressure tubes in HWR Fugen. *J. Nuclear Materials*, **159** (1988) 62–74.
3. Koike, M., Asada, T., Yuhara, S. & Kaneko, J., Mechanical properties change by irradiation for zirconium alloy pressure tube material. *Proceedings of Material Mechanics Lecture No. 870–12*, The Japan Society of Mechanical Engineers, Tokyo, 1987, pp. 32–3, in Japanese.
4. Honda, S., Fracture toughness of Zr–2·5wt%Nb pressure tubes. *Nuclear Engineering and Design*, **81** (1984) 159–67.
5. Newman, J. C., Fracture analysis of surface and through cracks in cylindrical pressure vessels. NASA Technical Note, NASA TN D-8325, December 1976.
6. Erdogan, F. & Kibler, J. J., Cylindrical and spherical shells with cracks. *Int. J. Fract. Mech.*, **5**(3) (1969) 229–37.

7. Folias, E. S., An axial crack in a pressurized cylindrical shell. *Int. J. Fract. Mech.*, **1**(2) (1965) 104–13.
8. Moan, G. D., Coleman, C. E., Price, E. G., Rodgers, D. K. & Sagat, S., Leak-before-break in the pressure tubes of CANDU reactors. *Int. J. Pres. Ves. & Piping*, **43** (1990) 1–21.
9. Tada, H., The effects of shell corrections on stress intensity factors and the crack opening area of a circumferential and longitudinal through-crack in a pipe. NUREG, CR-3464, 1983, pp. 71–81.
10. Akagawa, K., Two phase flow (in Japanese). Corona Co., Tokyo, 1976, p. 192.

Int. J. Pres. Ves. & Piping **43** (1990) 57–65

Development of USNRC Standard Review Plan 3.6.3 for Leak-Before-Break Applications to Nuclear Power Plants

K. Wichman & S. Lee

Office of Nuclear Reactor Regulation, US Nuclear Regulatory Commission, Washington, District of Columbia 20555, USA

ABSTRACT

In the United States, it is now permissible to eliminate the dynamic effects of postulated high energy pipe ruptures from the design basis of nuclear power plants using 'Leak-Before-Break' (LBB) technology. To provide review guidance for the implementation of LBB, a new Standard Review Plan (SRP) 3.6.3 was issued for public comment. Based upon public comments received and advances in fracture mechanics application, further development of SRP 3.6.3 is in progress.

SRP 3.6.3 will outline the review procedures and acceptance criteria for LBB licensing applications. A deterministic fracture mechanics evaluation accounting for material toughness will be required. Margins on load, crack size, and leakage will be specified and the load combination methods and leakage detection sensitivity will be described. Piping particularly susceptible to failure from potential degradation mechanisms will be excluded from the application of LBB. The design basis of containment, emergency core cooling systems, and environmental qualification of equipment in the context of LBB applicability will be clarified.

'LEAK-BEFORE-BREAK' LICENSING

General Design Criterion 4 (GDC-4) of Appendix A to Part 50 of Title 10 of the Code of Federal Regulations (10 CFR Part 50) of the United States regulations requires postulation of pipe breaks and provision of appropriate protection against associated dynamic effects. However, the US Nuclear Regulatory Commission (NRC) staff issued Generic Letter 84-04[1] accepting that the double-ended guillotine break (DEGB) of the pressurized water reactor (PWR) primary loop piping was unlikely to occur, provided it could

Int. J. Pres. Ves. & Piping 0308-0161/90/$03·50

be demonstrated by deterministic fracture mechanics analyses that postulated small through-wall flaws in plant-specific piping would be detected by the plant's leakage monitoring systems long before the flaws could grow to unstable sizes. Leakage exceeding the limit specified in plant Technical Specifications requires operator action or plant shutdown. The concept underlying such analyses is referred to as 'leak-before-break' (LBB). A detailed discussion of limitations and acceptance criteria for LBB used by the NRC staff is provided in NUREG-1061, Vol. 3.[2]

The NRC staff initiated a rule change to GDC-4[3,4] because of the extremely low probability of LBB pipe rupture and the potential for adverse safety implications resulting from the installation of protective devices. The placement of pipe whip restraints degrades plant safety when thermal growth is inadvertently restricted, reduces the accessibility for and effectiveness of in-service inspections, increases in-service inspection radiation dosages, and adversely affects construction and maintenance economics. On 11 April 1986, a final 'limited scope' rule[5] was published amending GDC-4 to permit the use of analyses to eliminate from the design basis the dynamic effects of postulated pipe ruptures of primary coolant loop piping in PWRs. On 27 October 1987, a final 'broad scope' rule[6] was published amending GDC-4 to permit the use of LBB analyses in all qualified high energy piping, i.e. pressure exceeding 1·9 MPa (275 psi) or temperature exceeding 93°C (200°F), in nuclear power units.

Currently, approximately two-thirds of the PWRs in the US have approval for the application of LBB in the primary coolant loop. There are also four PWRs which have LBB approval for their auxiliary lines, specifically, pressurizer surge, accumulator, and residual heat removal (RHR) lines. (One of the four plants also has approval for the safety injection lines and the reactor coolant loop bypass lines.) The approved auxiliary lines are all inside containment, fabricated from austenitic stainless steel, and at least 150 mm (6 in) in diameter. The application of LBB has not been approved as yet for any boiling water reactor (BWR).

STANDARD REVIEW PLAN 3.6.3

A new NRC Standard Review Plan (SRP) section, numbered 3.6.3[7] and entitled 'Leak-Before-Break Evaluation Procedures', providing review guidance for the implementation of the revised GDC-4 was published for public comment on 28 August 1987. The development of the final form of SRP 3.6.3 by the NRC staff is continuing. While this paper discusses the status of the development of the SRP 3.6.3, the final form of SRP 3.6.3 may differ from the discussion presented in this paper.

LBB LIMITATIONS

The application of LBB is limited to piping that is not likely to be susceptible to failure from various degradation mechanisms in service.[2] From the NRC staff's experience, a significant portion of any LBB review involves the evaluation of the susceptibility of the candidate piping to various degradation mechanisms.

The LBB approach cannot be applied to piping that can fail in service from such effects as water hammer, creep, erosion, corrosion, fatigue, and environmental conditions. The rationale is that these degradation mechanisms challenge the assumptions in the LBB acceptance criteria. For example, (1) water hammer may introduce excessive dynamic loads which are not accounted for in the LBB analyses, and (2) corrosion and fatigue may introduce flaws whose geometry may not be bounded by the postulated through-wall flaw in the LBB analyses. Adhering to the 'defense-in-depth' principle, piping susceptible to failure from these potential degradation mechanisms is excluded from LBB applications.

To demonstrate that the candidate piping is not susceptible to failure from these degradation mechanisms, the operating history and measures to prevent or mitigate these mechanisms must be reviewed.

One example is the application of LBB to the pressurizer surge line in PWRs. The pressurizer surge line was recently observed to have thermal stratification as discussed in NRC Bulletin 88-11.[8] Because thermal stratification loads were not considered in the original design of the pressurizer surge line and preliminary analyses indicated that the resulting thermal fatigue might be a concern, the utilities were requested to re-evaluate the integrity of the pressurizer surge line with thermal stratification according to Section III of the American Society of Mechanical Engineers (ASME) Boiler and Pressure Vessel Code requirements. Although some pressurizer surge line supports had to be modified, a number of utilities have completed the re-evaluation and found their pressurizer surge lines to have acceptable integrity, i.e. fatigue failure considering thermal stratification of the Code-acceptable pressurizer surge lines was unlikely. Thus, the application of LBB was permitted for these Code-acceptable pressurizer surge lines. A unique feature of the pressurizer surge line is that the stresses resulting from thermal stratification (gross bending) increase as shutdown occurs. Since these stresses are large relative to pertinent service limits, they cannot be ignored in LBB analyses. To ensure that the plant could be safely shutdown to allow for the repair of the pressurizer surge line, should a flaw occur in that line, a stability analysis was required for the forced shutdown situation with a postulated flaw in the pressurizer surge line.

Further, the NRC staff issued Bulletin 88-08[9] and Supplement 3[10] to it relating to the potential of thermal fatigue due to valve seat leakage and periodic valve packing leakage, respectively. Utilities would have to demonstrate that these particular thermal stratification mechanisms are unlikely to occur in their piping before LBB could be applied.

Piping in BWRs has a history of intergranular stress corrosion cracking (IGSCC). Although the broad scope rule[6] permits the application of LBB in BWRs provided the piping is treated by two IGSCC mitigating measures, the NRC staff is continuing to evaluate the effectiveness of various IGSCC mitigating measures as additional data become available. Mitigating measures under consideration are piping replacement with IGSCC resistant material, remedial residual stress improvement, and hydrogen water chemistry. Because of the IGSCC issue, LBB has not yet been applied to BWR piping.

LBB ACCEPTANCE CRITERIA

After the LBB candidate piping has been reviewed for degradation mechanisms as discussed previously and found to be acceptable, the piping is subjected to a rigorous fracture mechanics evaluation. The purpose of this evaluation is to show that there is flaw stability and the resulting leakage will be detected in the unlikely event that a flaw should develop. The current acceptance criteria established by the NRC staff are expected to remain the same. These are as follows:[2]

(1) The loading conditions should include the static forces and moments (pressure, deadweight, and thermal expansion) due to normal operation, and the forces and moments associated with the safe shutdown earthquake (SSE). These forces and moments should be located where the highest stresses, coincident with the poorest material properties, are induced for base materials, weldments, and safe ends. (A safe end provides a transition between dissimilar metals for welding.)

(2) A through-wall flaw should be postulated at the locations determined from criterion (1) above. The size of the flaw should be large enough so that the leakage from this postulated flaw is assured of detection. When the pipe is subjected to normal operational loads, it should be demonstrated that there is a factor of at least 10 between the leakage from the leakage-size flaw and the plant's installed leak detection capability. A leakage analysis which has been bench-marked against experimental or plant data is required. The margin on leakage is required to account for uncertainties such as the crack

opening area, crack surface roughness, two-phase flow, and leak detection capability.

(3) It should be demonstrated that the postulated leakage-size flaw is stable under normal plus SSE loads. It should be demonstrated that there is a factor of at least 1·4 between the loads that will cause flaw instability and the normal plus SSE loads. The factor of 1·4 stems from the factor of the square root of two on stress intensity for flaw evaluation under Service Level C and D Loadings in IWB-3610 of Section XI of the ASME Code. Service Level Loadings are defined in NCA-2140 of Section III of the ASME Code. However, the broad scope rule permits a reduction of the factor of 1·4 to 1·0 if the individual normal and seismic (pressure, deadweight, thermal expansion, SSE, and seismic anchor motion) loads are summed absolutely. This is because the NRC staff considers an absolute load summation sufficiently conservative to warrant a margin reduction.

(4) The flaw size margin should be determined by comparing the leakage-size flaw with the critical-size flaw. Under normal plus SSE loads, it should be demonstrated that there is a factor of at least two between the critical-size flaw and the leakage-size flaw. The factor of two stems from the factor of two on flaw size for flaw evaluation under Service Level C and D Loadings in IWB-3610 of Section XI of the ASME Code.

FLAW STABILITY ANALYSIS

A deterministic fracture mechanics evaluation accounting for material toughness is required. Utilities may propose any fracture mechanics evaluation method for NRC staff review. However, the utilities will have to demonstrate the accuracy of the method by comparing with other acceptable methods or with experimental data. Both circumferential and axial flaws are postulated; however, experience shows that the circumferential flaw is generally limiting because of the higher stress on the circumferential flaw as compared with that on the axial flaw.

In the proposed SRP 3.6.3,[7] the NRC staff included an acceptable fracture mechanics evaluation method for austenitic stainless steel piping, having a circumferential through-wall flaw, which was based on IWB-3640 of Section XI of the ASME Code. The method applied the results from an elastic–plastic fracture mechanics (EPFM) tearing modulus analysis to a limit load analysis. This was accomplished by increasing the applied loads by a 'Z factor'[11] for the limit load analysis. Thus, the load carrying capacity of a flawed pipe based on a limit load analysis is reduced by the 'Z factor' to account for material toughness. This method was evaluated by the ASME Code committees and approved for inclusion in Section XI of the ASME Code. The difference between IWB-3640 in Section XI of the ASME Code and the 'Z factor' approach for LBB analysis is that the former deals with surface flaws and the latter deals with through-wall flaws. However, it should be noted that 'Z factors' were developed from EPFM analyses for through-wall flaws. Thus, no inconsistency in methodology exists. Because cast stainless steel is subject to thermal aging, whereby the fracture toughness of the material degrades with time in service depending on material chemistry, cast stainless steel materials will be evaluated on a case-by-case basis.

The 'Z factor' method was extended to ferritic steel piping in ASME Code Case N-463[12] which was proposed to be incorporated into IWB-3650 of Section XI of the ASME Code. The NRC staff is developing this method for LBB applications for ferritic steel piping. Figure 1 shows draft curves being developed for determining the stable circumferential through-wall flaw length for axial stress P_a and bending stress P_b. The 'Z factor' has been generalized as Z' which applies to both austenitic and ferritic materials. The constant C is the margin associated with the load combination method selected.

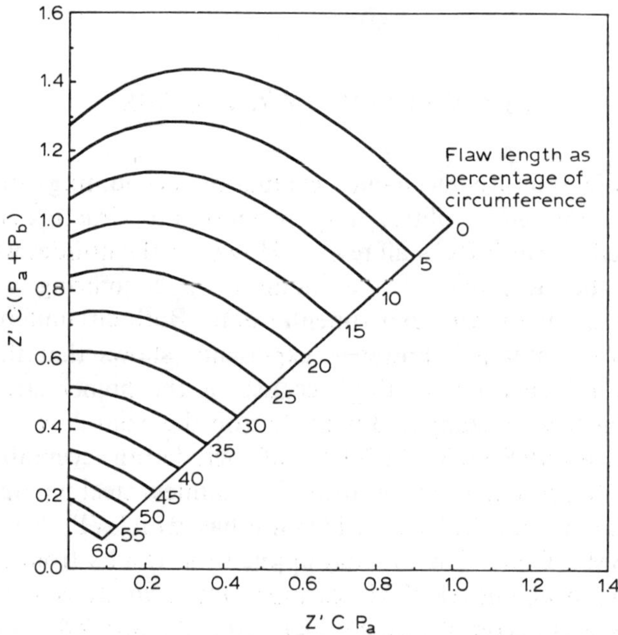

Fig. 1. Stable circumferential through-wall flaw size in austenitic and ferritic piping.

DESIGN BASIS WITH LBB APPLICATION

The broad scope rule[6] introduced an acknowledged inconsistency into the design basis by excluding the dynamic effects of postulated pipe ruptures while retaining non-mechanistic pipe rupture for containments, energency core cooling systems (ECCSs), and environmental qualification (EQ) of safety related electrical and mechanical equipment. The NRC staff subsequently clarified its intended treatment of the containment, ECCS, and EQ in the context of LBB application in a request for public comments on this issue that was published on 6 April 1988.[13] This section discusses in more detail the NRC staff's intent.

Effects resulting from postulated pipe ruptures can be generally divided into local dynamic effects and global effects. Local dynamic effects of a pipe rupture are uniquely associated with that particular pipe rupture. These specific effects are not caused by any other source or even a postulated pipe rupture at a different location. Examples of local dynamic effects are pipe whip, jet impingement, missiles, local pressurizations, pipe break reaction forces, and decompression waves in the intact portions of that piping or communicating piping. Global effects of a pipe rupture need not be associated with a particular pipe rupture. Similar effects can be caused by failures from sources such as pump seals, leaking valve packings, flanged connections, bellows, manways, rupture disks, and pipe ruptures of other piping. Examples of global effects are gross pressurizations, temperatures, humidity, flooding, loss of fluid inventory, radiation, and chemical condition.

The application of LBB technology eliminates the local dynamic effects of postulated pipe ruptures from the design basis because a LBB pipe does not rupture. However, global effects may still result from a source other than the postulated pipe rupture. Since the global effects from the postulated pipe rupture provide a convenient and conservative design umbrella, and the NRC staff is not prepared at this time to propose alternative criteria, the functional and performance requirements of containment, ECCS, and EQ are not affected by LBB applications. Thus, after LBB is demonstrated, (1) the containment must continue to be designed to withstand all global effects (such as pressure, temperature, and flooding) up to and including the rupture of the largest pipe in the reactor coolant system (RCS); (2) the heat removal and mass replacement capacity of the ECCS must continue to be designed to accommodate pipe ruptures up to and including the rupture of the largest pipe in the RCS; and (3) the EQ of equipment must continue to be based on all global effects (such as pressure, temperature, humidity, flooding, radiation, and chemical condition) resulting from pipe ruptures.

Although the broad scope rule[6] permits the utility to request an exemption from the EQ requirements based on LBB, the NRC staff is not allocating resources to such EQ exemption review until the safety benefits associated with such an action can be demonstrated. However, as published in a policy statement dated 2 May 1989,[14] the industry is still encouraged to develop justification which would allow serious consideration of extension of the scope of LBB application and to develop an appropriate substitute or replacement for the DEGB used in ECCS and EQ evaluations.

CONCLUSIONS

In the United States, the elimination of dynamic effects of postulated high energy pipe ruptures from the design basis of nuclear power plants using LBB fracture mechanics technology is permitted by the revised General Design Criterion 4 of Appendix A to Part 50 of Title 10 of the Code of Federal Regulations. A new Standard Review Plan section, numbered 3.6.3, providing review guidance for the implementation of the revised General Design Criterion 4 was published for public comment and will be issued in final form. The development of Standard Review Plan 3.6.3 is still continuing as described herein.

REFERENCES

1. Generic Letter 84-04. Safety evaluation of Westinghouse topical reports dealing with elimination of postulated pipe breaks in PWR primary main loops. US Nuclear Regulatory Commission, Washington, DC, 1 February 1984.
2. NUREG-1061, Vol. 3. Evaluation of potential for pipe breaks, Report of the US Nuclear Regulatory Commission Piping Review Committee. US Nuclear Regulatory Commission, Washington, DC, November 1984.
3. Modification of General Design Criterion 4 Requirements for Protection Against Dynamic Effects of Postulated Pipe Ruptures. *Federal Register*, Vol. 50, No. 126, published by the Office of the Federal Register, Washington, DC, 1 July 1985, pp. 27006–9.
4. Modification of General Design Criterion 4 Requirements for Protection Against Dynamic Effects of Postulated Pipe Ruptures. *Federal Register*, Vol. 51, No. 141, published by the Office of the Federal Register, Washington, DC, 23 July 1986, pp. 26393–9.
5. Modification of General Design Criterion 4 Requirements for Protection Against Dynamic Effects of Postulated Pipe Ruptures. *Federal Register*, Vol. 51, No. 70, published by the Office of the Federal Register, Washington, DC, 11 April 1986, pp. 12502–5.
6. Modification of General Design Criterion 4 Requirements for Protection Against Dynamic Effects of Postulated Pipe Ruptures. *Federal Register*, Vol. 52, No. 207, published by the Office of the Federal Register, Washington, DC, 27 October 1987, pp. 41288–95.

7. Standard Review Plan; Public Comment Solicited. *Federal Register*, Vol. 52, No. 167, published by the Office of the Federal Register, Washington, DC, 28 August 1987, pp. 32626–33.
8. NRC Bulletin 88-11. Pressurizer surge line thermal stratification. US Nuclear Regulatory Commission, Washington, DC, 20 December 1988.
9. NRC Bulletin 88-08. Thermal stresses in piping connected to reactor coolant systems. US Nuclear Regulatory Commission, Washington, DC, 22 June 1988.
10. NRC Bulletin 88-08, Supplement 3. Thermal stresses in piping connected to reactor coolant systems. US Nuclear Regulatory Commission, Washington, DC, 11 April 1989.
11. Case N-436-1. Alternative methods for evaluation of flaws in austenitic piping. *Cases of ASME Boiler and Pressure Vessel Code*. American Society of Mechanical Engineers, New York, 7 December 1987.
12. Case N-463. Evaluation procedures and acceptance criteria for flaws in Class 1 ferritic piping that exceed the acceptance standards of IWB-3514.2. *Cases of ASME Boiler and Pressure Vessel Code*. American Society of Mechanical Engineers, New York, 30 November 1988.
13. Leak-Before-Break Technology: Solicitation of Public Comment on Additional Applications. *Federal Register*, Vol. 53, No. 66, published by the Office of the Federal Register, Washington, DC, 6 April 1988, pp. 11311–2.
14. Policy Statement on Additional Applications of Leak-Before-Break Technology. *Federal Register*, Vol. 54, No. 83, published by the Office of the Federal Register, Washington, DC, 2 May 1989, pp. 18649–51.

Int. J. Pres. Ves. & Piping **43** (1990) 67–83

Leak-Before-Break Application in US Light Water Reactor Balance-of-Plant Piping

B. F. Beaudoin, D. F. Quiñones & T. C. Hardin

Robert L. Cloud and Associates, Inc., 125 University Avenue, Berkeley, California 94710, USA

ABSTRACT

This paper describes the criteria and methodology for a leak-before-break (LBB) program for high energy balance-of-plant (BOP) nuclear piping in the United States. LBB, the analytical demonstration that high toughness piping will leak detectably before catastrophic failure, can be applied to any operational or pre-operational light water reactor plant to minimize pipe rupture hardware and to discount pipe rupture dynamic effects.

The general methodology described herein, encompasses applicable US NRC regulatory requirements and incorporates experience gained in the licensing process of actual LBB programs. First, candidate piping systems must be carefully screened to verify that they are not subject to failure by phenomena that would adversely affect the accurate evaluation of flaws. Next, pipe stresses, material properties, and leak detection capabilities are gathered for the fracture mechanics and fluid mechanics analyses. At the piping locations which have the least favorable combination of material properties and stress, a crack is postulated which is of sufficient size that the resulting leakage will be detected by installed leak detection systems. Finally, LBB is demonstrated if the postulated crack remains stable even if a seismic event takes place before the crack is discovered and repaired. An LBB example is presented in this paper for a generic pressurizer surge line, and reflects the consideration of flow stratification on LBB analyses.

INTRODUCTION

Regulations for nuclear power plants in the United States have required that the plants be designed to assure safe shutdown in the event of a double-ended guillotine break (DEGB) in high energy piping systems. Protective

Int. J. Pres. Ves. & Piping 0308-0161/90/$03·50

measures which assure safe shutdown include physical separation of equipment, structures and components from postulated break locations, as well as the addition of hardware such as pipe whip rupture restraints and jet impingement shields at other locations.

In the last decade, several experimental and analytical evaluations have shown that the probability of the hypothetical DEGB, caused by either direct or indirect failure, is extremely low. Because nuclear high energy piping is made of high-toughness austenitic stainless and ferritic steel and is resistant to unstable crack growth, a crack should leak a detectable amount well in advance of any growth that could result in a sudden catastrophic break. This is the 'leak-before-break' (LBB) concept. Thus, if a conservative analysis shows that leaking pipes can be detected and repaired before a sudden rupture takes place, regulatory rule-making allows a utility to remove pipe whip restraints and jet impingement shields and to discount certain pipe rupture dynamic effects. LBB can also supplement certain piping analysis programs for use in snubber minimization applications.

This paper consists of two sections. The first section presents the general LBB program requirements for an LWR in the US as delineated in NUREG 1061, Vol. 3[1] and the proposed Standard Review Plan (SRP) 3.6.3.[2] The second section presents the application of LBB to a sample problem, using a step-by-step approach developed during earlier LBB programs successfully completed by the authors' firm.[3] The sample problem methodology also incorporates experience gained in the US licensing process.[4] The high energy line selected as the sample LBB application is a generic PWR pressurizer surge line subject to flow stratification. Flow stratification effects have recently become a major concern in pressurizer surge lines, and the sample problem demonstrates how it affects LBB for these lines.

OVERVIEW OF THE LBB APPROACH

The high energy fluid systems (temperature $> 93°C$ or pressure $> 1{\cdot}9$ MPa, or both) which are usually considered in a LBB program are ASME Classes 1 and 2 piping or their equivalent,[5] as well as some ANSI B31.1 non-nuclear safety piping. Only those systems that contain break locations that could potentially target structures, systems and components important to safety need to be analyzed; however, such systems must be analyzed anchor-to-anchor.

The first step in the LBB approach is to screen the candidate systems for the susceptibility to known degradation mechanisms. Next, for systems that potentially qualify for LBB, the fracture and material properties, piping loads, weld procedures and locations, and leak detection capabilities are gathered. At those locations in each system which have the least favorable combination of material properties and stress, a crack is postulated. Crack opening area and fluid mechanics calculations are performed to determine leak rate values. Finally, a fracture mechanics analysis is performed to demonstrate that the flaw will not cause pipe instability even if seismic loads are applied before the flaw is discovered and repaired.

The LBB Methodology is not usually applied to systems which have a history of excessive or unusual loads or degradation mechanisms present (e.g. main feedwater) because these phenomena are difficult to quantify. The excessive or unusual loads or degradation mechanisms of concern include water hammer, corrosion (particularly intergranular stress corrosion cracking), erosion, creep, brittle fracture, thermal aging and fatigue. Therefore, the plant piping service experience and design bases are screened. In addition, an evaluation is required to screen out those systems which could degrade or fail from indirect causes such as fires, missiles, equipment failures, and failure of systems and components which are in close proximity. To pass the evaluation, plant specific analyses supplemented by the service experience of related systems at other plants should show that these phenomena are unlikely to occur.

An assessment of leak detection capabilities is performed concurrently with the piping system screening program. A plant-specific evaluation of available leak detection systems both inside and (if applicable) outside containment is made to accurately establish the minimum leak rate that can be detected reliably under normal operating conditions. Leak detection systems typically employed inside containment include sump level and pump flow monitoring, atmosphere particulate and gaseous radioactivity monitoring, and pressure, temperature and humidity monitoring.

Specific plant leak detection capabilities, including both time and the ability to detect a leak, are a function of system sensitivity, monitoring frequency and trend monitoring. Outside containment, visual detection is the most effective method[6] even for systems with insulation. Visual leak detection strategy would be particularly effective for small diameter lines or auxiliary lines outside containment.

The minimum leak detection capability required by regulatory guides for inside containment is $63\,cm^3/s$ for PWR plants and $315\,cm^3/s$ for BWR plants, but experience has shown that in many cases leaks with flow rates as low as $6{\cdot}6\,cm^3/s$ can be found easily.[6] If the plant-specific evaluation demonstrates a leak detection sensitivity greater than that required in the regulatory guides, it may be used; a value of $32\,cm^3/s$ has been justified in at least one licensing situation for a PWR plant.

Another task that should be performed during the initial phases of the LBB assessment is base and weldment material data collection. The task of identifying materials and material specifications used for the base material and weldments in the LBB candidate high energy systems is one of the most time-consuming and potentially difficult tasks in the LBB methodology.

Material properties may be obtained from the literature[7] or from an archival materials testing program. Using data available in the literature is simpler but requires conservative bounding and a sometimes difficult confirmation of representativeness to the actual plant material. Unfortunately, material property data for some piping base metal, weldments and thermally-aged cast stainless steel are not readily available due to the proprietary treatment of the appropriate test data.

The primary material data requirements are multiple true stress–strain and *J*-resistance (*J–R*) curves at temperature for use in the LBB analyses. For this application, the *J*-parameter characterizes the elastic–plastic field around the crack tip. The stress–strain curve is first used to calculate crack opening area to establish leakage rates. A best-fit stress–strain curve at the appropriate operating temperature is used because a lower bound curve could overpredict crack opening and leakage. Crack stability calculations require both stress–strain and *J–R* curves if an elastic–plastic fracture mechanics approach is used. Because plant-specific material property data at the appropriate temperature are usually unavailable, lower bound stress–strain and *J–R* curves available in industry literature provide conservative crack stability analyses. This extra conservatism can be avoided if experimental data from actual plant piping material are available.

Welding specifications provide step-by-step procedures for fit-up, surface preparation, preheating, tack welding, root passes, cover gases, heat and power input, travel speed, number of passes and other appropriate weld quality items. The welding procedure, the filler material, and the welding technique may produce dramatic differences in the fracture resistance properties between the various types of weldments. Because industry weld material properties are used for plant welds, comparisons of plant and industry welding parameters are necessary to assure representativeness of weld material properties. Typical LWR welds are: gas metal arc weld (GMAW), shielded metal arc weld (SMAW), submerged arc weld (SAW) and gas tungsten arc weld (GTAW).

When considering carbon (ferritic) steel and some cast stainless steel (which has a duplex structure), the normal operating temperatures of the pipes also must be examined relative to the ductile-to-brittle transition temperature. Each material should be evaluated for the most critical service condition that can be experienced to assure that brittle fracture is not a concern.

For the candidate piping run, an anchor-to-anchor examination is made to identify the base metal and weld locations with the least favorable combination of stress and material properties. Normal plant operating conditions in combination with the applicable seismic conditions define the stresses.

Piping stresses generally are generated from the as-built analysis. Stresses of interest come from normal operation (pressure, deadweight, and thermal expansion), safe shutdown earthquake (SSE) and seismic anchor motion (SAM). Maximum piping stresses produced by a combination of these loads are then determined. In practice, various live methods of measuring stress values may be employed for lines subject to phenomena such as thermal stratification. The normal operating stresses are used to determine crack opening for the leak rate calculations. The normal plus seismic stresses are used for the crack stability calculations.

If the piping system is also undergoing a snubber optimization program, it should be completed before the LBB evaluation, because the seismic stresses may increase with dynamic snubber reduction. In addition, normal operating stresses may increase if some snubbers are replaced with rigid supports.

LBB ANALYSIS

To demonstrate that high energy fluid piping is unlikely to experience DEGB or its equivalent, LBB analyses are conducted to assure that: (1) leakage from a postulated through-wall flaw will be detected, and (2) any pre-existing leakage size crack will remain stable and will not cause a rupture even if an earthquake occurs. The leakage size crack (LSC) is the through-wall flaw length that yields detectable leakage.

Once the sensitivity of the leak detection system is established, the minimum detectable leak rate is multiplied by a margin of ten to yield the leakage used as the basis for crack postulation. For example, if the leak detection capability is $63\,cm^3/s$, then the LSC which yields $630\,cm^3/s$ ($63\,cm^3/s \times 10$) is postulated. The margin of ten on leak rate is necessary to conservatively compensate for physical uncertainties related to leak detection measurement and analytical uncertainties related to irregular crack shapes, crack surface roughness and crack plugging.

Standard leak rate computer codes can be used to determine a leak rate versus crack length relationship. For leak rate calculations, the LSC is determined using normal operating stresses and ‘best fit’ material properties. The computer output typically provides a data table of flow rate versus crack length and thus, for the specified detectable leak rate determined for that plant, the corresponding LSC can be identified.

A typical crack length versus leak rate curve for a circumferential crack in a PWR pressurizer surge line is shown in Fig. 1. Leakage rate versus crack

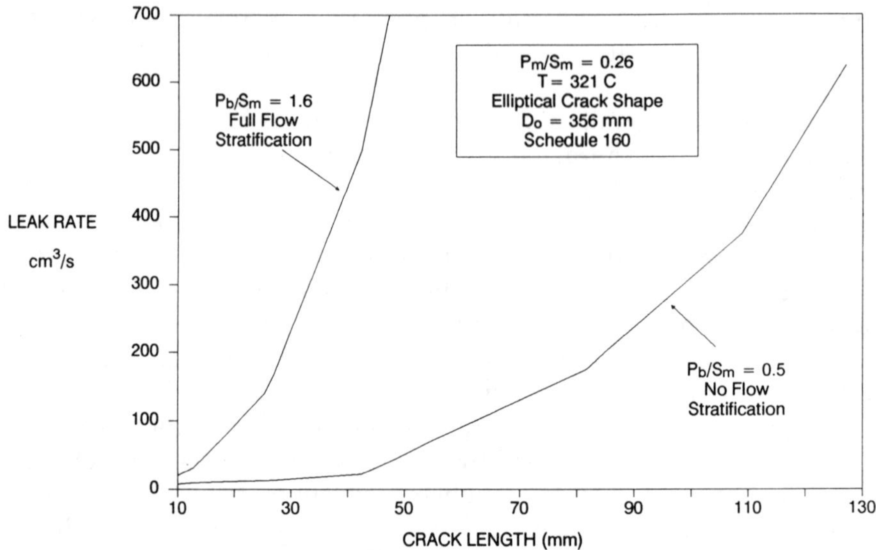

Fig. 1. Leak rate curve for circumferential crack in a generic pressurizer surge line.

length is shown for both full flow stratification and no flow stratification. This 356-mm diameter line is analyzed using a pressure of 16·5 MPa and a temperature of 321°C. The effects of flow stratification were addressed by including bending stresses up to 187·0 MPa. The bending stress without flow stratification was 82·0 MPa.

The final step in the LBB methodology is crack stability analysis to demonstrate that the LSC will not become unstable and result in a DEGB. There are two parts to the stability analyses: first, a check that the LSC is less than half the critical crack size under normal plus seismic stresses; second, a check that the LSC is less than the critical crack size even under excessively high (1·4 × the sum of the normal and seismic) stresses. The crack stability analysis is based either on a modified limit load approach or an elastic–plastic fracture mechanics (EPFM) approach, whichever is more appropriate.

For austenitic stainless steel base metal and weldments, the methodology shown in SRP 3.6.3[2] and the ASME Code Section XI IWB-3600 subsections[5] is based on the modified limit load approach. Further details regarding the modified limit load approach can be found in Refs 8 and 9.

SAMPLE LBB CALCULATIONS

This section presents a sample LBB calculation in a generic, parametric manner to illustrate the results calculated for a 356-mm diameter Type 304 stainless steel surge line. Pressurizer surge lines are good LBB candidates because they are large diameter BOP piping systems with numerous pieces of rupture mitigation hardware and large associated LOCA loads.

Any surge line LBB analysis should consider the stresses resulting from the flow stratification phenomena recently observed in operational PWR surge lines. The stresses resulting primarily from two effects—striping and the significant global bending moments resulting from some insurge and outsurge flow transients—must be added to the standard ASME Code analysis stresses so that the most limiting location on the line is properly identified.[10] Striping, the thermal oscillation of the hot and cold fluid layer interface, is a high frequency phenomenon and its effect on specific nodal piping stresses is difficult to quantify. The magnitudes of these stresses are a function of postulated break and hanger locations, surge line geometry, heat-up or cool-down rates, and plant operating conditions. For this sample problem, pipe bending stresses resulting from flow stratification ranged from zero to the maximum anticipated stress (105 MPa).

Previous parametric studies for other stainless and carbon steel lines are available in the literature.[11] These early studies did not incorporate the effects of thermal stratification, but did include separate fatigue crack growth analyses.

Leak rate estimates

The PICEP computer code[12] is a fracture mechanics program developed by the Electric Power Research Institute (EPRI) and is used to calculate the leakage for through-wall cracks in pipes, the crack opening area, and the limit load critical crack size. While flow rate estimates employ two-phase homogeneous non-equilibrium critical flow theory, the crack opening area is determined using elastic–plastic fracture mechanics methods similar to those used in the crack stability program, FLET,[13] another EPRI-developed fracture mechanics computer code. Both PICEP and FLET have been validated extensively using computer simulations, laboratory testing, and actual nuclear plant events involving fluid leakage. Good overall accuracy to actual test results for PICEP and FLET is verified in Refs 12 and 14.

For the highest normal plus SSE stress location corresponding to the worst material properties, leak rates as a function of crack size are determined. In order to identify the highest stress location, the stress state of each node is examined; for each node in the line, the crack opening forces and moments are combined algebraically (plus or minus) for normal operating conditions (i.e. combination of deadweight, thermal expansion and pressure) and then SSE moments and forces are added absolutely. Another case would combine thermal expansion moments with the SSE moments.

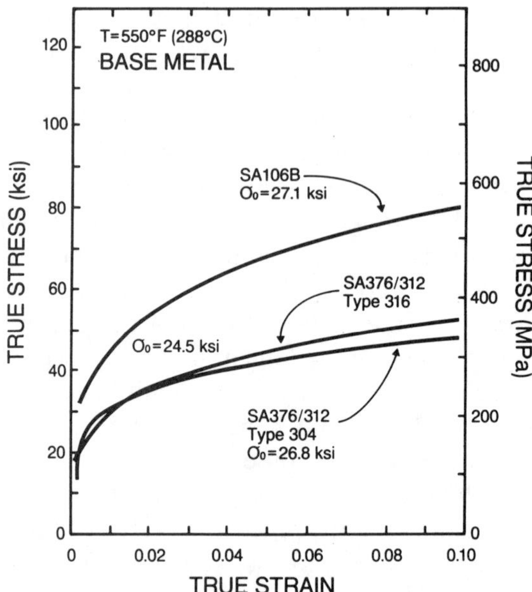

Fig. 2. Lower-bound true stress–true strain curves for base metal.

To alleviate the need for defining the stress/material property state at each node, the worst material properties can be postulated at the high stress location whether or not they exist at that node. This simplification conservatively bounds all possible stress/material property conditions and expedites the analysis. If the LBB calculations fail, then a more precise node-specific analysis may be required.

The PICEP code requires a stress–strain relationship (such as those shown in Figs 2 and 3) to be used for determining the crack opening area needed to calculate leakage. Recent EPRI results indicate that in the case of circumferential weld cracks, the base metal stress–strain properties more realistically (and conservatively) estimate the crack opening area.[15] Lower-bound true stress–strain properties do not produce lower-bound leak rate estimates; therefore, best fit base metal stress–strain properties slightly higher than those shown in Fig. 2 were used.

Piping stresses were resolved into two categories: membrane stress and bending stress. The membrane stress (P_m) results almost exclusively from pressure. The bending stress (P_b) values, resulting mostly from deadweight and thermal loading, were varied to provide a parametric study of thermal stratification effects.

For simplicity, the stresses were normalized to the ASME Section III Code S_m value of 117 MPa.[5] The stress values, ranging from normal operation with no stratification ($P_b = 0{\cdot}5\,S_m$ or 82·0 MPa) to normal

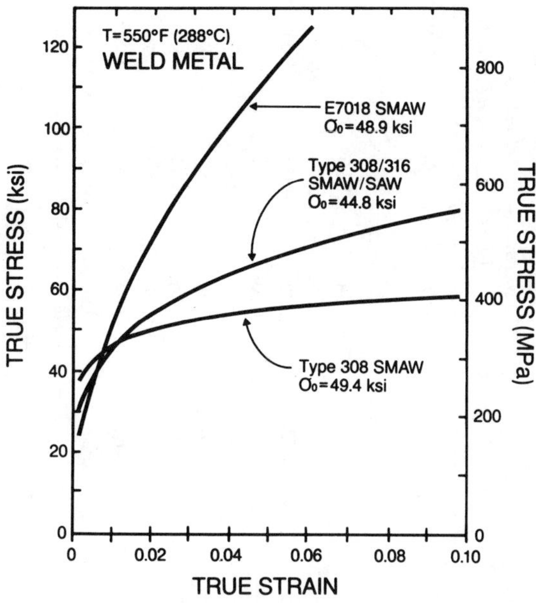

Fig. 3. Lower-bound true stress–true strain curves for weld metal.

operation with full stratification ($P_b = 1{\cdot}6\ S_m$ or 186 MPa), were used as input for PICEP leak rate calculations. The case for stresses resulting from thermal stratification combined with SSE stresses was not considered for the parametric studies described herein. Full flow stratification can be accompanied by a top-to-bottom pipe temperature spread of 150°C. From the PICEP output, the crack length corresponding to 315 cm^3/s (i.e. $10 \times 31{\cdot}5$ cm^3/s, the leak detection capability assumed for this sample problem) was determined for the generic pressurizer surge line. These results are shown in Table 1 and Fig. 1.

TABLE 1
Leak Rate Results for 356-mm Pressurizer Surge Line with a Circumferential Crack (Type 304 Stainless Steel)

Normal operating stresses		*315 cm^3/s COD (mm)*	*315 cm^3/s crack size (mm)*
Membrane P_m/S_m	*Bending P_b/S_m*		
0·26	0·5	0·014	117·0
0·26	0·7	0·158	97·8
0·26	1·0	0·190	77·0
0·26	1·3	0·227	52·3
0·26	1·6	0·295	33·5

TABLE 2
Stability Evaluation for 356-mm Stainless Steel Surge Line with Circumferential Crack

Normal operating stresses		*Normal + SSE stresses*	*Leakage size crack (LSC)*[a]	*Limit load* (l_c)	*Flaw size margin—limit load*[b]	*Flaw size margin—EPFM Normal + SSE loads and 2 × [LSC] using DPFAD*		
Membrane P_m/S_m	*Bending* P_b/S_m	*Bending* P_b/S_m	*(mm)*	*(mm)*	*Base*	*B/B*[c]	*B/SM*[d]	*B/SA*[e]
0·26	0·5	0·7	117·0	424	3·62	1·34	1·16	1·10
0·26	0·5	0·9	117·0	383	3·27	1·10	0·96	0·91
0·26	0·7	0·9	97·8	383	3·92	1·25	1·09	1·03
0·26	0·7	1·1	97·8	350	3·57	1·06	0·92	0·87
0·26	1·0	1·2	77·0	335	4·35	1·12	0·98	0·93
0·26	1·0	1·4	77·0	305	3·96	0·98	0·85	0·81
0·26	1·3	1·5	52·3	287	5·49	1·06	0·93	0·89
0·26	1·3	1·7	52·3	256	4·89	0·94	0·82	0·79
0·26	1·6	1·8	33·5	240	7·16	0·99	0·87	0·84
0·26	1·6	2·0	33·5	208	6·21	0·90	0·79	0·76

[a] Leakage size crack is the 315-cm^3/s crack size calculated for normal operating loads only.
[b] Ratio of the limit-load critical crack size for base metal (normal + SSE) loads to the leakage size crack (LSC).
[c] Using base metal (B) stress–strain and base metal (B) *J–R* curve properties; the DPFAD margin ≥ 1·0 for crack stability.
[d] Using base metal (B) stress–strain and SMAW weld metal (SM) *J–R* curve properties; margin ≥ 1·0 for crack stability.
[e] Using base metal (B) stress–strain and SAW weld metal (SA) *J–R* curve properties; margin ≥ 1·0 for crack stability.

Table 1 and Fig. 1 results also show that there is a significant difference in the crack length corresponding to 315 cm^3/s (i.e. the LSC), depending on the assumed magnitude of flow stratification effects. The LSC with no stratification is 117 mm; with full stratification, the LSC is 33·5 mm. These circumferential crack lengths will next be checked for stability. Longitudinal, through-wall cracks were also examined (not reported here); axial cracks are usually less limiting than circumferential cracks of similar length.

Crack stability under normal plus SSE loads

The purpose of this analysis is to demonstrate that the critical crack size under normal plus SSE loads is at least twice the length of the LSC (i.e. a margin of two exists on flaw size). While there are several possible methods one can use for these analyses, results are shown in Table 2 for two methodologies: limit load, and the EPFM approach employed by the EPRI FLET computer code.[13]

An unmodified limit load approach is not permitted by the US NRC, but is shown in this sample problem for general interest. The methodology will not be explained in detail. In the limit load approach, the critical crack size under normal plus seismic stresses is calculated and the resultant critical crack size is compared with the LSC to determine the crack size margin (l_{crit}/LSC).

To assess crack stability in the elastic–plastic regime, the deformation plasticity failure assessment diagram (DPFAD) method[16] is used. The

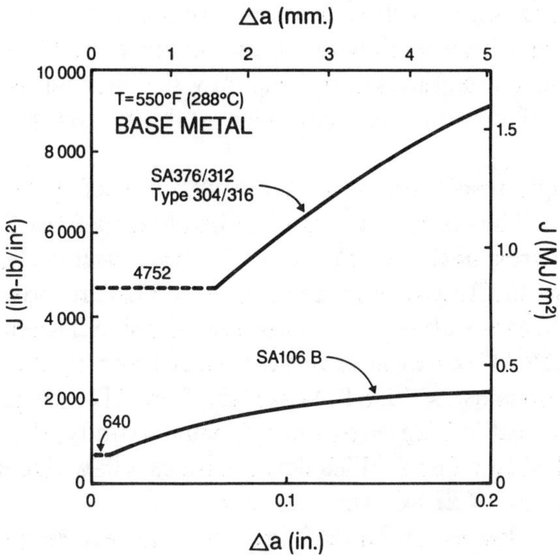

Fig. 4. Lower-bound *J–R* curves for base metal.

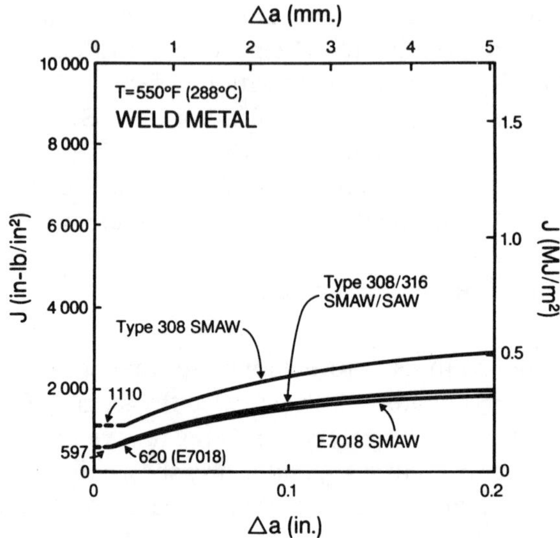

Fig. 5. Lower-bound *J–R* curves for weld metal.

applied loads, leakage size crack length, and material stress–strain and *J–R* curve properties such as those shown in Figs 2–5, are used. Because normal plus SSE inertia stresses are used to assess stability, the SSE forces and moments are combined absolutely with the normal operating forces and moments to define conservative normal plus SSE stress conditions. The resultant normal plus SSE stress condition is thus defined for each degree of assumed flow stratification.

The FLET program uses these varying stress input values to assess the stability of a crack with a crack length equal to twice the LSC from Table 1. The resulting ratio, sometimes called the flaw size margin, is l_{cit}/2LSC and must be $\geq 1{\cdot}0$ to demonstrate adequate stability for the given stress condition.

Flaw size margin results from the stability analyses for the surge line are shown in Table 2. The column in Table 2 identified as 'Flaw Size Margin—Limit Load' corresponds to the results from unmodified limit-load calculations using the flow stress as the net section failure stress; a margin of two or more indicates stability. The columns in Table 2 identified as 'Flaw Size Margin—EPFM' correspond to the results from elastic–plastic crack stability calculations using FLET. When the DPFAD margin (for normal plus SSE loads) in Table 2 is greater than or equal to unity, the crack is stable based upon EPFM (see Fig. 6). This demonstrates a flaw size margin of two because a crack size of 2LSC was analyzed.

The last four columns of Table 2 indicate that elastic–plastic fracture generally dictates crack stability. The flaw size margins (l_{crit}/2LSC)

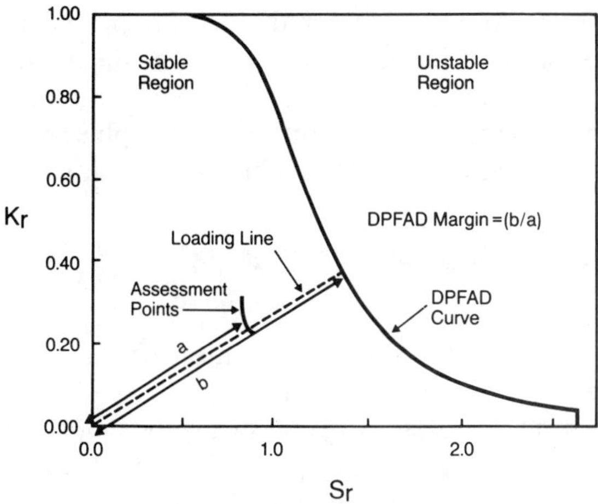

Fig. 6. Sample DPFAD calculation showing stable crack growth.

calculated using EPFM methods indicate a higher degree of instability than the unmodified limit-load flaw size margins for the same stress condition.

The DPFAD method presents stability results in a safety/failure plane and the DPFAD margin is illustrated in Fig. 6. The ordinate of the DPFAD diagram K_r is the elastic stress intensity factor normalized by the crack driving force, and the abscissa is the applied stress normalized by the reference stress, which is related to the limit stress. The DPFAD or evaluation curve generated from calculated J values separates a region of crack stability from a region of crack instability. The assessment points (Fig. 6) are based on applied loading and material fracture resistance values. Additional details regarding the DPFAD method can be found in Refs 13 and 16.

The results in Table 2 also address the combinations of materials used when evaluating welds. Stated a different way, the base metal stress–strain properties can be matched with the base metal J–R curve properties for a base metal break location. When a weld is present, the bulk stress–strain is best represented by the base metal properties,[15] while the weld metal J–R curve, because it is more limiting locally than the base metal J–R curve, is appropriate for use in the fracture mechanics analysis. To illustrate the effects of various material property assumptions, three separate groups of results are presented in Table 2. These are as follows: (1) base metal stress–strain and base metal J–R, (2) base metal stress–strain and SMAW J–R, and (3) base metal stress–strain and SAW J–R.

Note that in some cases for the stainless steel line (Table 2) the DPFAD margin is $< 1{\cdot}0$ indicating instability; this means that a flaw which is twice the size of the LSC is unstable and the required margin to failure does not exist. It is important to note that for this sample problem however, flow stratification stresses were conservatively added to the highest normal plus SSE stresses. In reality, the point of highest stress due to flow stratification may not be the point of highest stress due to normal plus SSE loads, and this line would probably pass if a node-specific evaluation was performed.

Crack stability under excessively high loads

The final step involves crack stability verification after an additional margin is applied to the normal plus SSE loads.

Assuming that a stable leakage size crack exists in a pipe, it must be shown that, if a faulted loading condition occurs, this through-wall crack is stable under loads in excess of normal plus SSE. Thus, for this stability check, a margin of 1·414 (i.e. $\sqrt{2}$) is placed on normal plus SSE loads, but the margin of two is removed from the LSC flaw size. If the LSC is stable even when these high loads are imposed, the DPFAD load margin is greater than 1·0 and the line passes the final acceptance check for LBB.

Load margin results from the stability analyses for the surge line are shown in Table 3. The column in Table 3 identified as 'Load Margin—Limit Load' corresponds to the results from unmodified limit-load calculations using the flow stress as the net section failure stress. The columns in Table 3 identified as 'Load Margin—EPFM' correspond to the results from elastic–plastic crack stability calculations using the DPFAD methodology.

As before, Table 3 illustrates various combinations of material properties (stress–strain and *J–R* curves). Again, some cases have DPFAD margins $<1{\cdot}0$, indicating instability for the overload condition due to the conservative assumptions described earlier.

LBB assessment

The DPFAD results from Tables 2 and 3 show that generally for situations in which stratification causes P_b/S_m to be $>1{\cdot}0$, the surge line does not meet the required flaw size and load margins. A formal LBB submittal for this surge line would require a more specific identification of the location or worst combination of stress and material properties. This would provide relief from the assumption made in these sample analyses that the highest stress location is also the site of worst material properties. Alternatively, the stresses resulting from flow stratification could be reduced or limited by changing plant operating procedures, for example, to limit maximum permitted loop-to-pressurizer differential temperature.

TABLE 3
Excessive Load Stability Evaluation for 356-mm Stainless Steel Surge Line with Circumferential Crack

Normal operating stresses		*Normal + SSE stresses*	*Leakage size crack [LSC]*[a]	*Limit load* (l_c)	*Load margin—limit load*[b]	*Load margin–EPFM for 1·414 × (Normal + SSE) loads and 1 × [LSC] using DPFAD*		
Membrane P_m/S_m	*Bending* P_b/S_m	*Bending* P_b/S_m	*(mm)*	*(mm)*	*Base*	*B/B*[c]	*B/SM*[d]	*B/SA*[e]
0·26	0·5	0·7	117·0	368	3·14	1·48	1·30	1·24
0·26	0·5	0·9	117·0	319	2·73	1·19	1·04	1·00
0·26	0·7	0·9	97·8	319	3·26	1·26	1·10	1·06
0·26	0·7	1·1	97·8	280	2·86	1·04	0·91	0·88
0·26	1·0	1·2	77·0	256	3·32	1·02	0·90	0·86
0·26	1·0	1·4	77·0	216	2·80	0·88	0·77	0·74
0·26	1·3	1·5	52·3	192	3·67	0·89	0·78	0·75
0·26	1·3	1·7	52·3	152	2·91	0·79	0·69	0·66
0·26	1·6	1·8	33·5	128	3·82	0·80	0·71	0·68
0·26	1·6	2·0	33·5	88	2·62	0·72	0·64	0·61

[a] Leakage size crack is the 315-cm^3/s crack size calculated for normal operating loads only.
[b] Ratio of the limit-load critical crack size for base metal loads [1·414 × (normal + SSE)] to the leakage size crack (LSC).
[c] Using base metal (B) stress–strain and base metal (B) *J–R* curve properties; the DPFAD margin ≥ 1·0 for crack stability.
[d] Using base metal (B) stress–strain and SMAW weld metal (SM) *J–R* curve properties; the margin ≥ 1·0 for crack stability.
[e] Using base metal (B) stress–strain and SAW weld metal (SA) *J–R* curve properties; the margin ≥ 1·0 for crack stability.

SUMMARY AND CONCLUSIONS

This paper has briefly reviewed the general requirements and methodology used during the application of LBB to nuclear BOP piping in the US. Regulatory guidelines established in NUREG-1061, Vol. 3 and SRP 3.6.3 remain in effect and have been augmented by additional technical demands as industry experience dictates. In general, the LBB methodology consists of the following:

- *Screening:* determine the applicability of LBB.
- *Information gathering:* assemble the necessary material property data and leak detection capabilities.
- *Leakage size crack determination:* calculate the LSC at the location of least favorable combination of stress and material properties.
- *Flaw size margin determination:* demonstrate an adequate margin (l_{crit}/LSC $\geq$ 2) under Normal plus SSE loads, and
- *Load margin determination:* demonstrate an adequate margin (l_{crit}/LSC $\geq$ 1) under 1·414 × (Normal plus SSE) loads.

A sample LBB analysis for a generic pressurizer surge line has demonstrated how industry experience with flow stratification may affect LBB results. While the sample problem showed unsatisfactory results for some degrees of flow stratification due to the conservatisms employed, the surge line could still pass all applicable LBB criteria, even for the case of full stratification. However, a more rigorous examination in order to quantify and locate specifically the highest stress due to flow stratification would be required.

ACKNOWLEDGEMENTS

The authors wish to thank Dr Robert L. Cloud of Robert L. Cloud & Associates, Inc. for his guidance and review. The Electric Power Research Institute authorized the use of PICEP and FLET for this paper.

REFERENCES

1. US Nuclear Regulatory Commission. *Report of the US Nuclear Regulatory Commission Piping Review Committee*: *Evaluation of Potential for Pipe Breaks.* Washington, DC, November 1984, NUREG-1061, Vol. 3.
2. US Nuclear Regulatory Commission. *Standard Review Plan for the Review of Safety Analysis Reports for Nuclear Power Plants*, Washington, DC, 1981, NUREG-0800.

3. Server, W. L., Beaudoin, B. F. & Quiñones, D. F., Applying Leak-Before-Break to high energy piping, Nuclear Safety Analysis Center, Electric Power Research Institute, Palo Alto, November 1987, NSAC-114.
4. Server, W. L., Beaudoin, B. F. & Quiñones, D. F.,Lead plant application of Leak-Before-Break to high energy piping, Nuclear Safety Analysis Center, Electric Power Research Institute, Palo Alto, January 1989, NSAC-141.
5. ASME boiler and pressure vessel code. American Society of Mechanical Engineers, New York, July 1989.
6. Bausch, H. P., Leak detection in nuclear piping outside containment. Electric Power Research Institute, Palo Alto, March 1987, NSAC-110.
7. Hiser, A. L. & Callahan, G. M., Users guide to the NRC's piping fracture mechanics data base (PIFRAC). US NRC, Washington, DC, March 1987, NUREG/CR-4894.
8. Section XI Task Group for Piping Flaw Evaluation. Evaluation of flaws in austenitic steel piping. Electric Power Research Institute, Palo Alto, CA, Special Report, July 1986, NP-4690-SR.
9. Section XI Task Group for Piping Flaw Evaluation. Evaluation of flaws in ferritic piping. Electric Power Research Institute, Palo Alto, CA, October 1988, NP-6045.
10. Hirschberg, P. & Antaki, G. A., Measurement of stratification in the pressurizer surge line. The 1989 ASME Pressure Vessels and Piping Conference —JSME Co-Sponsorship, Honolulu, *Design and Analysis of Piping and Components—1989, PVP-Vol. 169*, ed. Q. N. Truong. pp. 85–90.
11. Mukherjee, S. K., Szy Slow Ski, J. J., Chexal, V., Norris, D. M., Goldstein, N. A., Beaudoin, B. F. Quiñones, D. F. & Server, W. L., Rupture hardware minimization in pressurized water reactor piping. *ASME Journal of Pressure Vessels, Technology*, **111** (1989) 64–71.
12. Norris, D. M. & Chexal, V., PICEP: Pile crack evaluation computer program. Electric Power Research Institute, Palo Alto, CA December 1987, NP-3596-SR, Revision 1.
13. Okamoto, A. & Norris, D. M., FLET: Pipe crack instability program. Electric Power Research Institute, Palo Alto, CA, February 1990, NP-6717-CCML.
14. Kishida, K. & Zahoor, A., Crack-opening area calculations for circumferential through-wall pipe cracks. Electric Power Research Institute, Palo Alto, CA, August 1988, NP-5959-SR.
15. Ganta, B. R. & Ayres, D. J., Analysis of cracked pipe weldments. Electric Power Research Institute, Palo Alto, CA, February 1987, NP-5057.
16. Bloom, J. M. & Malik, S. N., Procedure for the assessment of the integrity of nuclear pressure vessels and piping containing defects. Electric Power Research Institute, Palo Alto, CA, June 1982, NP-2431.

Int. J. Pres. Ves. & Piping **43** (1990) 85–93

Application of Leak-Before-Break to Primary Loop Piping to Eliminate Pipe Whip Restraints in a Spanish Nuclear Power Plant

M. Rodríguez & A. Esteban

Consejo de Seguridad Nuclear (CSN), 11 Justo Dorado, 28040 Madrid, Spain

ABSTRACT

The Spanish plant described in this study is a 982 MWe PWR plant with a three-loop primary circuit of piping made from centrifugally-cast stainless steel SA351 CF8A.

The licensee requested an exemption to GDC-4, from CSN, so as to avoid the postulation of guillotine rupture of the primary loop piping. The request was based on the generic work performed for a US PWR plant group to have such exemption to GDC-4. As the piping material in the Spanish plant is different from that in the plants included in the generic work, CSN performed a review of the applicability of the generic results to the Spanish plant. Also, aspects such as fatigue evaluation, net section collapse, crack growth and leak detection, specifically analyzed for the Spanish plant, were reviewed.

CSN found that fracture toughness test results from generic work are applicable to the Spanish plant; sufficient margin exists against unstable crack extension, and adequate leak detection capability exists with the leakage detection systems available in the plant.

Exemption to GDC-4 was approved and CSN authorized the licensee to remove protection devices against dynamic loads from guillotine breaks in the primary coolant loops.

INTRODUCTION

The Spanish plant is a 982 MWe PWR plant with a three-loop primary piping circuit manufactured with piping made from centrifugally-cast stainless steel SA351 CF8A.

The design basis includes the 10CFR50 General Design criterion (GDC-4) requiring that, 'Structures, systems and components important to safety

Int. J. Pres. Ves. & Piping 0308-0161/90/$03·50

shall be designed to accommodate the effects associated with the postulated accidents, including loss of coolant accidents.'

Early in 1986 the owner submitted to Consejo de Seguridad Nuclear (CSN) an application to approve an exemption from a portion of the requirements of GCD-4, in order to eliminate the need to postulate circumferential and longitudinal pipe breaks in the primary loop of the Reactor Coolant System (RCS) and consequently:

(1) eliminate pipe whip restraints and jet impingement shields;
(2) eliminate the consideration of dynamic effects and loading conditions associated with such breaks in the design of primary system components, piping, access platforms and connected piping, including their supports.

REGULATORY APPROACH

As a guide for the technical approach, the applicant followed the NRC Generic Letter (GL) 84·04.[1] The GL 84·04 presents NRC evaluation results of technical reports submitted by the main supplier of an American PWR group, eliminating the need to postulate breaks in primary loop piping:

—Acceptable technical bases have been provided to eliminate such breaks in the plants included in the proposal.
—Other PWR licensees or applicants may also request exemptions on the same basis from the requirements of GDC-4 if they can demonstrate the applicability to their plants of modeling and conclusions contained in the reports used for the evaluation.

The NRC evaluation criteria established in GL 84·04, and followed by CSN, are as follows:

—Use the *J*-integral based tearing stability concept considering:

- static loads + Safe Shutdown Earthquake (SSE);
- a postulated circumferential through-wall crack with a length larger than either twice the wall thickness or that corresponding to a calculated leak rate of 37·85 kg/min (10 gpm) at normal operating conditions.

—The occurrence of phenomena such as water hammer, stress corrosion cracking (SCC) or unanticipated cyclic stresses shall be excluded.
—Material fracture toughness should be based on experimental *J–R* curves for both base and weld material. The condition $T_{mat} \geq 3T_{app}$ should be satisfied at the applied *J*-value.

—ASME code limits shall be satisfied for faulted conditions in an uncracked section. Cracked sections shall have adequate margins against net section plastic failure.

Enclosed with the application were reports[2,3] prepared by the main supplier having NRC exemption for the first group of plants included in GL 84·04. Moreover, a plant-specific report[4] for the Spanish plant, also prepared by the supplier, was enclosed to demonstrate applicability of the generic evaluation.

EVALUATION

The evaluation was performed in two steps:

(1) It was checked that all parameters used in the generic evaluation covered those of the Spanish plant.
(2) Aspects specifically analyzed for the plant were reviewed, following a different approach from that in the generic evaluation.

Generic evaluation review

Stress corrosion cracking (SCC) on primary loop piping was excluded by the main supplier because the three conditions for SCC occurrence (corrosive environment, susceptible material and high tensile stresses) are not present in this piping. Also water hammer was excluded based on considerations of the system design.

Crack size selection for generic evaluation was made by the main supplier assuming probability of detection-crack depth (per cent through wall) curves similar to those obtained from a Round Robin Test.[5] Values for the probability of detection were selected considering:

—experience from pre-service and in-service inspections obtained by the non-destructive examination (NDE) personnel about crack sizes that have been missed and detected;
—crack sizes that must be repaired following ASME Boiler and Pressure Vessel Code Section XI requirements for pre-service and in-service inspection;
—ASME III Appendix G requirements for brittle fracture evaluation in Class I piping.

Material tensile properties for generic evaluation were:

—minimum yield strength (σ_y) both at room and operation temperature from quality assurance files;
—minimum yield and ultimate strength (σ_y and σ_u) from tests on six heats of wrought stainless steel.

Values for J-integral initiation toughness, J_{IN}, and material tearing modulus, T_{mat}, were obtained for generic evaluation by testing compact tension specimens following acceptable published procedures.[6]

To account for the resistance to faulted loading conditions of the cracked section, both static and dynamic analyses were carried out in the generic evaluation. Umbrella loads for the group of plants included in the generic evaluation were considered; these included pressure, weight, thermal and SSE. The axial load and bending moment were obtained by algebraically combining the static loads and then absolutely adding SSE loads. As result, an umbrella axial force and bending moment were obtained.

Finite element analyses were reviewed. Both static and dynamic calculations were performed with a section containing 190·5 and 88·9 mm through-wall cracks. Allowable stress and strain results were found; a lower bound plastic instability load was obtained. Dynamic calculation results were found to be higher than those for static calculation.

Also by finite element analysis, the applied J (J_{app})-values were obtained by the main supplier for both static and dynamic cases in the generic evaluation. Calculations were performed considering a 190·5 mm crack length. Values obtained for dynamic calculation were higher than those for static calculation.

The tearing modulus for the J_{app} value (T_{app}) was obtained by the main supplier for generic evaluation by the equation proposed by Tada included in NUREG CR-0838.[7]

Applicability to weld material of the above results, which were obtained for the base material, was established, by the main supplier by obtaining J_{IN} and crack growth values by testing several weld materials. Shielded metal-arc welding (SMAW) and submerged-arc welding (SAW) were considered as the main welding processes and tests were performed on both as-welded and thermally-treated materials. No significant differences exist between the results from base and weld materials.

To account for the thermal aging effect on fracture toughness of centrifugally-cast stainless steel, results from five different testing programs were considered by the main supplier for the generic evaluation.[8] J_{IN} and T_{mat} were obtained for aged material using ASTM testing procedures.[9] Two of these programs, conducted in the USA,[10–12] showed differents Charpy-V toughness and J_{IC} reductions with aging for 316 CF8M stainless steel. The fourth program, conducted in France,[13] included a heat with high sensitivity for aging. The fifth program was conducted by the main supplier, using the sensitive heat considered to be a lower bound for the generic evaluation (which is confirmed by the French results).

An empirical model was used to predict end-of-life Charpy-V toughness.

Values for the most sensitive heat conservatively enveloped those anticipated for SA351 CF8A at the end of plant life.

The crack opening area for a 190·5 mm long through-wall crack was obtained, by the main supplier, for the generic evaluation by finite element calculations. With the crack opening area results, leak rates were calculated for normal and upset conditions by the Fauske[14] two-phase flow model taking account of pressure and friction losses. The calculational model was applied to data from leak rate tests having good agreement between calculated and measured flow rates, using a friction factor value derived from test data. Application of the model to cracks experienced in plant operation showed the model accuracy when a proper description of the crack (length and width) is given.

The leak rate obtained for a 190·5 mm long through-wall crack can be easily detected with the leak detection systems normally available in the plants, assuming these systems comply with NRC RG 1·45[15] requirements.

Plant-specific evaluation review

Fatigue exclusion as a failure cause was performed in the design of the primary system by analysis following the ASME III requirements. This analysis was reviewed by CSN as part of the primary system stress report. Additional finite element analysis of low cycle fatigue crack growth was performed by the main supplier to determine the primary system sensitivity to the presence of small cracks ($<12·7$ mm), using a crack growth rate law for stainless steel.

The global failure mechanism was reviewed by the main supplier in a different way from that in the generic evaluation using experimentally-verified, net section collapse theory with the limit moment calculated by an analytical model as described in EPRI NP-192.[15] Material tensile properties were obtained from ASME III-W/1975.

The axial force and bending moment of the cracked section for the faulted load condition were derived from stress analysis. The bending moment value obtained for generic evaluation covered this value for the Spanish plant, but the axial force was higher for the Spanish plant.

No dynamic testing was performed to determine fracture toughness of SA351 CF8A. Static tests were performed by the main supplier to obtain J_{IN} for SA351 CF8A without aging. Values of J_{IN} for aged SA351 CF8A were adopted from experimental data used in the generic evaluation.[8]

Charpy-V tests were carried out for the Spanish material; results were interpolated in the T_{mat}-Charpy-V energy relationship derived from tests on SA351 CF8M.

J_{app} values were derived from J_{app}-static moment curves developed by finite element calculations for generic evaluation, considering loads and geometry of the Spanish plant. T_{app} calculations were performed with the same methodology as for the generic evaluation (NUREG CR838).[7]

Leak rate calculations for the generic evaluation were assumed to apply to the Spanish plant. The five primary leak detection systems described in RG 1.45 are available. A specific report to demonstrate the compliance with RG 1.45 was submitted by the utility.

EVALUATION FINDINGS

From the generic evaluation

Sufficient evidence was obtained to assure that the following assumptions can be applied to the Spanish plant:

—Stress corrosion cracking and waterhammer can be excluded as mechanisms of failure of primary loop piping.
—Cracks longer than those supposed detectable are not critical.
—Umbrella loads for faulted conditions will not produce failure of sections containing cracks, with adequate safety margins.
—A large amount of experimental work has been performed to obtain the fracture toughness properties J_{IN} and T_{mat} for different materials. When not directly applicable to the Spanish plant, the lower bound values are used.
—Methods to obtain J_{app} and T_{app} are adequate.
—Leak rates for crack sizes below the critical value for instability can be detected by the leak detection systems normally available.

From the specific evaluation

—Material tensile properties used are more conservative than the generic evaluation values and cover the values from history dockets of the plant material.
—Although the axial force for the Spanish plant is higher than the value of generic evaluation, the generic demonstration of resistance to faulted load conditions of the cracked section is considered acceptable as the direct stress on the section is lower for the Spanish plant. Sufficient margin exists for the global failure of the cracked section.
—Weld materials SS-308 and SS-303C used in the primary loop for the Spanish plant were included in tests for generic evaluation to

demonstrate that they have fracture toughness comparable to the base material.

—Sufficient design margin exists to account for fatigue. Fatigue crack growth of small cracks is very small.

—No aging tests were carried out for SA351 CF8A and consequently, because of the approximative methods used, not too much credit can be taken from J_{IN} and T_{mat} estimated values. Lower bound values for J_{IN} and T_{mat} obtained in the referenced tests for the most sensitive heat[8] conservatively cover those anticipated for SA351 CF8A and can be used for evaluation.

—Acceptable methods have been employed to obtain J_{app} and T_{app} for the Spanish plant.

—If $J_{IN} > J_{app}$ then no crack growth is expected for a 190·5 mm crack. If lower bound values for J_{IN} are assumed, $J_{IN} < J_{app}$.

—The criterion $1/3\ T_{mat} \geq T_{app}$ is verified, both for the best estimated T_{mat} values for the Spanish plant and the lower bound experimental values for aged material.

—The critical flaw size is greater than 190·5 mm for the Spanish plant.

—Sufficient leak detection capability exists within the systems available in the plant. Three of those systems (sump level, condensate flow rate from coolers and radioactive particulate detectors) have a detection sensitivity of 3·78 kg/min (1 gpm) in one hour, as required by the RG 1.45.

CONCLUSIONS

The technical information furnished is sufficient to demonstrate the compliance with the requirements stated in GL 84-04. From a documents review and audits, CSN concluded that double-ended guillotine breaks of the primary loop piping will not occur in the Spanish plant. Exemption from the requirements of GDC-4 can be authorized to exclude guillotine breaks of primary coolant loop piping as a design basis, subject to the following:

—Other primary components, including connecting piping are excluded from the exemption.

—Guillotine primary loop break shall be maintained as a design basis for containment, emergency core coolant systems and safety systems design, for compartment pressurization analysis and for equipment qualification.

As a result of the exemption, only structures that protect against dynamic effects associated with primary loop breaks can be dismantled and loads from such effects can be eliminated from design. Any other modification related to the exemption shall be approved by CSN.

Primary loop pipe whip restraints were dismantled except those embedded in the biological shield wall. Impingement shields were also dismantled. Considerable benefits were obtained in terms of personal radiation exposure during maintenance and in-service inspections and optimization of primary loop piping thermal isolation with a reduction in thermal load to containment.

REFERENCES

1. Eisenhut, D. G., Safety evaluation of Westinghouse reactors: topical reports dealing with elimination of postulated pipe breaks in PWR primary main loop. Generic Letter 84-04, US Nuclear Regulatory Commission, February 1984.
2. Palusamy, S. S. & Hartmann, A. J., Mechanistic fracture evaluation of reactor coolant pipe containing a postulated circumferential through-wall crack. WCAP 9558, Rev. 2 Westinghouse Proprietary Class 2, May 1981.
3. Palumsay, S. S., Tensile and toughness properties of primary piping weld metal for use in mechanistic fracture evaluation. WCAP 9787, Westinghouse Proprietary Class 2, May 1981.
4. Monette, P., Swamy, S., Lee, Y. S. & Calvin, E. R., Technical bases for eliminating large primary loop pipe rupture as a structural design basis for Vandellos II. WENX/85/37, Westinghouse Proprietary Class 2, June 1985.
5. Doctor, S. R., Becker, F. L., Heasler, P. G. & Selby, G. P., Pacific Northwest Laboratory (PNL). Paper presented to specialist meeting on Defect Detection and Sizing. CSNI Report No. 75, 3–6 May 1983, pp. 669–78.
6. Clarke, G. A., Andrews, W. R., Begley, J. A., Donald, J. K., Embley, G. T., Landes, J. D., McCabe, D. E. & Underwood, J. H., A procedure for the determination of ductile fracture toughness values using *J*-integral techniques. *Journal of Testing and Evaluation*, **7** (1979) 49–56.
7. Tada, H., Paris, P. & Gamble, R., Stability analysis of circumferential cracks in reactor piping systems. US Nuclear Regulatory Commission report NUREG/CR-0838, June 1979.
8. Bamford, W. H., Kunka, M. K., Spitznagel, J. & Witt, F. J., The effects of thermal aging on the structural integrity of cast stainless steel piping for Westinghouse nuclear steam supply systems. WCAP-10456, Westinghouse Proprietary Class 2, November 1983.
9. Standard test method for J_{IC}. A measure of fracture toughness. ASTM E 813–81, 1981.
10. Beck, F. H., Schoefer, E. A., Flowers, J. W. & Fontana, J. W., New cast high strength alloy grades by structure control. *ASTM STP 369*, American Society for Testing and Materials, Philadelphia, PA, 1965, pp. 159–74.
11. Landerman, E. I. & Bamford, W. H., Fracture toughness and fatigue characteristics of centrifugally cast type 316 stainless steel after simulated thermal service conditions. Paper presented at the Winter Annual Meeting of the ASME, San Francisco, CA, MPC-8, ASME, December 1978.

12. Bamford, W. H. Landerman, E. I. & Diaz, E., Thermal aging of cast stainless steel and its impact on piping integrity. Paper presented at ASME Pressure Vessel and Piping Conference, Portland, Oregon, June 1983.
13. Slama, G., Petrequin, P., Masson, S. M. & Mager, T. R., Effect of aging on mechanical properties of austenitic stainless steel casting and welds. Paper presented at SMIRT 7 Post-conference Seminar 6, Assuring Structural Integrity of Steel Reactor Pressure Boundary Components. Monterey, CA, 29–30 August, 1983.
14. Fauske, H. K., Critical two-phase, steam water flows. Proceedings of the Heat Transfer and Fluid Mechanics Institute, Stanford University Press, Stanford, CA, 1961.
15. Reactor coolant pressure boundary leakage detection systems. Regulatory Guide 1.45, US Atomic Energy Commission, May 1973.
16. Kanninen, M. F., Broek, D., Marschall, C. W., Rybicki, E. F., Sampath, S. G., Simonen, F. A. & Wilkowski, G. H., Mechanical fracture predictions for sensitized stainless steel piping with circumferential cracks. EPRI NP-192, September 1976.

Int. J. Pres. Ves. & Piping **43** (1990) 95–111

Development of Criteria for Protection against Pipe Breaks in LWR Plants

Y. Asada

Tokyo University, 3-1 Hongo 7-Chome, Bunkyo-ku,
Tokyo 113, Japan

K. Takumi

Nuclear Power Engineering Test Center, Shuwa Kamiyacho Building,
3–13 Toranomon 4-Chome, Minato-ku, Tokyo 105, Japan

H. Hata

Mitsubishi Atomic Power Industries Inc., 4-1 Shibakouen 2-Chome,
Minato-ku, Tokyo 105, Japan

&

Y. Yamamoto

Toshiba Corp., 8 Shinsugitacho, Isogo-ku,
Yokohama 230, Japan

ABSTRACT

A proving test for the structural integrity of safety-related carbon steel piping components in light water reactor plants was conducted in NUPEC as a four-year project, in which the applicability of the Leak-Before-Break (LBB) concept to protect against a postulated pipe break was reviewed in parallel with the clarification of fracture behavior. The comprehensive review of LBB applications consists of applicable piping systems, premise for evaluations, procedure and evaluation findings. The review concluded that present practice for design, fabrication, installation and operation can ensure structural integrity and moreover postulated that instantaneous pipe break as a basis for structural design is unrealistic if certain conditions are met. Fatigue is the only failure mechanism to be considered to affect the piping system.

Int. J. Pres. Ves. & Piping 0308-0161/90/$03·50

1 INTRODUCTION

An application of the Leak-Before-Break (LBB) concept to protect against a postulated pipe break in a light water reactor (LWR) plant has been reviewed in Japan to achieve the rationalization of component design in the context of design improvements and standardization.

This report presents the results of the second step of the review on LBB evaluation for the safety-related carbon steel piping systems in both PWR and BWR plants in Japan, as NUPEC's verification test program. The report consists of the following:

(1) piping systems applicable to LBB
(2) potential piping failure mechanisms that may limit LBB application
(3) a sequence of evaluation to verify LBB and its premise
(4) evaluation findings

2 BACKGROUND

2.1 Design criteria on postulated pipe break

Historically, pipe breaks up to double-ended guillotine breaks (DEGB) of maximum pipe size in the reactor coolant pressure boundary (RCPB) in LWR plants have been postulated as the design basis of the containment vessel and emergency core cooling system (ECCS) from the standpoint of reactor safety. Later on, structures, systems and components important to reactor safety have been required to protect against dynamic effects such as pipewhip and jet impingement resulting from postulated pipe break.

Recently, research and development (R&D) to investigate the fracture mechanisms of pipes has advanced worldwide sufficiently to convince regulators that LBB can be applied. As a result, some countries have already introduced LBB into the design criteria and others have been continuing to review the design applicability. This is because the postulation of an instantaneous pipe break is unrealistic. The application of LBB will enhance reactor safety owing to improved maintainability and inspection, and the resulting reduction of occupational radiation exposure by eliminating massive structures, like pipe whip restraints. But it is recognized at present that the application of LBB to such reactor safety evaluation, design of containment vessel and ECCS is a future issue since many related regulatory guidelines and design criteria must be reviewed, the potential failure mechanism of pressure boundaries, other than piping, must be clarified, and moreover international consensus is necessary.

Under such circumstances, Japan started to develop the rationalized protection criteria in consideration of the LBB concept by taking a two-step approach. First of all, research and development on austenitic stainless steel piping belonging to RCPB was completed technically and is under review for implementation. Then the verification of LBB for carbon steel piping belonging to Class 1 and 3 piping (equivalent to ASME code Sec. III, Class 1 and 2) was started in 1985 as a four-year project by NUPEC under a contract with the Ministry of International Trade and Industries and completed at the end of fiscal 1988.

2.2 Development of fracture criteria of carbon steel piping

To develop fracture criteria applicable to safety-related carbon steel piping, fracture behavior tests, LBB verification tests and analyses were performed. The test matrix is shown in Table 1.

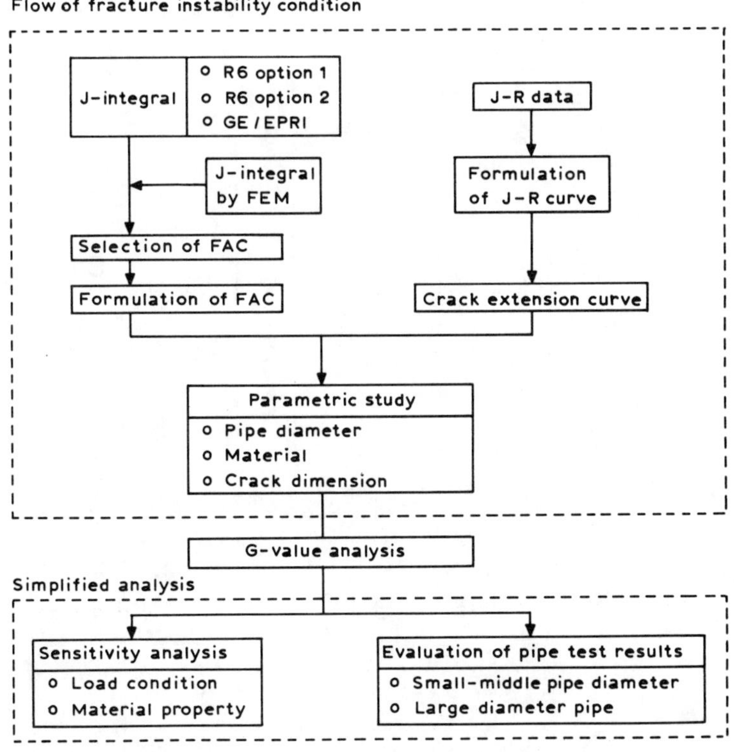

Fig. 1. Evaluation flow of fracture analysis. J–R = J-resistance; FAC = failure assessment curve; R6 = fracture evaluation method developed by the CEGB; *G*-Value = modification factor to net section collapse criterion.

TABLE 1
Matrix of the Verification Test

Material	*Fracture behavior test (straight pipe)*									*LBB verification test*		
	Low comp. bending		*High comp. bending*				*Low comp. tension*			*High comp.*		
	Quasi-static		*Quasi-static*				*Quasi-static*		*Dynamic*	*Cyclic load*		
	6B	*16B*	*6B*		*16B*		*6B*		*6B*	*Straight pipe*	*Elbow*	*Tee*
	RT	*RT*	*RT*	*HT*	*RT*	*HT*	*RT*	*HT*	*HT*	*HT*	*HT*	*HT*
STS42	○	○	○	○	○	○	○	○	○	○		
SFVC2B											○	○
Weld joint	○									○		

RT = Room temperature; HT = high temperature (300°C).

The tests yielded data on fracture behavior for nominal 6-inch and 16-inch pipes (representative of base and weld metals). Although the data are limited, they showed that a net section stress criteria dominates the fracture behavior. A fracture evaluation was conducted to develop a comprehensive method applicable to all the materials and large diameter pipes under consideration.

The study aimed at developing a simple but conservative engineering method. The fracture criteria under various conditions were screened by means of the R6 procedure developed by the CEGB in the UK embracing the *J*-integral, tearing modulus and plastic collapse criteria. The evaluation procedure is shown in Fig. 1. The major factors influencing fracture behavior were analyzed by the finite element method (FEM) and various simplified methods to determine which method would yield the most appropriate failure assessment curves (FAC) in the R6 procedure.[1] Evaluation in terms of simplicity, conservativeness and accuracy concluded that the R6 procedure, Option 2, with modification, yielded the most appropriate FAC. From results of this study,[2] the modification factor, *G*, to the critical bending moment in the net section stress formula was established as the following:

G factor = plastic collapse load/maximum fracture load

$$G = 1 \qquad D < 6$$

$$G = \{0{\cdot}692 - 0{\cdot}0115D\} + \{0{\cdot}188 + 0{\cdot}0104D\}\log_{10}(\theta) \qquad 6 \leq D \leq 30$$

Where D denotes the nominal pipe diameter (inch), and θ, the half crack angle (degrees).

3 PIPING SYSTEM APPLICABLE TO LBB

3.1 Premise

When the LBB concept is introduced to the design of plant safety features, its validity must be assured by some premise which includes design which will maintain the structural integrity of the piping system, limiting the conditions of potential failure mechanisms that may jeopardize the overall structural integrity during service, and early detection of defects. The premise, in addition to conservative fracture mechanics analysis for the postulated conditions, must give additional protection when applying LBB to design. The following conditions should be satisfied when applying the LBB:

(1) *Highly qualified piping.* Material selection, design, fabrication,

inspection and testing should be performed in accordance with the applicable regulations, codes, standards and guidelines. It is especially important to maintain welding reliability to secure sufficient fracture toughness of the piping components. Therefore, potential piping systems for the application of LBB should be highly qualified Class 1 and 3 piping, by the Notification No. 501, 'Technical Standard for Nuclear Power Plant Equipment Structure', which belongs to reactor safety facilities equivalent to ASME code Classes 1 and 2. This requirement gives additional confidence for the application of LBB.

(2) *Potential failure mechanism.* The piping systems shall not be susceptible to any failure mechanism so far experienced during operation.

(3) *Leak detection capability.* Leak detection capability shall be provided to monitor fluid leakage continuously. Sump water level and condensed water monitoring systems, which are installed inside the containment vessel at present and are capable of detecting 3·8 kg/min (1 gpm) leakage, are judged effective to ensure LBB.

(4) *In-service inspection (ISI) requirement.* Although no credit of ISI is taken to verify LBB, the ISI requirement will not be changed even when LBB is verified.

3.2 Assessment of potential failure mechanism

The initial step is a review that will identify all possible failure mechanisms that might lead to a catastrophic break in the piping. The review consists of past experiences in other industries, and any other potential failure mechanisms. The potential failure mechanisms identified were stress corrosion cracking, general corrosion, erosion and corrosion, cavitation, fatigue, aging, water hammer, vibration and fretting.

These failure mechanisms were closely evaluated when they affected the application of LBB. The results of the review concluded that most of the potential failure mechanisms could be avoided by proper design, fabrication, installation and operating control implemented in nuclear power stations at present. Although erosion with corrosion is an important failure mechanism for carbon steel piping, present water chemistry control, oxygen control for BWRs and pH control for PWRs, was judged to be sufficient to avoid degradation in the piping systems. The only failure mechanism to be taken into account is fatigue, although even fatigue failure is protected against by design.

4 LBB EVALUATION

4.1 Sequence of evaluation

The evaluation method used to verify the LBB concept is based on the behavior of fatigue cracks. Fatigue cracks propagate more quickly in the through-thickness direction of a pipe than in circumferential or axial directions. The flaw was evaluated in the following sequence (as shown in Fig. 2):

(1) Postulate a single initial flaw on the inside surface of the pipe.
(2) The crack size for stability evaluation is based on the crack growth analysis with margins related to periods of the plant life and to the imposed loads. The final crack shapes, which is through-wall or part-through-wall, is defined as the postulated crack size. The maximum through-wall crack length is based on the leak detection capability and its margin.
(3) Crack stability analysis is performed for design loading conditions using the method described in Section 2.2.
(4) The postulated design basis event is defined as the result of the analysis.

4.2 Method of evaluation

4.2.1 Fundamental conditions

4.2.1.1 Location for the evaluation. The existence of significant defects in nuclear grade piping is unlikely because the material is inspected during fabrication and pre-service inspection (PSI). Also the structural integrity of the pressure boundary will be maintained by strict quality control and planned operational control. When a defect is postulated, it is more likely to be found in a weld than in the base metal. Accordingly, the weld is the location to be evaluated.

4.2.1.2 Size of initial flaw. A single, semi-elliptical flaw is postulated. The probability that many significant flaws or a flaw covering the pipe circumference will not be detected by PSI is very low. Therefore, using the detectability of ultrasonic testing used for PSI multiplied by a factor of 2, the size of the flaw is set to half an ellipse of 10% of wall thickness in depth by 50% of wall thickness in length, with a minimum size of 2×10 mm. Although the crack size is more conservative in large diameter piping, this procedure was considered prudent because of the greater consequences should a large diameter pipe fail. Crack orientation is circumferential or axial, depending on the direction of the weld.

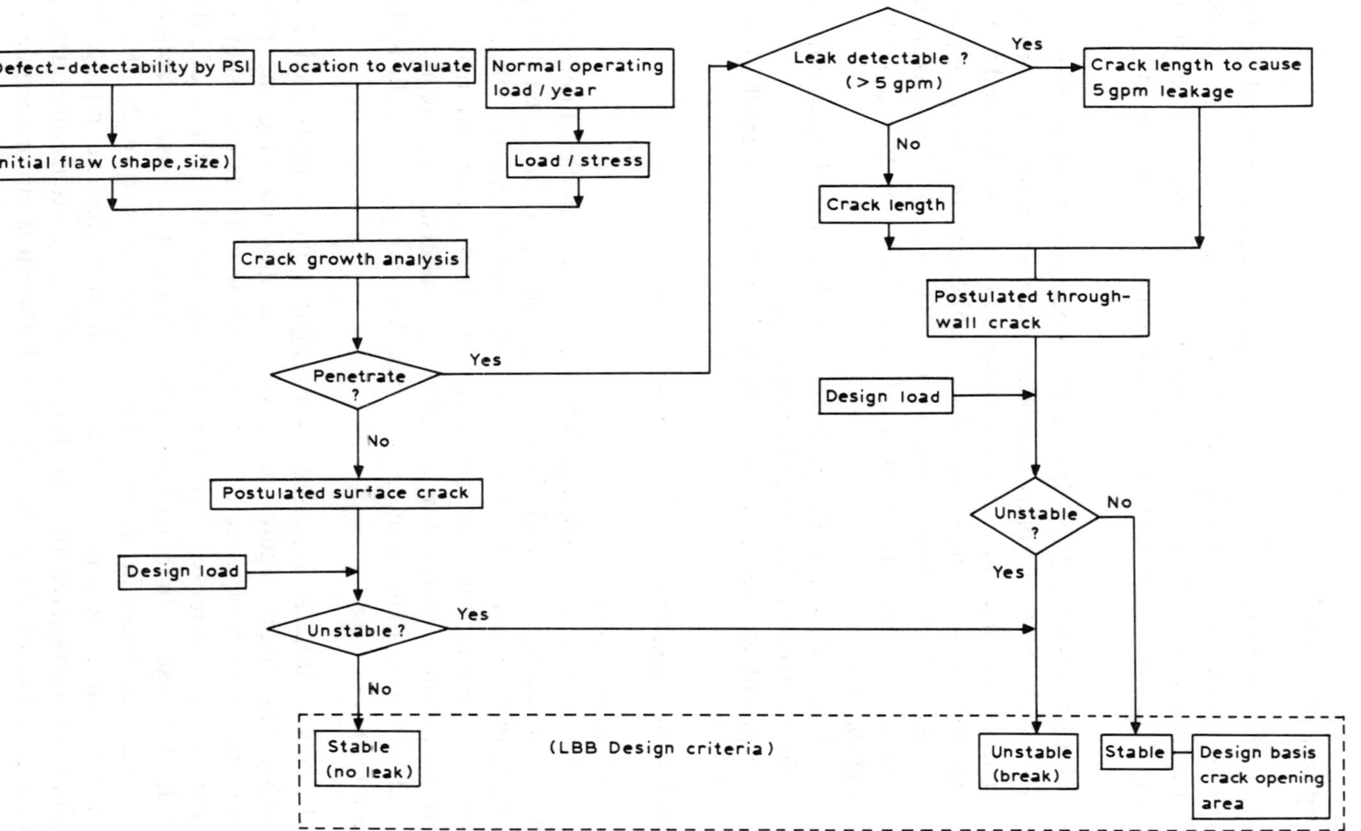

Fig. 2. Evaluation flow to establish LBB design criteria. PSI = Pre-service inspection.

4.2.1.3 Load and stress. Fracture mechanics analyses consist of crack growth analysis and crack stability analysis. Crack growth analysis is performed by conservatively setting the load based on stress reports, using twice as many cycles as required by the design and loading for operating conditions I and II, including equivalent seismic events. For crack stability analysis, operating conditions I, II and III, including equivalent seismic events, were used. The operating conditions I, II and III are equivalent to levels A, B and C in ASME Code Section III, respectively.

The stress cycles for crack growth analysis are shown in Fig. 3 for a PWR and in Fig. 4 for a BWR for a circumferential flaw.

4.2.2 Crack growth analysis

The pipe is assumed to behave like a flat plate. The crack growth rate was determined from corrosion fatigue experimental results under simulated LWR environmental conditions.

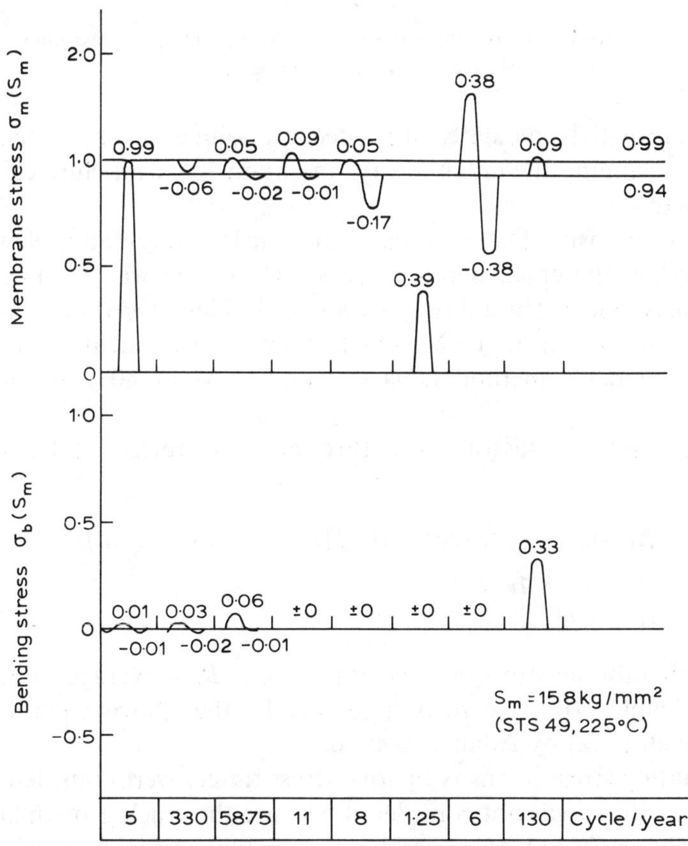

Fig. 3. Stress cycle for crack growth analysis in PWR.

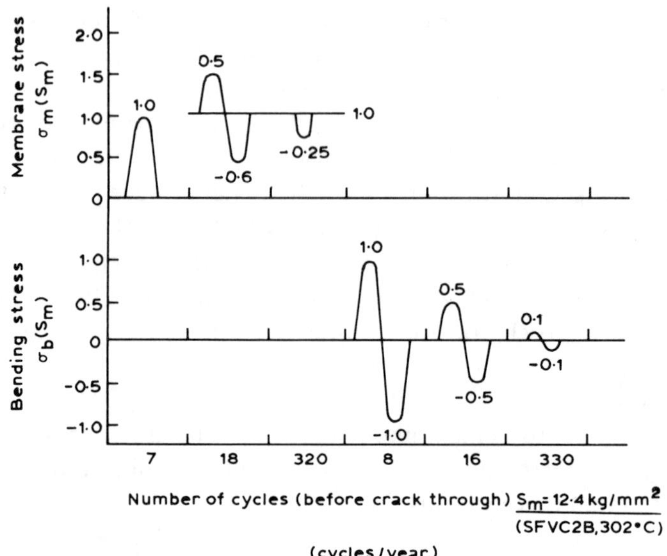

Fig. 4. Stress cycle for crack growth analysis in BWR. S_m = Design stress intensity; σ_m = membrane stress; σ_b = bending stress.

The crack growth is calculated at the deepest point and at the crack tip on the surface, assuming the crack always maintains its semi-elliptical shape during growth.

The stress intensity factor is calculated using Raju and Newman's solution.[3] When the crack depth reaches 80% of the wall thickness, the surface crack is replaced by a through-wall crack. This procedure is based on the results of analysis using FEM, which show that the calculated value by Raju and Newman's method departs from FEM at 80% of the wall thickness.

The stress intensity factor for a through-wall crack is calculated as follows:

$$K_\mathrm{I} = 0{\cdot}31\sigma\sqrt{\pi c \sec \pi c/W}\,.f(\lambda) \qquad (\mathrm{MPa}\sqrt{\mathrm{m}})$$
$$\sigma = \sigma_m + \sigma_b$$
$$W = \pi R_0$$

where σ_m = membrane stress, σ_b = bending stress, R_0 = average radius, c = half crack length, $f(\lambda)$ = correction factor. In the above equation, the correction factor $f(\lambda)$ by Adams[4] is used.

For estimating stress intensity factors, stress ranges were considered. The effect of mean stress was not considered because the crack growth law was conservatively set based on experiments enveloping the data related to parameters of load ratios, materials and environments.

4.3 Postulated crack

4.3.1 Determination of postulated crack

The postulated crack is determined as follows:

(1) Unless the crack propagates through-wall, the crack size resulting from crack growth analysis is regarded as the postulated crack.
(2) If the length of the as-grown Crack 'B' exceeds the length of the through-wall Crack 'A', which leaks at 5 gpm (19 litre/min) under normal plant operation, Crack 'A' is regarded as the postulated crack. This method is used because such a large amount of leakage can be easily detected by the leak detection system.

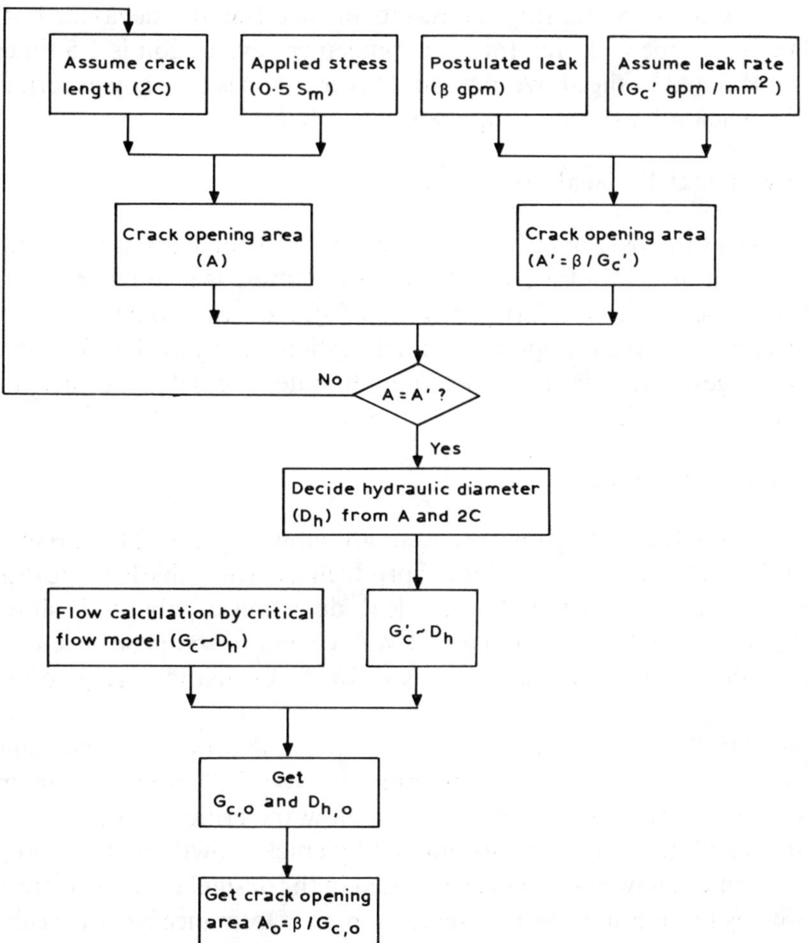

Fig. 5. Method of deciding crack opening area. A, A′, Ao = Crack opening area; $2C$ = crack length; Sm = design stress intensity; β = detectable leak; G_c, G_c', $G_{c.o}$ = leak rate; D_h, $D_{h.o}$ = hydraulic diameter.

(3) If the as-grown crack is smaller than Crack 'A', Crack 'B' is regarded as the postulated crack.

4.3.2 Determination of crack opening area
The crack opening area, which produces a detectable leakage, can be determined according to the sequence shown in Fig. 5.

Leakage from a crack is calculated using the critical flow model, Henry's model for subcooled water,[5] and Moody's slip model for saturated water and steam,[6] based on experimental results.

The crack opening area is obtained by Tada and Paris's method,[7] with loading of 0·5 S_m (S_m = design stress) membrane stress, which will be mainly produced by internal pressure. The estimated crack opening displacement was reviewed by comparing the results of pipe fracture behavior tests.

The crack opening area for jet impingement evaluation is 1·5 times the area calculated by the above method. This criterion is based on experimental results when a large load is applied.

4.4 Crack stability analysis

The stresses (loads) applied in the crack stability analyses are the primary stress components, but a thermal expansion stress is also included.

The net section stress formula, with a *G*-factor using pipe diameter and crack configuration, is applied for crack stability analysis. The flow stress is the average of the design yield and ultimate strengths specified in the Notification No. 501.

4.5 Results of analysis

The results of the crack growth analysis are shown in Table 2 for a BWR and Table 3 for a PWR. The crack length producing a detectable leakage (5 gpm), which is five times greater than the leak detection capability, is shown in Table 4 for a BWR and Table 5 for a PWR. Chemical compositions, strength and toughness of piping materials used for the evaluation are described in Ref. 2.

The critical stress determined from the postulated crack size and using the method described in Section 2.2 is approximately 3 S_m, while the maximum anticipated stress is about 1·6 S_m; well below the critical stress.

The postulated crack size is obtained by crack growth analysis from the assumed initial flaw and stress cycles used in the design. If the realistic crack allowed by radiographic testing specified in the Ordinance No. 81 'Technical Standard for Welding' is employed as an initial crack, the period to reach the postulated crack size is estimated to be as long as 15 to 60 times the design life. This safety margin is judged sufficient.

TABLE 2
Results of Crack Growth Analysis (BWR)

Pipe size (1 inch)	Material	Outside diameter (mm)	Thickness (mm)	Initial crack (mm)		After 80 years (mm)		Critical stress (S_m)	Initial crack (mm)		After N years (mm)		Period (N)	Safety margin (N/40)
				a	*c*	*a*	*c*		*a*	*c*	*a*	*c*		
26 (MS)	STS 49	660·4	33·3	3·33	8·33	6·38	10·62	3·34	3·00	3·00	6·38	10·62	>1 000	>25
16 (MS)	STS 42	406·4	26·2	2·62	6·55	5·30	8·44	3·39	3·00	3·00	5·30	8·44	>1 000	>25
10 (MS)	STS 42	267·4	18·2	2·00	5·00	4·33	6·45	3·38	2·00	2·00	4·33	6·45	>1 000	>25
4 (MS)	STS 42	114·3	11·1	2·00	5·00	4·41	6·64	3·30	2·00	2·00	4·41	6·64	>1 000	>25
20 (FDW)	STS 49	508·0	32·5	3·25	8·13	5·26	8·67	3·34	3·00	3·00	5·26	8·67	>1 000	>25
16 (FDW)	STS 42	406·4	26·2	2·62	6·55	3·55	6·74	3·40	3·00	3·00	3·55	6·74	640	16
10 (FDW)	STS 42	267·4	18·2	2·00	5·00	2·38	5·06	3·40	2·00	2·00	2·38	5·06	>1 000	>25
4 (FDW)	STS 42	114·3	11·1	2·00	5·00	2·41	5·07	3·37	2·00	2·00	2·41	5·07	>1 000	>25
20 (ECCS)	STS 49	508·0	32·5	3·25	8·13	4·57	8·74	3·34	3·00	3·00	4·57	8·74	860	21
16 (ECCS)	STS 42	406·4	30·9	3·09	7·73	4·36	8·26	3·40	3·00	3·00	4·36	8·26	820	20
10 (ECCS)	STS 42	267·4	18·2	2·00	5·00	2·54	5·11	3·40	2·00	2·00	2·54	5·11	>1 000	>25
4 (ECCS)	STS 42	114·3	11·1	2·00	5·00	2·54	5·12	3·36	2·00	2·00	2·54	5·14	>1 000	>25

MS, Main steam system; FDW, Feed water system; ECCS, Emergency core cooling system. Material: STS 42, STS 49, carbon steel.

TABLE 3
Results of Crack Growth Analysis (PWR)

Pipe size (inch)	*Material*	*Outside diameter (mm)*	*Thickness (mm)*	*Initial crack (mm)*		*Stress cycle*	*After 80 years (mm)*		*Critical stress*	*Initial crack (mm)*		*After N years (mm)*		*Period (N)*	*Safety margin (N/40)*
				a	*c*		*a*	*c*		*a*	*c*	*a*	*c*		
16 (FWS)	STS 49	406·4	21·4	2·14	5·35	A	2·19	5·37	3·04	2·0	2·0	2·19	2·27	920	23
						B	5·91	9·09	3·02			5·91	7·96	780	19
28 (MSS)	SGV 42	711·2	31·0	3·1	7·75	A	3·27	7·78	3·32	3·0	3·0	3·27	3·42	780	19
						B	3·38	7·84	3·32			3·38	3·54	950	23
			34	3·4	8·5	A	3·63	8·54	3·32	3·0	3·0	3·36	3·55	>1 000	>25
						B	3·74	8·61	3·32			3·40	3·57	>1 000	>25
30 (MSS)	SGV 42	762	33	3·3	8·25	A	3·51	8·28	3·32	3·0	3·0	3·36	3·55	>1 000	>25
						B	3·62	8·35	3·32			3·40	3·57	>1 000	>25

FWS, Feed water system; MSS, main steam system.
Material: STS 49, SGV 42—carbon steel.

TABLE 4

Leak Detectable Crack Length (BWR) (Stress: 0·5*Sm*; leak rate: 1·9 kg/min (0·5 gpm))

Pipe size (inch)	*Fluid*	*Material*	*Outside diameter (mm)*	*Wall thickness (mm)*	*Crack angle (θ degrees)*
20		STS 49	508·0	32·5	44·4
16	Subcooled water		406·4	30·9	56·0
10		STS 42	267·4	18·2	72·4
4			114·3	11·1	113·2
26		STS 49	660·4	33·3	64·4
16	Saturated steam		406·4	26·2	90·8
10		STS 42	267·4	18·2	109·2
4			114·3	11·1	149·6

4.6 Evaluation for conditions beyond design basis

The above evaluation is based on the consideration of events within design. The inital crack size is based on postulated fabrication defects although crack initiation from in-service defects can be avoided by proper quality and operating control. Also, the fatigue crack growth analysis is very conservative. However, it may be necessary to consider unanticipated events which the engineer could not expect from past operating experiences, such failure might be caused by fatigue. A review was performed of fatigue crack growth behavior.

From Raju and Newman's equation for a stress intensity factor, fatigue crack growth of a surface crack, which is not affected by initial crack size, is characterized by a specific load ratio, L (bending stress/tension stress) and an exponent of stress intensity factor m from Paris's law.[3] The final crack length on the inside surface of the pipe at penetration is six times the wall thickness

TABLE 5

Leak Detectable Crack Length (PWR) (Stress: 0·5*Sm*; leak rate: 1·9 kg/min (5 gpm))

Pipe size (inch)	*Fluid*	*Material*	*Outside diameter (mm)*	*Wall thickness (θ degrees)*	*Crack angle*
16 (FWS)	Subcooled water	STS 49	406·4	21·4	38·6
28 (MSS)	Saturated steam	SGV 42	711·2	31·0 34·0	72·1 71·6
30		SGV 42	762·0	33·0	69·8

TABLE 6
Results beyond Design Basis (BWR)

System pipe size (inch) thickness (mm)	*Crack angle (degrees)*			*G-factor*	*Critical stress (S_m)*
	5 gpm leakage crack	*6 t crack*	*Postulated crack*		
MS-26-33·3	64·4	36·5	64·4	1·14	1·96
MS-16-26·2	90·8	47·4	90·8	1·14	1·62
MS-10-18·2	109·2	50·2	109·2	1·09	1·22
MS-4-11·1	149·2	74·0	149·2	1·00	0·62
FDW-20-32·5	44·4	47·0	47·0	1·07	2·36
FDW-16-26·2	54·8	47·4	54·8	1·07	2·19
FDW-10-18·2	72·4	50·2	72·4	1·04	1·91
FDW-4-11·1	113·2	74·0	113·2	1·00	1·21
ECCS-20-32·5	44·4	47·0	47·0	1·07	2·36
ECCS-16-30·9	56·0	56·6	56·6	1·07	2·16
ECCS-10-18·2	72·4	50·2	72·4	1·04	1·91
ECCS-4-11·1	113·2	74·0	113·2	1·00	1·21

when $L = 1{\cdot}0$ and $m = 5{\cdot}95$; these values are bounding conditions. Tables 6 and 7 show the crack lengths that are detectable by their leaks, fatigue crack length of 6 t (wall thickness, 6 mm) and critical stresses of both BWR and PWR plants. The results of the review are summarized as follows:

(1) A fatigue crack is detected by the leak detection system either when the pipe wall is penetrated, or in the course of slow crack propagation after complete wall penetration.
(2) Under the actual stresses in piping, these cracks are stable, except in small diameter piping, nominal diameter of 4 inch or less for actual use.
(3) The LBB condition is still satisfied during events beyond design, so long as the events are caused by fatigue.

TABLE 7
Results beyond Design Basis (PWR)

System pipe size (inch) thickness (mm)	*Crack angle (degrees)*			*G-factor*	*Critical stress (S_m)*
	5 gpm leakage crack	*6 t crack*	*Postulated crack*		
MSS-28-34·0	71·6	34·5	71·6	1·14	1·82
MSS-28-31·0	72·1	31·3	72·1	1·14	1·81
MSS-30-33·0	69·8	31·1	69·8	1·16	1·82
FWS-16-21·4	38·6	38·2	38·6	1·00	2·34

5 SUMMARY

A review of LBB for carbon steel piping systems in both PWR and BWR plants was conducted as part of NUPEC's verification test program. The following conclusions were obtained:

(1) The premise for an application of LBB and a comprehensive evaluation procedure was clarified.
(2) The structural integrity of safety-related carbon steel piping systems ensures the applicability of LBB.
(3) A fracture mechanics assessment found that LBB conditions could be satisfied with an ample margin of safety and a great degree of reliability, even if fatigue-caused events beyond design occurred.
(4) The LBB concept can be introduced to design against dynamic effects resulting from a postulated pipe break, except for small diameter piping.

REFERENCES

1. Harrison, R. P., Loosamore, K., Milne, I. & Dowling, A. R. Assessment of the integrity of structures containing defects. CEGB Report R/H/R6—Rev. 2, 1980.
2. Asada, Y., Takumi, K., Gotoh, N., Umemoto, T. & Kashima, K., Leak-Before-Break verification test and evaluations of crack growth and fracture criterion for carbon steel piping. *J. Pres. Ves. & Piping*, **43** (1190) 379–97.
3. Newman, J. C. Jr & Raju, I. S., Analyses of surface cracks in finite plates under tension or bending loads. NASA TP-1578, December 1979.
4. Adams, N. J. I., The influence of curvature on stress intensity at the top of a circumferential crack in a cylindrical shell. *ASTM STP 486*, American Society for Testing and Materials, Philadelphia, PA, 1979, p. 39.
5. Henry, R. E., The two-phase critical discharge of initially saturated or subcooled liquid. *Nuclear Science and Engineering*, **41** (1970) 336–42.
6. Moody, F. J., Maximum two-phase vessel blowdown from pipes. *Journal of Heat Transfer*, **88** (1966) 285–95.
7. Paris, P. C. & Tada, H., The Application of fracture proof design methods using tearing instability theory to nuclear piping postulating circumferential through wall cracks. NUREG/CR-3464, September 1983.

Int. J. Pres. Ves. & Piping **43** (1990) 113–127

Ontario Hydro's Leak-Before-Break Approach to Darlington NGS Heat Transport System Piping

J. S. Nathwani & J. D. Stebbing

Nuclear Safety Department, Design and Development Division—Generation, Ontario Hydro, 700 University Avenue, Toronto, Canada M5G IX6

ABSTRACT

The primary objective in our Leak-Before-Break studies is to show how a rational and comprehensive approach can provide an adequate measure of confidence in the assessment of piping integrity such that provision of design features (viz. pipewhip restraints, jet impingement shields) to protect against the dynamic effects of pipe rupture is not necessary. This study is one component of the overall Leak-Before-Break approach adopted at Ontario Hydro.

The results of a review undertaken to evaluate the system transients or events sequences which may subject the piping to a potentially significant increase in loadings are reported. The focus in this paper is to show the approach used in deriving loadings for use in the elastic–plastic fracture mechanics (EPFM) analyses required to demonstrate crack stability.

1 INTRODUCTION

The Leak-Before-Break (LBB) approach which has been developed for Darlington NGS A is being specifically applied to the large diameter heat transport piping: the 21 in. (533 mm) heat transport pump discharge, the 22 in. (559 mm) steam generator inlet, and the 24 in. (610 mm) heat transport pump suction.

A systematic review and critical evaluation of the failure mechanism and causes which could jeopardize integrity of the specific piping is considered the first important step in establishing role and acceptability of the LBB

Int. J. Pres. Ves. & Piping 0308-0161/90/$03·50

concept. The other important aspects are those related to a specific evaluation of material properties, demonstration of crack stability using elastic–plastic fracture mechanics (EPFM) methods, modelling of leak rates and evaluation of the design and operational capability to detect leakage. The results of these studies have been reported elsewhere.[1-8]

In general, the approach as applied to the specified piping, provides assurance that: (a) the largest credible inaugural flaw will not grow through the pipe wall (or to a critical size) for the designed service life of the piping system and, (b) if a crack were to develop through the wall, then it would be detected in a timely manner and that the crack would remain stable even when subjected to the largest credible abnormal loading that the piping system may experience.

This paper briefly describes the overall safety and licensing requirements which define the scope of application of LBB to the large diameter piping at Darlington NGS A. The paper focuses on the results of a review undertaken to evaluate the conditions in the heat transport system which may subject the piping system to a potentially significant increase in loadings. These pressure, thermal and mechanical loadings are used in the EPFM analyses which demonstrate that the readily detectable through-wall crack is stable.

2 LICENSING REQUIREMENTS—BACKGROUND

A licensing requirement imposed on Darlington NGS A was the provision of adequate protection in the design to ensure that no unacceptable consequences result from postulated catastrophic ruptures of HT piping and their associated dynamic effects (i.e. pipewhip or jet impingement). Break locations, for which protection was to be provided, were selected on the basis of stress and geometry-related criteria and included consideration of both circumferentially- and longitudinally-oriented ruptures. The stress-related criterion required postulating breaks at those locations where stresses exceeded a given value, that is, $\frac{2}{3}$ of the appropriate allowable stress. Guillotine ruptures at girth butt-welds and longitudinal ruptures at the sides of those elbows which exceeded the stress-related criterion were to be postulated. The geometry-related criterion required postulating guillotine ruptures at piping anchors, nozzles, changes in geometry and also required postulating longitudinal ruptures at the crotch of branch/tee connections. In Canada, these criteria were developed by a multi-disciplinary joint Ontario Hydro-AECL working group.[9]

For Darlington NGS A large diameter HT piping, difficulties were encountered for some break locations in predicting and quantifying accurately the extent and nature of consequential damage.

The same recognition of the difficulties and undesirability of installing pipewhip restraints had previously prompted American vendors and utilities, the US Nuclear Regulatory Commission and their West German counterparts (the German Reactor Safety Commission, RSK) to reassess the likelihood of large pipe breaks, particularly in nuclear piping systems. In the United States, significant progress has been made in assessing the behaviour of cracks in piping by using the EPFM analysis methods. The US NRC has since reviewed and accepted, on a case by case basis, applications for exemptions from providing pipewhip restraints on the basis of this approach.

The German Basis Safety concept was adopted in principle by the German Reactor Safety Commission (RSK) in 1977. When applied, this concept precluded the previously perceived need for installation of pipewhip restraints. The essence of the Basis Safety Concept is that through optimized materials (as determined by full-scale testing), optimized design and quality in manufacture, supplemented by a number of independent redundancies such as inspection, sufficient assurance against catastrophic pipe failure is provided. The approach is used for both primary and secondary side piping. A comparison of these international approaches is considered in detail in Ref. 10.

An Ontario Hydro review in 1983/84 of the developments in EPFM techniques formed the basis for the application. The method used was based on the *J*-Integral/Tearing Modulus approach. The viability of the approach was confirmed when it was applied to the Darlington NGS A 24-inch heat transport pump suction line; the analysis provided assurance that large margins exist between crack sizes which could leak at rates for which immediate shutdown action was already required and for those crack sizes which could grow unstably causing catastrophic guillotine pipe rupture.

In early 1985, Ontario Hydro obtained agreement in principle for use of EPFM in lieu of pipewhip restraints from the Atomic Energy Control Board (AECB), the Canadian Regulatory Authority.

3 DARLINGTON NGS A—DESIGN FEATURES AND LBB APPROACH

3.1 Design features

Darlington NGS A is a four unit CANDU, a pressurized heavy water cooled, heavy water moderated and natural uranium fuelled nuclear generating station. The reactor is contained in a cylindrical, horizontal, single-walled stainless steel vessel called the calandria. The calandria houses the heavy water moderator and reflector (at a low pressure, 24 kPa(g), and low temperature 65°C) and is axially penetrated by 480 calandria tubes. The calandria tubes contain and support the in-core portions of the fuel channels and isolate the pressure tubes (which contain the fuel and heavy water coolant) from direct contact with the moderator.

The heat transport system circulates pressurized heavy water (10 MPa; 310°C) through the reactor fuel channels. The heat transport main circuit consists of two loops each containing two pumps, two steam generators, two reactor inlet headers, two reactor outlet headers, and one inlet feeder and one outlet feeder for each of the 240 reactor fuel channels. Each loop removes the heat from half of the core. The two main circuits provide bi-directional flow through the core such that flow is in opposite directions in adjacent channels and individual feeder pipe sizes are selected and channel flows trimmed to give the same exit quality from all outlet feeders at full reactor power. Heavy water is pumped through the large diameter heat transport pump discharge lines to the reactor inlet header (RIH) and from there fed to each of the fuel channels through individual inlet feeder pipes,

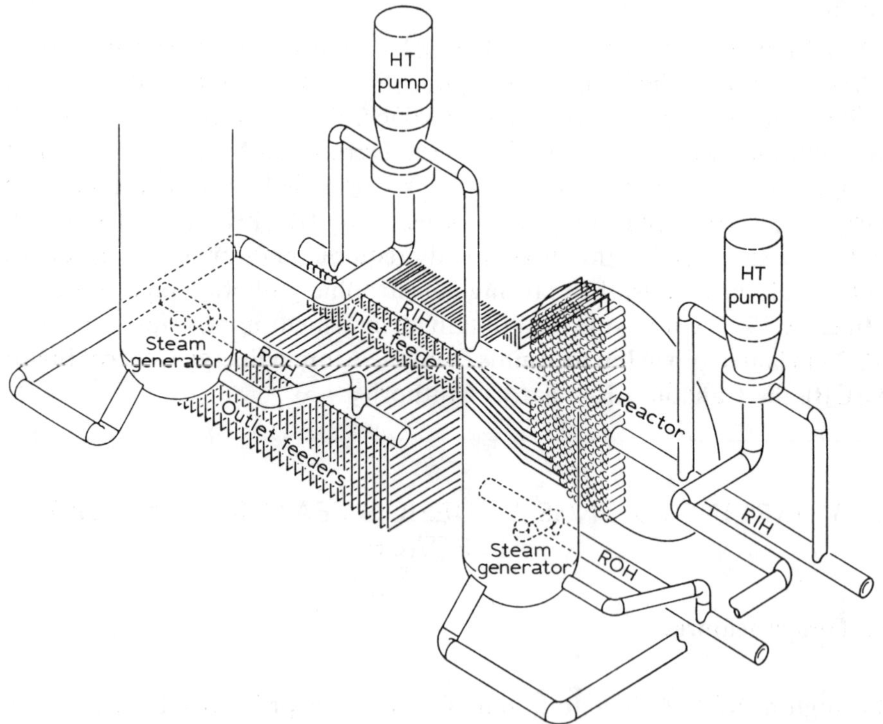

Fig. 1. Darlington NGS A—main heat transport circuit large diameter piping.

and from each fuel channel through individual outlet feeder pipes to the horizontal reactor outlet headers (ROH).

From the reactor outlet headers, the coolant flows through two 559 mm (NPS 22 Schedule 100) lines to the steam generator where the heat is transferred to the secondary side. Each steam generator is connected to the suction of the heat transport pump by one 610 mm (NPS 24 Schedule 100) line. Each heat transport pump delivers heavy water to one reactor inlet through two 533 mm (NPS 21) lines. The LBB approach is applied to the large diameter piping (greater than or equal to NPS 21) as illustrated in Fig. 1.

3.2 LBB approach

Generally, the elements of the LBB approach for Darlington NGS A can be grouped into two areas: (a) those related to the demonstration of crack stability utilizing fracture mechanics methods; and (b) those related to leakage rate predictions and assessment of detection capability. Details of the overall approach are provided in Refs 1 and 2.

The essential results of a review undertaken to identify the possible events which may subject the piping to increased loadings are presented below. These loadings are to be then used in the EPFM analyses to demonstrate crack growth stability under such conditions. Only the top-end of the fault tree is shown to illustrate the ways in which an increase in loadings can occur in the system. The detailed fault trees are documented in Ref. 11.

4 TYPES OF LOADINGS

The loadings on any piping system can be considered in terms of three basic elements: thermal, mechanical and pressure loadings.

4.1 Thermal loadings

Heat can induce two different types of stresses in piping system:

(a) thermal expansion of the piping and thermal anchor movement which gives rise to net piping section loads and induces secondary stresses; and
(b) thermal gradients through the wall which induce stresses that can be classified in terms of peak or secondary loadings as used in the ASME context.

The effects of heat on piping can be characterized in terms of three parameters as follows:

(a) the steady state temperature (T_1);
(b) the magnitude of temperature change (ΔT_1); and
(c) the rate of temperature change ($\Delta T_1/s$).

4.2 Mechanical loadings

Mechanical loadings involve mechanical restraint or excitation of a component during operation and are normally defined as forces and displacements applied at given locations to a system or component. For ease of identification, mechanical loadings in this study are categorized as being generated either externally or internally.

An example of an externally applied mechanical loading is a design basis earthquake (DBE) where inertia and anchor point motions are applied to the heat transport system piping and components. These are superimposed on loadings arising from normal operating conditions and deadweight. Internally generated mechanical loadings are those associated with acceleration or deceleration of fluid flow conditions within the system. An example of this type of loading is the mechanical loading applied by the heat transport pump under two-phase flow conditions.

4.3 Pressure loadings

Internal pressure leads to primary stresses. These are not self-limiting and are, therefore, considered in any summation of net piping section loadings.

4.4 Initiating events and sequences

Identification of initiating events and sequences which cause a change in the applied loadings on the piping is the first step in the evaluation process. Although the abnormal loadings arising from any sequence of events are expected to comprise a combination of the three basic loading elements (namely, pressure, thermal and mechanical), the objective at the first step was to identify the event sequences which could cause the maximum loading on the piping due to a specific loading element (for example, pressure or mechanical or thermal). Hence, each loading element was examined separately. For example, sequences were identified which can give rise to increased pressure loadings; then other sequences were considered for mechanical loadings and, subsequently, the sequences for thermal loadings.

By following a process of deductive reasoning, a fault tree was developed

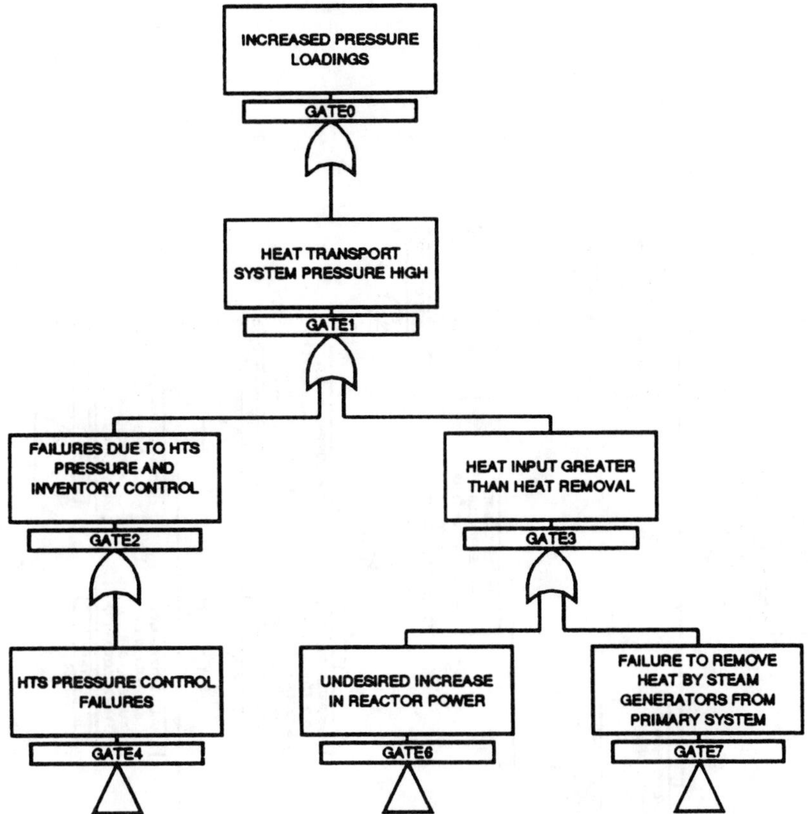

Fig. 2. Darlington NGS A—LBB pressure loadings.

for each basic loading element. The results are shown in Figs 2, 3 and 4. The purpose of the fault trees is to show qualitatively all the ways in which piping can be subjected to increased loadings. These possible ways involve either the initiating events or event sequences.

Once all the initiating events and event sequences had been identified, they were examined to identify the sequences with the maximum value for each loading element. Details of the evaluations for each loading element are described in Ref. 11.

The loading elements were then combined. The objective in combining the maxima for each of the basic loading elements (i.e. pressure, thermal, mechanical) is to provide a set of loadings for input to the EPFM analyses which conservatively envelopes all identified credible events that can give rise to one or more of the basic loading types. In specifying the loadings for EPFM analyses in this way, it was recognized that temperature and pressure are related, as may be pressure or temperature to a specific type of mechanical loading. Hence, this is recognized when specifying the

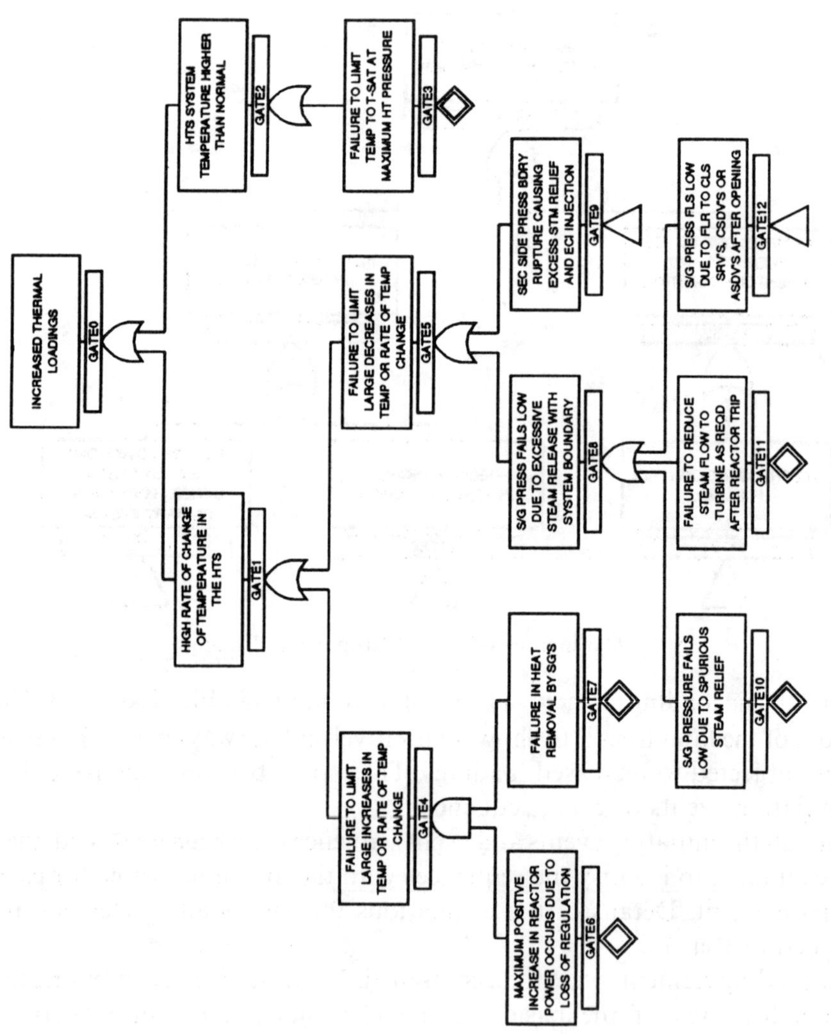

Fig. 3. Darlington NGS A—LBB thermal loadings.

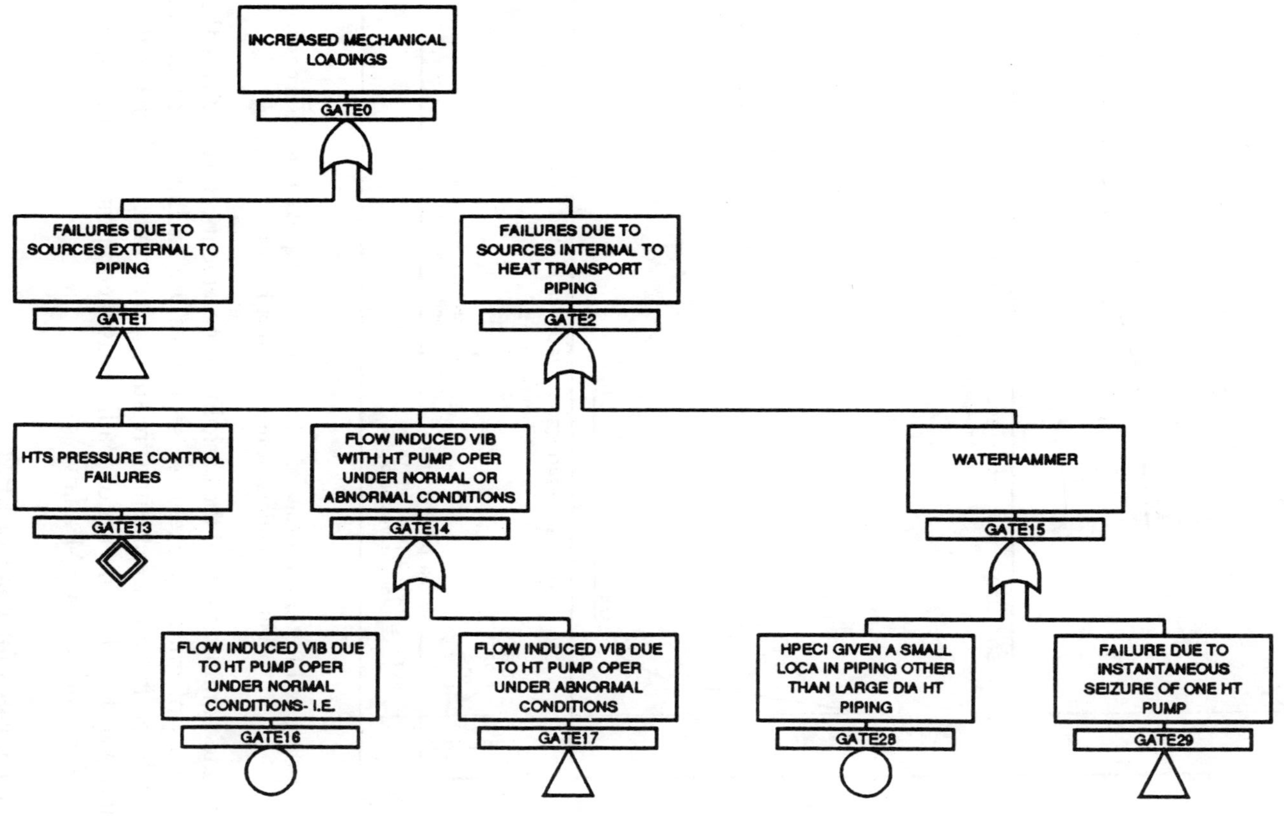

Fig. 4. Darlington NGS A—LBB mechanical loadings.

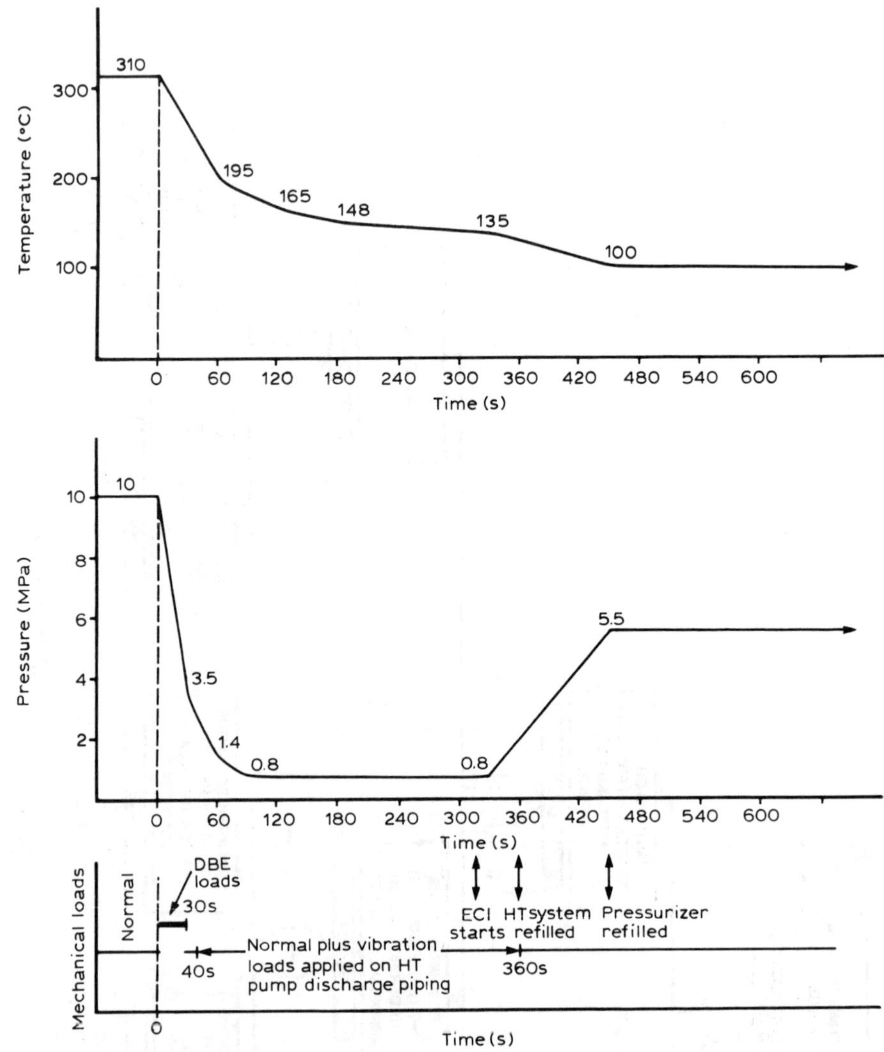

Fig. 5. Darlington LBB/EPFM—Case (d): design basis earthquake plus seismic induced failure conditions at reactor outlet headers.

'appropriate' combinations of the basic loading elements, thus ensuring that combinations of mutually exclusive sequences are excluded. The combination of the basic loading elements (Pressure$_{max}$, Thermal$_{max}$, Mechanical$_{max}$) is discussed in Ref. 11. This set of loading cases comprise the appropriate transient for input to EPFM analyses.

4.5 Loading cases for input to EPM

The cases which were further developed to enable pressure, thermal and mechanical loadings to be quantified are as follows:

(a) rapid cooldown
(b) rapid pressurization
(c) design basis earthquake (DBE) at normal operating conditions
(d) DBE plus seismically induced failures
(e) failures causing waterhammer and vibration effects

Cases (d) and (e) above are discussed in detail below.

4.5.1 Case (d)—DBE plus seismically induced failures

The composition case (Case (d)) is shown in Fig. 5.

The early portion of the temperature transient is similar to that experienced in the rapid cooldown. However, the later portions are more severe, due to overcooling of the heat transport system, on the assumption of good heat transfer coupling within the steam generator, and with rapid heat rejection to the secondary side through the assumed broken steam balance header. The main steam isolating valves are assumed to remain open throughout. Because there is no reactor building high pressure signal, on which the Emergency Cooling Injection (ECI) is conditioned, the ECI signal is conditioned out for 5 min at about the 20 s mark, on the basis of low heat transport pressure plus a 5 min delay. High pressure ECI is not assumed to start, therefore, until approximately 320 s. Thereafter, the temperature decreases more quickly than it would on steam cooling only.

The pressure decrease depicted in the transient is derived from the thermohydraulic computer code simulation enveloping this case. Pressure drops quickly because of overcooling of the heat transport circuit, void creation, the assumed loss of D_2O makeup feed, and marginal loss of D_2O due to the failed fuelling machine catenary. Pressure remains at a low level for approximately 5 min, and recovers as the system and pressurizer refill once ECI starts.

Mechanical loading from the DBE is applied over the first 30 s of the vent. This is consistent treatment of DBE both in design, and in the assumed seismically induced failures. Loadings due to pump induced vibration were also considered (details are given in Ref. 11).

While the Case (d) addresses seismically induced secondary failures, random secondary pipe failure cannot be discounted entirely. In the secondary side of the plant, random pipe failures may be postulated in any piping system. Thermal effects on the main heat transport circuit could be transmitted through the steam generator. The three main piping connections on the secondary side of the steam generator are the main feedwater nozzle, the main steam outlet nozzle, and the re-heat condensate nozzle. As the re-heat condensate nozzle and associated piping are significantly smaller in size than the feedwater connection, only the feedwater and steam nozzle failures are considered as design basis events.

Piping failures at each nozzle respectively are postulated to occur independently (at the feed pipe nozzle weld to the steam generator, or at the steam piping nozzle weld to the steam generator nozzle). Single random pipe failure events are assumed to occur on only one vessel at any specific time.

The feedwater line break at the steam generator feedwater nozzle imposes mechanical loadings on the steam generator and its supports. Analyses performed for the steam generator internals, which are exposed to hydraulic and internal mechanical loading during the event, demonstrate that primary system integrity (i.e. pressure boundary shell, tubes and tubesheet) is maintained throughout the transient. For the heat transport system, the loss of feedwater to all steam generators causes heat transport temperature and pressure to rise. Heat transport high pressure, steam generator low level, and feedwater line low pressure trips are credited to trip the reactor; this results in a short-term primary side perturbation in both temperature and pressure. However, since the other three steam generators within the reactor unit continue to reject heat, the overall effect on the heat transport system is minimal.

The steam line break, postulated to occur at the steam generator steam outlet nozzle weld to the steam piping, is more significant than that of the feedwater line break in terms of its effect on the heat transport system. Heat rejection from this steam generator is rapid, and local effects in its piping quadrant are as seen in Case (d) (Fig. 5). The general effect on the other three quadrants of the heat transport system, remote from the steam generator with the break approach that of the rapid cooldown transient. Completed analyses indicate that steam generator primary pressure boundaries (tubing, tubesheet, etc.) survive loadings from this event also.

4.5.2 Case (e)—failures causing waterhammer and vibration effects

Flow induced vibration with HT pump operating under abnormal conditions is a source internal to piping which may result in increased mechanical loadings. For the loading case for input to EPFM analyses, the pump induced vibration forcing function is applied to conditions consistent with those found during testing. Figure 6 illustrates Case (e).

The cooldown rate is controlled by secondary side steam rejection; the rate of temperature change is only slightly faster than the rapid cooldown due to the effect of injection of cooler emergency coolant injection (ECI) fluid. In each of Cases (d) and (e), assuming Class IV power remains available to power the pumps, conditions were found in which pump induced piping vibration is plausible. Superimposed, therefore, on Cases (d) and (e), are the vibration loads over the time periods where hydraulic and temperature

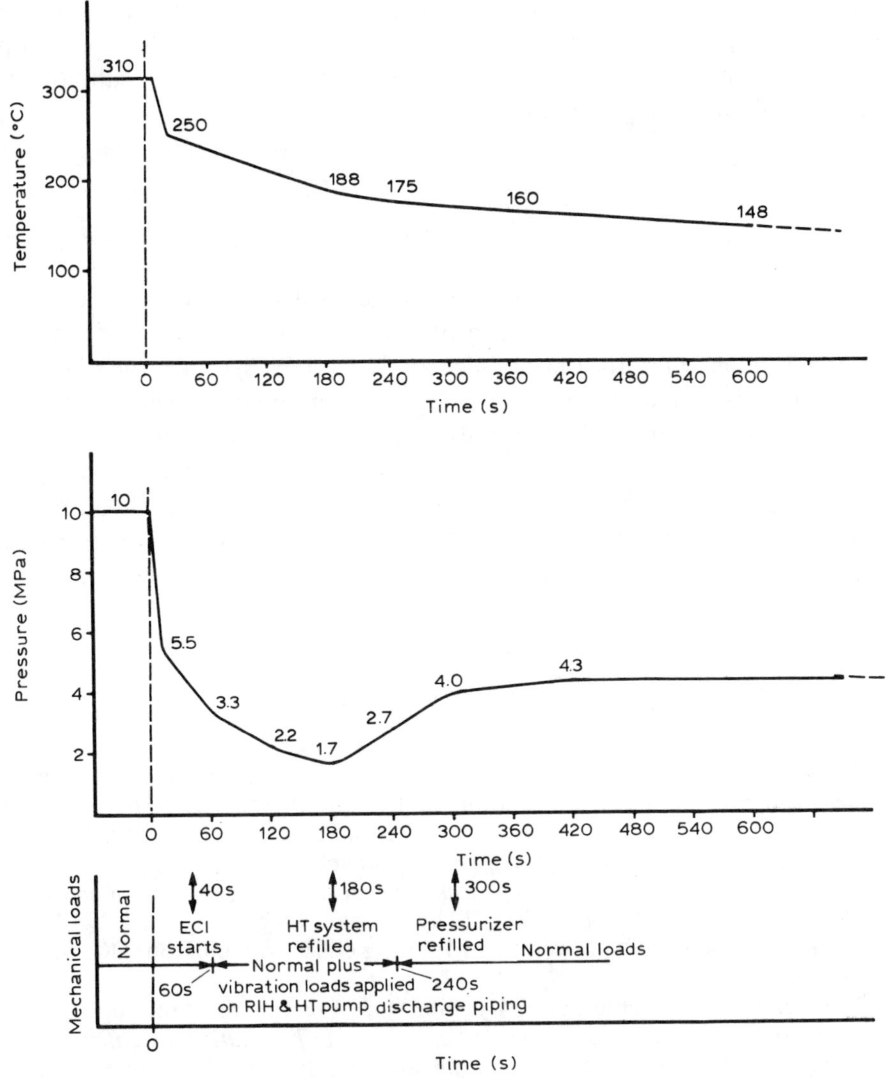

Fig. 6. Darlington LBB/EPFM—Case (e): failure causing waterhammer and vibration effects.

conditions permit pump vibration to occur. The magnitude of the forcing function assumed is the maximum experienced during the pump testing. While data are available indicating lesser magnitudes of forcing function in specific temperature ranges, the maximum is used as a measure of conservatism, regardless of the temperature. The time periods over which the vibration is applied are assumed to start and finish with the onset of insufficient NPSH, rather than strictly at saturation conditions (as observed in testing).

In summary, the loading Case (e) incorporates the thermal, pressure and mechanical loadings which consider the waterhammer and flow or pump induced vibration effects following a feeder size break and subsequent initiation of emergency coolant injection.

5 CONCLUSIONS

The objective of this paper was to present the results of a review undertaken to identify and generate a set of appropriate loading conditions for input to the elastic–plastic fracture mechanics analyses (EPFM) for large diameter heat transport system piping. The composite loading cases were developed for transient pressures, temperatures and mechanical loadings.

It concluded that the cases evaluated for the Darlington heat transport system provide appropriate 'upper bound' loading conditions for input to EPFM analyses which are required to demonstrate Leak-Before-Break behaviour of the specified large diameter heat transport system piping.

REFERENCES

1. Nathwani, J. S., Kee, B. L., Kim, C. S. & Kozluk, M. J., Ontario Hydro's heat before break approach: application to the Darlington (CANDU) Nuclear Generating Station A. *Nuclear Engineering and Design*, **111** (1989) 85–107.
2. Nathwani, J. S., Ontario Hydro's LBB approach for Darlington NGS A. In *Proceedings of the Seminar on 'Leak-Before-Break Progress in Regulatory Policies and Supporting Research'*, Tokyo, Japan, 14–15 May 1987. US NRC NUREG/CP-0092, US Government Printing Office, Washington, DC, March 1988 pp. 113–50.
3. Mukherjee, B., Carpenter, D. & Kozluk, M. J., Fracture resistance of SA106B seamless piping and welds. *Transactions of the 9th International Conference on Structural Mechanics in Reactor Technology*, Volume A. American Society of Mechanical Engineers, Pittsburgh, PA, 1987, pp. 155–62.1.
4. Kozluk, M. J., Aggarwal, M. L. & Vijay, D. K., Demonstrating Leak-Before-Break for CANDU heat transport piping. *Int. J. Pres. Ves. & Piping*, **34** (1988) 255–63.
5. Kozluk, M. J., Manning, B. W., Misra, A. S., Lin, T. C. & Vijay, D. K., Linear–elastic solutions for long radius piping elbows with curvilinear throughwall cracks. In *Advanced Topics in Finite Element Analysis*, ASME PVP-Vol. 143; 19–23 June, American Society of Mechanical Engineers, Pittsburgh, PA, 1988, pp. 23–8.
6. Aggarwal, M. L., Kozluk, M. J., Lin, T. C., Manning, B. W. & Vijay, D. K., A Leak-Before-Break strategy for CANDU primary piping systems. *Int. J. Pres. Ves. & Piping*, **25** (1986) 239–56.

7. Kozluk, M. J., Lin, T. C., Manning, B. W., Misra, A. S., Scarth, D. A. & Vanderglas, M. L., Effects of pressure loading on throughwall cracks. In *Transactions of the 9th International Conference on Structural Mechanics in Reactor Technology*, Volume G, American Society of Mechanical Engineers, Pittsburgh, PA, 1987, pp. 481–8.
8. Mukherjee, B., Observations on the effect of post-weld heat treatment on T-resistance curves of SA106B seamless piping welds. *Nuclear Engineering and Design*, **111** (1989) 63–75.
9. Ontario Hydro, Guidelines for selecting location at which to postulate ruptures in CANDU reactor piping. Report No. 80062, Ontario Hydro Design and Development Division, January 1980.
10. Kee, B. L. & Nathwani, J. S., Darlington NGS A Leak-Before-Break approach for large diameter HT piping. Design and Development Division Report No. 85371, Ontario Hydro, December 1985.
11. Nathwani, J. S., Loading conditions for LBB analysis of the Darlington NGS A large diameter HT piping. D&D Report 88583, Ontario Hydro, July 1988.

Int. J. Pres. Ves. & Piping **43** (1990) 129–149

Analysis of Two Simplified Methods (R6 & GE-EPRI) for Circumferential Crack Stability in Leak-Before-Break Applications

Ph. Taupin, Ph. Gilles & S. Bhandari

Framatome, Tour Fiat, La Defense, 92084 Paris, Cedex 16, France

ABSTRACT

The Ainsworth method (R6 Option 2) and the GE-EPRI method are compared and analysed. These J-estimation schemes allow us to assess at low cost, crack stability in piping systems, which is an important step in Leak-Before-Break applications. The reliability, flexibility and self-consistency of the two methods are examined by focusing on the effect of geometry, nature of loading and superposition of tension and bending. The GE-EPRI method, based on Finite Element results, is used to check the validity of the hypothesis on which the simplifications in the R6 method for through-wall circumferentially cracked pipes rely. The R6 method appears to be easier to use for the treatment of secondary loads. It is shown that, the Ainsworth Failure Assessment Line, derived in the single load case, is still valid for combined proportional tension and bending loads, provided the appropriate limit load formula is chosen.

1 INTRODUCTION

One of important issues in the application of the Leak-Before Break (LBB) concept on nuclear piping is the development of simplified methods for a systematic evaluation of a large range of pipe geometries and loading conditions. Any approach based on the Finite Element Method would be too costly in time and money, linear elastic fracture mechanics methods (LEFM methods such as fracture loading estimation based on plastic-zone

Int. J. Pres. Ves. & Piping 0308-0161/90/$03·50

size-corrected stress-intensity factors) would be unable to account for non-linear phenomena, such as initiation and stable crack growth, involved in LBB analysis.

Framatome has selected two simplified methods meeting these two requirements of low cost and efficiency; the RH-R6 method developed by the CEGB,[1] and the GE-EPRI method.[2] This paper analyses and compares these two methods by focusing on three key problems encountered in LBB applications performed by Framatome.

The first problem concerns the domain of applicability of these methods. Are they valid for any type of crack geometry? How do they account for the material characteristics (strain hardening effects, welded joints)? What is their accuracy?

The second point deals with the effect of the nature of loading: non-linear analysis is applied only to the cracked component, thus the loads provided by the linear elastic analysis of the pipe system have to be clarified in secondary or primary load as to whether or not they are self-limited.[3]

Once the cracked body is characterized and the boundary conditions are fixed, loading conditions must be considered. These two methods, the R6 method (or more precisely, the Failure Assessment Line of this method) and the GE-EPRI handbook method have been initially developed for cracked components subjected to a single uniaxial load. The third part of this paper examines some different loading combination rules and points out their limitations.

2 STUDY OF THE EFFECT OF SPECIMEN GEOMETRY AND MATERIAL CHARACTERISTICS

To analyse the first problem, one has to go back to the basis of the development of the present double-criteria of the CEGB technique[4] to assess how the material characteristics and the geometrical behaviour have been incorporated into this method.

2.1 The basis of the double-criteria of the CEGB method

This method, like the EPRI method, is based on the J-integral. Three options have been proposed in Ref. 1. We shall mainly refer to Option 2, the basis of which is described below.[4]

The Failure Assessment Diagram (FAD) requires the evaluation of the parameter

$$K_r = \sqrt{J^e/J} \tag{1}$$

as a function of

$$L_r = \frac{\text{applied load}}{\text{limit load}} = \frac{P}{P_0} \tag{2}$$

where J^e and J represent the elastic and the total value of the J-integral, respectively and are functions of load and geometry. The limit load P_0 is a function of geometry.

Using the EPRI document,[5,6] for a material following a Ramberg-Osgood type stress–strain ($\sigma - \varepsilon$) curve

$$\frac{\varepsilon}{\varepsilon_0} = \frac{\sigma}{\sigma_0} + \alpha\left(\frac{\sigma}{\sigma_0}\right)^n \tag{3}$$

(where σ_0 is arbitrarily chosen, α and n are fitting parameters, $\varepsilon_0 = \sigma_0/E$, E is the Young's modulus) one can write

$$J = K^2(a_e)/E' + \alpha\sigma_0\varepsilon_0 c h_1(a/w, n)\left(\frac{P}{P_0}\right)^{n+1} \tag{4}$$

where

a is the crack length in a section of width $w = a + c$;

h_1 is a coefficient obtained from Finite Element computations and tabulated in Ref. 5;

$E' = E$ in plane-stress
$= E/(1 - \nu^2)$ in plane-strain (ν is Poisson's ratio);

K represents the stress intensity factor for an effective crack length a_e of

$$a_e = a + \frac{1}{\beta\pi}\left(\frac{n-1}{n+1}\right)\left(\frac{K}{\sigma_0}\right)^2\left[\frac{1}{1 + (P/P_0)^2}\right] \tag{5}$$

where $\beta = 2$ for plane-stress and 6 for plane-strain.

The use of this method, with the J-integral as defined above, poses the problem of FAD dependence on the parameter n through two terms

(i) term $K^2(a_e)$ as a_e depends on n
(ii) term $h_1(P/P_0)^{n+1}$

Ainsworth[4] has proposed some modifications so that the FAD becomes independent of the parameter n. In the following, these modifications are analysed.

2.2 The use of the reference stress concept

The reference stress is defined as

$$\sigma_{ref} = \frac{P}{P_0} \times \sigma_0 \tag{6}$$

Using the stress–strain curve (σ_{ref}, ε_{ref}) defined by eqn (3), the J-integral thus becomes

$$J = K^2(a_e)/E' + ch_1(a/w, n) \times \sigma_{ref}\left(\varepsilon_{ref} - \sigma_{ref}\frac{\varepsilon_0}{\sigma_0}\right) \tag{7}$$

The plastic part of J for $n = 1$ corresponds to the elastic solution for $v = 0{\cdot}5$ which gives

$$J = J^e\left\{\frac{K^2(a_e)}{K^2(a)} + \mu\left(\frac{E'}{E}\right)\frac{h_1(n)}{h_1(1)}\left(E\frac{\varepsilon_{ref}}{\sigma_{ref}} - 1\right)\right\} \tag{8}$$

where $\mu = 1{\cdot}0$ for plane-stress and 0·75 for plane-strain conditions. The second term in the above equation still depends on n such that

$$h_1(n)/(P_0)^{n+1} = \text{constant}$$

Ainsworth proposed to allow

$$\frac{h_1(n)}{h_1(1)} = 1 \tag{9}$$

through the choice of P_0 such that the dependence of h_1 on n is minimized.
The resulting equation is

$$J = J^e\left\{\frac{K^2(a_e)}{K^2(a)} + \mu\left(\frac{E'}{E}\right)\left(E\frac{\varepsilon_{ref}}{\sigma_{ref}} - 1\right)\right\} \tag{10}$$

2.3 Plasticity correction

The first term in eqn 10 still depends on n through a_e. For small values of $\Delta a = a_e - a$ one can write

$$\frac{K^2(a_e)}{K^2(a)} = 1 + \gamma\frac{L_r^2}{(L_r^2 + 1)} \tag{11}$$

where

$$\gamma = \left(\frac{2}{\mu\beta\pi}\right)\left(\frac{n-1}{n+1}\right) ch_1(1)\frac{\partial K/\partial a}{K}$$

Ainsworth then proposed that the plastic zone corresponds to the case of an infinite plate under tension in plane stress conditions for $n \to \infty$ which gives $\gamma = 0{\cdot}5$, $\mu = 1{\cdot}0$. Besides, he considered $\mu E'$ equal to E.

Thus

$$K_r = \left\{ E\frac{\varepsilon_{ref}}{\sigma_{ref}} + 0{\cdot}5\frac{L_r^2}{1+L_r^2} \right\}^{-1/2} \tag{12}$$

which is the Failure Assessment Line (FAL) of the double-criteria R6 method, Option 2.

2.4 Analysis of the two hypotheses of Ainsworth

To appreciate the difference between the GE-EPRI method and the R6 criterion, one has to evaluate the two hypotheses of Ainsworth:

(i) $\dfrac{h_1(n)}{h_1(1)} = 1$

(ii) $\gamma = 0{\cdot}5$

The first one is non-conservative if $\dfrac{h_1(n)}{h_1(1)} > 1$

The second is conservative if the plastic zone is smaller than that obtained for an infinite plate under tension for $n \to \infty$ in plane stress conditions.

A systematic study was carried out to compare the Ainsworth's FAL with that obtained using the EPRI results. It is observed that Ainsworth's criterion, although giving conservative results in compact tension specimens

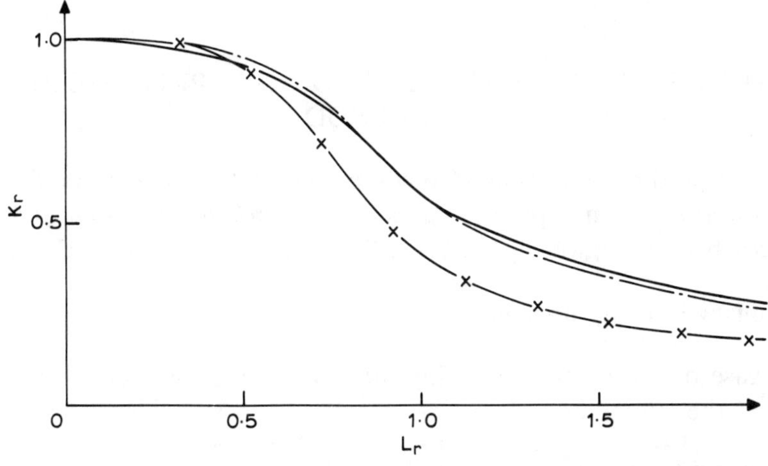

Fig. 1. FAD for a circumferentially through-wall cracked cylinder under bending ($R/t = 20$, $a/b = 0{\cdot}0625$). ——, Ainsworth; —·—, EPRI No. 2; —×—, EPRI No. 7.

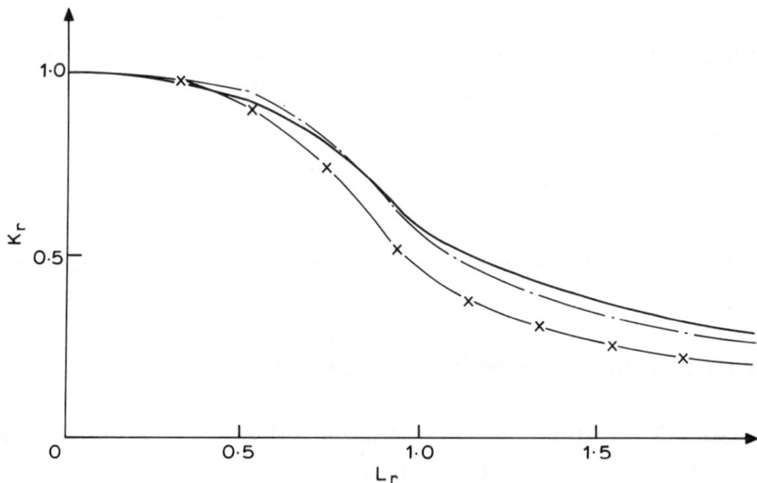

Fig. 2. FAD for a circumferentially through-wall cracked cylinder under axial tension ($R/t = 20$, $a/b = 0{\cdot}0625$). Key as Fig. 1.

and a center-cracked tension specimen in plane-strain conditions, can lead to unconservative results in some cases; in particular, for a circumferential through-wall crack in a cylinder under tension or bending, for some values of n (Figs 1 and 2). This case is of special importance in LBB analysis.

Thus the simplification in the method is at the cost of unconservatism for some cases of the strain hardening coefficient n. The authors feel that it is important to underline this point to avoid any misinterpretation in the results. For low values of n (between 3 and 5), generally met in the materials for LBB demonstrations, the results obtained by both methods are quite similar.

3 NATURE OF LOADING EFFECTS ON GE-EPRI SCHEME AND R6-METHOD

In this section the methods used to compute initiation of a circumferential *through-wall crack* in a pipe in cases of pure *primary* bending and pure *secondary* bending following the GE-EPRI and R6 methods are developed.

3.1 Primary bending moment

In the case of the primary bending moment, the GE-EPRI scheme is as simple as the R6 method.

3.1.1 GE-EPRI (primary load)

The J-integral is calculated as a function of crack and pipe geometries, primary bending moment M_P and Ramberg–Osgood parameters α and n with the following relation:

$$J = J^e(a_e, M_P) + J^P(a, M_P) \tag{13}$$

Equations of $J^e(a_e)$ and J^P, are the following[6]:

$$J = f_1\left(a_e, \frac{R}{t}\right)\frac{M_P^2}{E} + \alpha\sigma_y\varepsilon_y(b-a)\frac{a}{b}h_1\left(\frac{M_P}{M_y}\right)^{n+1} \tag{14}$$

where h_1 and f_1 are values tabulated in Ref. 6 as functions of a/b, R/t and n; $2b$ is the pipe circumference; a is the half crack $[(a/b) = (\gamma/\pi)]$; the other coefficients are defined in Section 2.1, σ_0 being taken as σ_y. M_y is the limit load of the crack pipe computed with σ_y:

$$M_y = 4\sigma_y R^2 t\left(\cos\frac{\gamma}{2} - \frac{1}{2}\sin\gamma\right) \tag{15}$$

3.1.2 R6 method (primary load)
The R6 Failure Assessment Line is computed (see eqn (12) with $L_R = \sigma_{ref}/\sigma_y$) using the method proposed by Ainsworth.[4]

The point (K_r, L_r) corresponding to the crack pipe under a primary bending moment is computed as follows:

$$K_r = \frac{K_I}{K_{JC}}(M_P) \tag{16}$$

where $K_I(M_P)$ is the Stress Intensity Factor computed with appropriate weight function of the cracked pipe and stress profile through the thickness corresponding to M_P, and $K_{JC} = \sqrt{J_{IC}E}$ in plane stress, and L_r represents the ratio of the elastic applied load M_P to the limit load of the cracked pipe M_y.

3.2 Secondary bending moment

In the case of 100% secondary bending moment, the GE-EPRI scheme is more complex to use than the R6-method.

3.2.1 GE-EPRI (secondary load)
For a given secondary bending moment M_s, we have to compute, as a first step, the corresponding elastic rotation of the uncracked pipe in pure bending, considered as a cantilever beam:

$$\theta = \frac{M_s L}{EI} \tag{17}$$

The value θ is the effective rotation at the end of the pipe induced by thermal expansion or anchor movement, L is the half pipe length and I the moment of inertia of cross-section.

The pipe is only loaded by this rotation θ. The elastic and plastic rotation Φ_c^e and Φ_c^p due to the presence of the crack as a function of a true (or effective) bending moment M_{eff} are given in Ref. 5.

$$\Phi_c^e = 4\frac{R}{I}V_3\frac{M_{eff}}{E}$$

$$\Phi_c^p = \alpha\varepsilon_y h_4\left(\frac{M_{eff}}{M_y}\right)^n$$

where V_3 and h_4 are tabulated in an EPRI report as functions of a/b, R/t and n.

The elastic and plastic rotation of the uncracked pipe are given by the following two relations:

$$\Phi_{nc}^e = \frac{M_{eff}}{EI}L$$

$$\Phi_{nc}^p = \frac{L\alpha\varepsilon_y}{R}\left[\frac{M_{eff}}{4\sigma_y R^2 t\beta}\right]^n$$

where $\beta(n)$ is given in Ref. 5 and the total rotation:

$$\Phi = \Phi_c^e + \Phi_{nc}^e + \Phi_c^p + \Phi_{nc}^p \quad (18)$$

With the value of θ computed previously, the solution of

$$\Phi = \theta \quad (19)$$

by an iterative technique, gives M_{eff}, the effective bending moment supported by the cracked pipe. This value can be simply obtained graphically as shown in Fig. 3. To compute the J-integral, we consider this value of M_{eff} as a primary bending moment, as in Section 3.1.

3.2.2 R6 method (secondary load)

In the case of a pure secondary load, L_r is simply equal to 0. K_I is computed using a plastic zone correction:

$$K_{cp} = K_I(a_e, M_s) \qquad a_e = a + r_y \qquad K_r = K_{cp}/K_{JC} \quad (20)$$

3.2.3 An alternative method

An alternative method is proposed to obtain more conservative results:

(a) if $M_s < M_y$

$$L_r = M_s/M_y \qquad \text{and} \qquad K_r = K_I(M_s)/K_{JC} \quad (21)$$

this formulation is identical to the case of primary load.

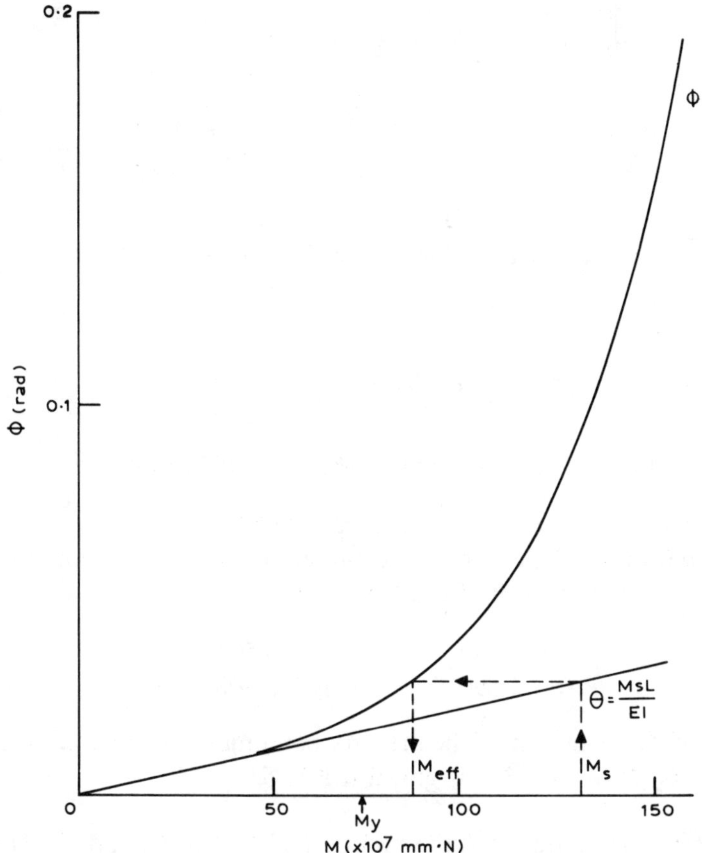

Fig. 3. Iterative scheme giving the effective bending moment.

(b) if $M_s > M_y$

The effect of secondary bending moment is limited to the limit load:

$$L_r = 1 \qquad \text{and} \qquad K_r = K_I(M_s)/K_{JC}$$

3.3 Application and comparison in a simple case

In this section, the authors propose to study the basic case of a pipe with a through-wall circumferential crack loaded by a primary and a secondary bending moment using these two methods.

Crack and pipe geometries (Fig. 4)

$$R = 300\,\text{mm} \qquad \gamma = 45^\circ$$
$$t = 30\,\text{mm} \qquad a/b = \tfrac{1}{4}$$
$$L = 10\,\text{m} \qquad a = 236\,\text{mm}$$

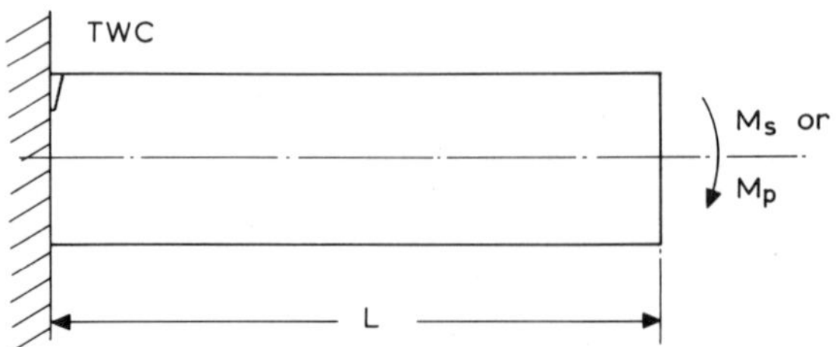

Fig. 4. Through-wall cracked (TWC) pipe in bending.

Material characteristics

$$E = 175\,000\ \text{MPa} \qquad \alpha = 1$$
$$\sigma_y = 120\ \text{MPa} \qquad n = 5$$

The limit load of the cracked pipe is $M_y = 7{\cdot}40\,10^8$ N mm.

J-integral results with a primary bending moment
EPRI: the parameters to compute the *J*-integral and the total rotation of the cracked pipe Φ are given in the tables of Ref. 5:

$$h_1 = 3{\cdot}969 \qquad V_3 = 0{\cdot}504$$
$$f_1 = 1{\cdot}599 \qquad h_4 = 2{\cdot}308$$

The value of *J* as function of the primary bending moment, computed with the equations in Section 3.1, is given in Fig. 5.

R6: the FAL is calculated with the equation of Section 3.1 and the Ramberg-Osgood parameters: α, n and σ_y.

To express these results in terms of *J*, we perform the following transformation:

with $L_r = M_p/M_y$, we obtain K_r from eqn (12), and:

$$J^{\mathrm{P}}_{\mathrm{R6}} = J^{\mathrm{e}}(M_{\mathrm{P}})/K_r^2$$

where J^{e} is computed without the plastic zone correction.

Figure 5 indicates that the values of the *J*-integral obtained with these two methods are almost the same for primary bending.

J integral results with a secondary bending moment
EPRI: the curve of the rotation Φ is plotted in Fig. 5 with the method developed previously. From the secondary bending moment M_s, computed with an elastic analysis of the complete structure including the uncracked pipe, we deduce simply the value of the elastic rotation imposed on this uncracked pipe using eqn (17).

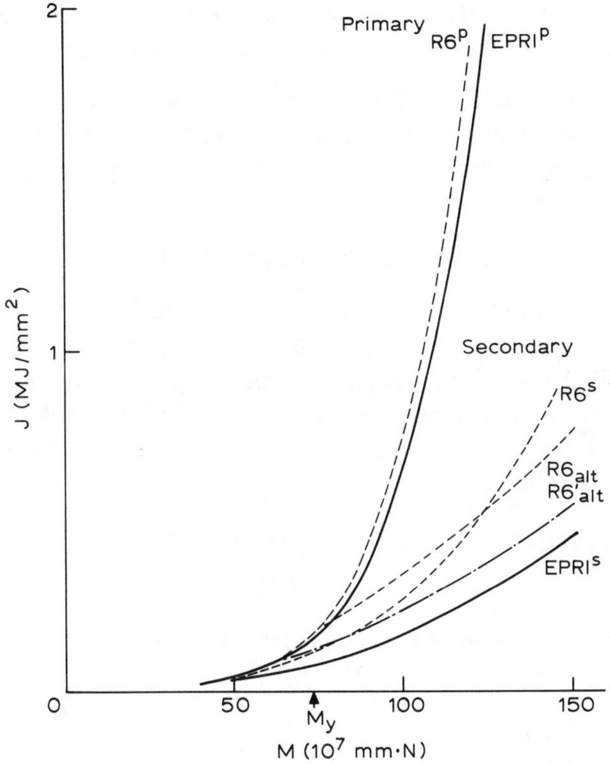

Fig. 5. *J* as a function of primary and secondary bending moment.

With: $\Phi = \theta$, we obtain graphically from Fig. 4 the effective bending moment M_{eff} to be applied to the pipe. Then, J is calculated with the EPRI relations of Section 2.1 considering M_{eff} as a primary bending moment.

The results in Fig. 5 (EPRI^{S}) indicate a strong effect of the type of loading. The J-integral is clearly greater with the primary load than with the secondary load.

R6: in the conventional application, $L_{\text{r}} = 0$, $J^{\text{e}}(a_{\text{e}})$ is computed with the effective crack length taking into account the plastic zone correction. For comparison with the EPRI method, the authors have used the same plastic zone size formula:

$$r_{\text{y}} = \frac{1}{2\pi}\frac{n-1}{n+1}\left(\frac{K_{\text{I}}}{\sigma_{\text{y}}}\right)^2$$

and

$$a_{\text{e}} = a + \Phi r_{\text{y}}$$

(Φ is considered equal to 1.)

We note that $J^e(a_e)$, R6^S is the same as the elastic part of the J-integral computed with the EPRI methodology for primary loading (Fig. 5).

R6 alternative method (R6$_{alt}$): for $M_s < M_y$, we obtain the same curve for J as for primary loading. For $M_s > M_y$, $L_r = 1$ and the corresponding value of K_r, from eqn (16) is 0·67. Then, for values of $M_s > M_y$, J is given by the following relation:

$$J(M_s) = J^e(M_s)/K_r^2 = 2{\cdot}26 J^e(M_s) \tag{23}$$

where J^e is calculated with no plastic zone correction.

Figure 5 indicates that the J-integral following the R6 Option 2 method and GE-EPRI are similar for secondary bending moments smaller than the limit load; smaller values being obtained for J EPRI, (EPRI^S < R6^S). With the alternative technique, J (R6$_{alt}$) is greater than values obtained with the two previous methods.

R6 second alternative method (R6$'_{alt}$): taking:

$$L_r = M_s/M'_y \tag{24a}$$

where M'_y is the bending moment at the onset of plasticity in the pipe:

$$M'_y = \pi/4 \times M_y \tag{24b}$$

And, for

$$M_s > M'_y, J(M_s) = J^e(M_s)/K_r^2 \tag{24c}$$

where

$$K_r = f\left(L_r = \frac{\pi}{4}\right) = 0{\cdot}79$$

we obtain:

$$J(M_s) = 1{\cdot}60 J^e(M_s)$$

which is approximately the result as obtained with the GE-EPRI method.

4. SIMPLIFIED METHOD FOR CRACK STABILITY ASSESSMENT UNDER COMBINED LOADING

4.1 General considerations

In the treatment of combined loadings by simplified methods, two types of difficulties are encountered.

(1) Firstly, there are the theoretical problems of non-proportional loadings and crack growth to consider. These two problems are linked since during crack growth, unloading takes place on the newly created faces. The GE-EPRI method, as well as the R6-method, is based on the deformation theory of plasticity using parameters (J, K_{cp} and the FAL) which are only valid when no unloading occurs. Thus, strictly speaking, these methods should be applied only to monotonically increasing loadings up to crack initiation. In other cases, more sophisticated methods should be used such as local approaches of fracture or improved crack tip integral techniques (see the general presentation[7] and examples of applications in crack stability[8,9] or fatigue analysis.[10]

However, Sönnerlind and Kaiser[11] have shown that, for a single-edge notch (SEN) specimen, the effect of loading path could be neglected. Also, Hutchinson and Paris[12] established that a limited amount of crack growth could be analysed using J (see also Kanninen *et al.*[13]). The authors assume these conclusions may be extended to piping containing a through-wall crack under pressure (or tension) and bending.

(2) The second difficulty is the non-linear behaviour of elastic–plastic fracture parameters as a function of the loading. The superposition principle, which is so useful in LEFM, is obviously not applicable to J and limit load estimates. In the following, the authors present and compare different simple ways to solve this problem; then more general conclusions on selection, use and improvement of these formulations will be drawn.

4.2 Comparison of the different approaches

4.2.1 The GE-EPRI method

The GE-EPRI method has been recently extended to the analysis of circumferential through-wall and surface flaws in cylinders subjected to combined tension and bending.[2] The scheme is defined by the following equations:

(a) Proportional loading parameter

$$\lambda = \frac{M}{P \times R} \tag{25}$$

where M is the bending moment; P, the axial force, and R, the pipe mean radius.

(b) Ramberg-Osgood stress–strain law (eqn (3)).

(c) J-separation formula

$$J = J^{e} + J^{p} \tag{26}$$

(d) Elastic component of J for combined tension and bending

$$J^{e} = \frac{K_{cp}^{2}}{E} \tag{27}$$

where $K_{cp} = K(a_e)$

$$K(a) = \left[F_1^{b} \times \frac{M}{Z^{b}} + F_1^{t} \times \frac{P}{Z^{t}} \right] \times \sqrt{\pi \times a} \tag{28}$$

The values F_1^b and F_1^t are tabulated functions of the relative crack size $[(a/b) = (\gamma/\pi)]$ and the R/t ratio.

$Z^{b} = \left(\frac{I}{R}\right)$ is the bending uncracked section modulus

$Z^{t} = 2\pi Rt$ is the tension uncracked section modulus

$$a_e = a + \frac{r_y}{1 + \left(\frac{P}{P'_0}\right)^2} \tag{29}$$

$$r_y = \frac{1}{2\pi}\left[\frac{n-1}{n+1}\right]\left(\frac{K}{\sigma_0}\right)^2 \tag{30}$$

$$P'_0 = \frac{1}{2}\left[\frac{-\lambda P_0^2 R}{M_0} + \sqrt{\left(\frac{\lambda P_0^2 R}{M_0}\right)^2 + 4P_0^2}\right] \tag{31}$$

$$P_0 = 2\sigma_0 Rt\left[\pi - \gamma - 2\,\mathrm{Arc\,sin}\left(\frac{1}{2}\sin\gamma\right)\right] \tag{32}$$

$$M_0 = 4\sigma_0 R^2 t\left[\cos\frac{\gamma}{2} - \frac{1}{2}\sin\gamma\right] \tag{33}$$

(e) Plastic component of J for combined tension and loading

$$J^{p} = \alpha\sigma_0\varepsilon_0 R(\pi - \gamma)\frac{\gamma}{\pi} h_1 \left[\frac{P}{P'_0}\right]^{n+1} \tag{34}$$

where h_1 is a tabulated function of $\frac{\gamma}{\pi}$, n, λ, $\frac{R}{t}$

and P'_0 is given by eqn (31).

These formulae have been presented because the methodology for the plastic zone size correction in J_e is not explicitly given in Ref. 2.

The three limit load formulae are derived from a global beam analysis. Equation (31) is obtained from the uncracked beam limit load equation:

$$\frac{M}{M_0}+\left(\frac{P}{P_0}\right)^2=1 \tag{35}$$

If we replace M by λPR in eqn (35) and then multiply all terms by $M_0 p_0^2$, we obtain a second degree equation whose solution gives the load P'_0.

The GE-EPRI method allows one to compute J for non-proportional loading by interpolating linearly using λ. This may be cumbersome for loading cases where λ varies, and somewhat limited since only a few λ values are considered in Ref. 2. In the present paper, the authors interpolate linearly using λ between the given h_1 values, and beyond the limits use the closest h_1 value (in this example, $h_1 = 6{\cdot}063$ for $\lambda < 0{\cdot}5$ in combined tension and bending).

4.2.2 The R6 method

Concerning the R6 method, we first check if the FAL proposed by Ainsworth[2] remains valid for combined loading, then we compare two L_r formulae. This FAL is very close to the R6, Rev. 3, Option 2 Curve. The difference in the example presented here is less than 1·3%. Option 2 retains a less severe plastic zone size correction which is probably more realistic, but gives a less conservative FAL.

The first step is achieved by using the above described GE-EPRI method, Option 3 of the R6 method and GE-EPRI limit loads. The case presented in Figs 6–9 is defined by the following basic data: the stress–strain curve for the

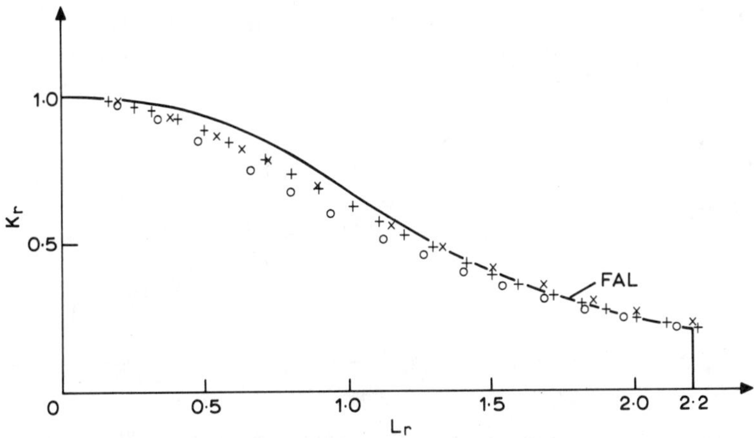

Fig. 6. FAL for proportional loadings using GE-EPRI limit loads. +, Pure tension; ×, pure bending; ○, tension and bending proportional ($\lambda = 0{\cdot}6$).

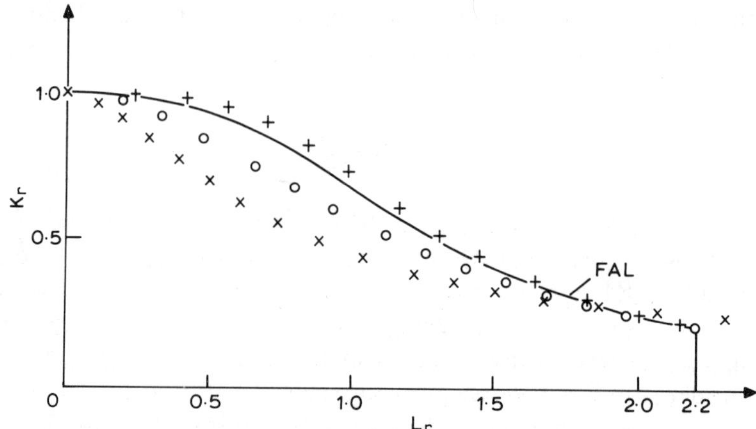

Fig. 7. FAL for combined tension and blending proportional ($\lambda = 0{\cdot}6$). +, J_{total} without plastic zone correction, GE-EPRI L_r; ○, J with plastic zone correction, GE-EPRI L_r; ×, J with plastic zone correction, L_r from Kanninen formula.

material follows the Ramberg–Osgood law; $n = 5$; $\alpha = 1$; $\sigma_0 = 120$ MPa; $E = 175\,000$ MPa ($\varepsilon_0 = \sigma_0/E$); the pipe is defined by $R = 0{\cdot}3$ m, $t = 0{\cdot}03$ m and a circumferential crack half angle of 45°. In pure bending the limit load equals 0·7391 MN m, and in pure tension 3·528 MN.

Figure 6 compares the FAL for three different proportional loadings and shows that combined tension and bending with fixed λ ($\lambda = 0{\cdot}6$) gives results slightly lower than for pure tension or pure bending; all these three curves being lower than Ainsworth's FAL. This difference is due to the plastic zone

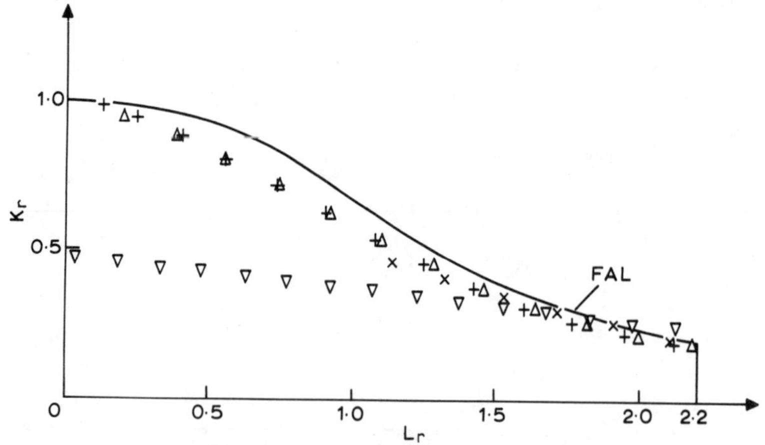

Fig. 8. FAL for constant pressure and varying bending moment. +, $P = 0{\cdot}4$, L_r from GE-EPRI; ×, $P = 4{\cdot}0$, L_r from GE-EPRI; △, $P = 0{\cdot}4$, L_r from Kanninen; ▽, $P = 4{\cdot}0$, L_r from Kanninen.

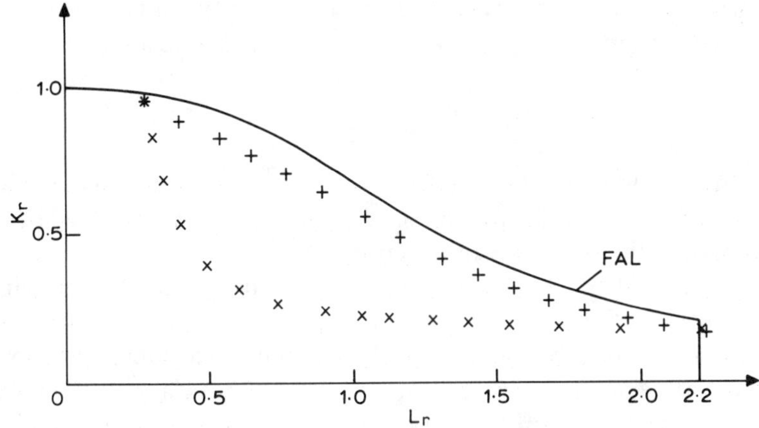

Fig. 9. FAL for constant bending moment and varying pressure. $M = 2$. +, GE-EPRI; ×, Kanninen.

correction; without this correction Ainsworth's FAL would be inside the FAD defined by the FAL computed for $\lambda = 0{\cdot}6$ (see Fig. 7). For large cracks, Finite Element computations have shown[14] that the J-integral estimates by using the GE-EPRI method would be better without applying the plastic zone size correction.

For L_r formulae, the CEGB report[1] refers to the limit load equations recommended by Miller,[15] who, for through-wall circumferential cracks under bending and pressure, recommends the net-section collapse formula derived by Kanninen *et al.*[16]

$$M_0 = 4\sigma_1 \times R^2 t\left(\cos\beta - \frac{1}{2}\sin\gamma\right) \tag{36}$$

where

$$\beta = \frac{\gamma}{2} + \frac{\pi}{2}\frac{P}{\sigma_2 \times Z_t} \tag{37}$$

$$\sigma_1 = \sigma_2 = \sigma_y$$

In pure bending this equation reduces to the GE-EPRI formula if σ_0 is substituted for σ_y. This formula has been extended to surface cracks under tension and bending by Ranganath *et al.*[17]

To apply such a formula to L_r, one has to distinguish between the two material characteristics σ_1 and σ_2. By definition of L_r, σ_1 is taken as σ_y, but setting σ_2 to the same value would give too conservative results. The shift of the neutral axis due to the presence of an axial force P, is better estimated by eqn (37) if σ_2 is equal to the flow stress σ_f. The validity limit of this L_r formula is obtained by setting M_0 to zero in eqn (36), which leads to

$$P = Z_t\sigma_2\left[-1 - \frac{\gamma}{\pi} + \frac{2}{\pi}\mathrm{Arc\,sin}\left(\frac{1}{2}\sin\gamma\right)\right] \tag{38}$$

This formula equals the GE-EPRI limit load for $\sigma_2 = \sigma_0$. Again, choosing $\sigma_2 = \sigma_y$ would give some reference load but not a collapse load which would correspond to the real limit of acceptability for P.

Taking $L_r = M/M_0$ from eqn (36) and $\sigma_2 = \sigma_F$ generates a much lower FAL than the EPRI solution as is shown in Fig. 7. Therefore we conclude that this formula may be used only if the moment is varying and the pressure being kept at a *low level*. This is clearly shown in Figs 8 and 9: $P = 0{\cdot}4$ MN is much lower than the limit load in tension (3·528 MN). Otherwise, choosing the Ainsworth's FAL and eqn (36) for L_r would be unconservative and changing the FAL as in Figs 8 or 9 (L_r from Ref. 16) would be too conservative.

4.3 Recommendation for R6 Option 2 use under combined primary loading

The problem of loading superposition has been most studied for the case of combined thermal and mechanical loading[18] (see other references in Ref. 19 p. 13). Ainsworth *et al.* analysed in a CSNI workshop[20] a part-through-wall circumferential crack in a large diameter pipe subjected to a constant axial load ($\sigma_y/3$) and a monotonically increasing bending moment. The J results obtained from the R6 method, Option 1, were conservative compared with finite element results.

However, this welded pipe is a special case and the L_r formula used has not been validated. This formula corresponds to a modified version of eqn (36) given by Ranta-Maunus and Achenbach[21] where L_r is applied to both σ_t and σ_b:

$$\frac{\pi}{4}\sigma_b = L_r\sigma_y\left[\cos\beta - \frac{1}{2}\sin\gamma\right] \tag{39}$$

where

$$\beta = \frac{\gamma}{2} + \frac{\pi}{2}\frac{\sigma_t}{L_r\sigma_y} \tag{40}$$

These two equations differ from eqns (36) and (37) only by σ_2 being replaced by $L_r\sigma_y$. This modification tends to reduce L_r, but the workshop results for J are increased by using Option 1 instead of Option 2. For general use it is better to make σ_2 higher, by substituting σ_f for σ_2; very good correlations have been obtained between predictions based on such a rule and experimental results of the Degraded Piping Program.[22]

More research should be done in this area of combined mechanical loading; as an example, work is underway at Framatome to examine if the effects of pressure and tensile loading are similar in thick shells.

At the present time the authors would recommend, for combined mechanical loading, to use the GE-EPRI results within the (short) range of tabulated values or the R6 Option 2 method in connection with the GE-EPRI limit load formula (eqn (31)). Under constant pressure, one may compute the initiation moment or the moment at instability using the R6 method and Kanninen formula for L_r (eqns (36) and (37)) if σ_2 is replaced by σ_f. But this formula may produce much larger margins (between applied and critical loads) than the real ones, thus it should be avoided when evaluating safety factors.

5 CONCLUSIONS

Two simplified methods for estimating crack initiation load and stable crack growth are analysed: the Ainsworth criterion, referred to here as the R6 method, and the GE-EPRI method.

First, it is recalled that the R6 method is a simplification of the GE-EPRI J-estimation scheme, but the criterion derived for plane geometries is not always conservative when applied to through-wall circumferentially cracked pipes and compared with the GE-EPRI method. However, the differences between estimates of the critical load are small for crack stability analyses. Moreover, Finite Element computations have shown, at least for large cracks, that the plastic zone correction can be disregarded in the GE-EPRI method, which suppresses the previously mentioned discrepancy.

Second, we explain that the treatment of secondary loading is much simpler using the R6 method. For a pipe with a through-wall circumferential crack subjected to a secondary bending moment, the R6 method provides much higher J-integral values than the GE-EPRI method. A simple alternative method $R6'_{alt}$, which reduces the differences, is proposed.

For combined loading, it is shown that the two methods give similar results, if in the R6 approach the GE-EPRI limit load equation is used rather than the recommended one. The GE-EPRI method, based on Finite Element results, may be considered as reliable and accurate, but the R6 method is easier to use and not restricted by the Ramberg-Osgood stress–strain law representation. The R6 method may be very conservative for high secondary loading, and unconservative under combined loading. Recommendations are given for removing these drawbacks.

REFERENCES

1. Milne, I., Ainsworth, R. A., Dowling, A. R. & Stewart, A. T., Assessment of the integrity of structures containing defects. Internal report R/H/R6, Rev. 3, CEGB, UK, 1986.
2. Kumar, V. & German, M. D., Elastic–plastic fracture analysis of through-wall and surface flaws in cylinders. EPRI Report NP-5596, Electric Power Research Institute, Palo Alto, CA, 1988.
3. Roche, R. L., Modes of failure—primary and secondary stresses. *J. Pres. Vessel Tech.*, **110** (1988) 234–9.
4. Ainsworth, R. A., The assessment of defects in structures of strain hardening materials. *Engng. Fract. Mech.*, **19** (1984) 633–42.
5. Kumar, V., German, M. D. & Shih, C. F., An engineering approach for elastic–plastic fracture analysis. EPRI Report NP-1913, Electric Power Research Institute, Palo Alto, Ca, 1981.
6. Kumar, V., German, M. D., Wilkening, W. W., Andrews, W. R., de Lorenzi, H. G. & Mowbray, D. F., Advances in elastic–plastic fracture analysis. EPRI Report NP-3607, Electric Power Research Institute, Palo Alto, CA, 1984.
7. Brust, F. W., Nakagaki, M. & Gilles, Ph., Comparison of elastic–plastic fracture mechanics techniques. *Proc. of the 21st ASTM, National Symposium of Fracture Mechanics*, June 1988, Annapolis, MA. *ASTM STP 1074*, American Society for Testing and Materials, Philadelphia, PA, 1990.
8. Brust, F. W., McGowan, J. J. & Atluri, S. N., A combined numerical/experimental study of ductile crack growth after a large unloading using T*, J and CTOA criteria. *Engng. Fract. Mech.*, **23**(3) (1986) 537–50.
9. Billardon, R., Adam, C. & Lemaitre, J., Study of non-uniform growth of a plane crack in a three dimensional body subjected to non-proportional loadings. *Int. J. Solids Struc.*, **22**(7) (1986) 677–92.
10. Leis, B. N. & Laflen, J. M., Problems in damage analysis under non-proportional cycling. *J. Engng. Mat. Tech.*, **102** (1980) 127–34.
11. Sönnerlind, H. & Kaiser, S., The *J*-integral for a SEN specimen under nonproportionally applied bending and tension. *Engng. Fract. Mech.*, **24**(5) (1986) 637–46.
12. Hutchinson, J. W. & Paris, P. C., Stability analysis of *J*-controlled crack growth. In *Elastic–plastic fracture*, *ASTM STP668*, ed. J. D. Landes, J. A. Begley & G. A. Clarke. American Society for Testing and Materials, Philadelphia, PA, 1979, pp. 37–64.
13. Kanninen, M. F., Hahn, G. T., Broek, D., Stonesifer, R. B., Marschall, C. W., Abou-Sayed, I. S. & Zahoor, A., Development of a plastic fracture methodology, Final Report EPRI NP-1734, Electric Power Research Institute, Palo Alto, CA, 1981.
14. Faidy, C. & Coustillas, F., Leak before break and through-wall cracked pipe under pure bending. *Proc. of the International Symposium on Pressure Vessel Technology & Nuclear Codes and Standards*, Seoul, Korea. Korean Society of Mechanical Engineers, Seoul, Korea, April 1989, pp. 15–35.
15. Miller, A. G., Review of limit loads of structures containing defects (3rd Edn). CEGB, UK, TPRD/B/093/N, 1987.
16. Kanninen, M. F., Broek, D., Marschall, C. W., Rybicki, E. F., Sampath, S. G., Simonen, F. A. & Wilkowski, G. M., Mechanical fracture predictions for

sensitized stainless steel piping with circumferential cracks. EPRI Report NP-192, Electric Power Research Institute, Palo Alto, CA, 1976.
17. Ranganath, S. & Metha, H. S., Engineering methods for the assessment of ductile fracture margin in nuclear power plan piping. In *Proc. Elastic–Plastic Fracture Second Symposium, Volume II, Fracture Resistance Curves and Engineering Applications, ASTM STP 803,* ed. C. F. Shih & J. P. Gudas, American Society for Testing and Materials, Philadelphia, PA, 1983 pp. II-309–II-330.
18. Muscati, A., Elastic–plastic fracture analysis of a thick cylinder subjected to combined thermal and mechanical loading, CEGB Report SWR/SSD/0626/N/85, 1985.
19. Milne, I., Ainsworth, R. A., Dowling, A. R. & Stewart, A. T., Background to and validation of CEGB Report R/H/R6—Rev. 3. CEGB Report R/H/R6. Rev. 3 Validation, CERL, Leatherhead, UK, 1987.
20. Ainsworth, R. A., Chell, G. G. & Milne, I., Solution to the workshop problem using the CEGB fracture assessment procedure [R6]. *Proc. of the CSNI/NRC workshop on ductile piping fracture mechanics*, ed. M. F. Kanninen. Southwest Research Institute, San-Antonio, TX, 1984.
21. Ranta-Maunus, A. K. & Achenbach, J. D., Stability of circumferential through-cracks in ductile pipes. *Nucl. Engng. Design*, **60** (1980) 339–45.
22. Wilkowski, G. M. *et al.*, Degraded piping program phase II, NUREG Report CR-4082, BMI-2120, 8, USNRC, Washington, DC, 1989.

Int. J. Pres. Ves. & Piping **43** (1990) 151–163

Leak-Before-Break in French Nuclear Power Plants

C. Faidy

EDF-SEPTIN, 12–14, Avenue Dutrievoz, 69628 Villeurbanne Cedex, France

S. Bhandari

Framatome Tour Fiat, La Defense, 92084 Paris, Cedex 16, France

&

P. Jamet

CEA-DEMT, Centre d'Etudes Nucléaires de Saclay,
91191 Gif-sur-Yvette, France

ABSTRACT

Practical applications of the leak-before-break concept at the present stage are quite limited in French nuclear power plants. However, discussions with safety authorities have included leak-before-break arguments for the different types of reactors: pressurized water reactors (PWRs), liquid metal fast-breeder reactors (LMFBRs) and gas graphite reactors (GGRs).

At present, the fracture mechanics part of the studies are complete for the following components:

—pipes in GGRs;
—primary and auxiliary lines in PWRs;
—steam-generator tubes in PWRs;
—pipes and main vessels in LMFBRs.

The different approaches are consistent but some specific problems have to be taken into account, depending on the plant, such as creep regime, thin shell components, in-service inspection or the issue of design safety.

A large research and development program, realized in different cooperative agreements (national or international), completes the general approaches. It comprises different topics, such as material properties, elasto-plastic fracture mechanics, leak-area determination and leak-detection devices.

Although practical applications are limited at present, EDF, in conjunction with Framatome, Novatome and CEA, will define a complete validated methodology to be used on a variety of cases.

1 INTRODUCTION

Present French regulations do not require the application of the leak-before-break (LBB) concept on French pressurised water reactors (PWRs). However, much research and development (R&D) activity is going on,[1] especially under three-party agreements between EDF (Electricité de France), CEA (French Atomic Energy Commission) and Framatome (designer and manufacturer), to show the applicability of leak-before-break and to develop the means and methods for its applications, particularly for primary (including auxiliary) piping, secondary piping and steam-generator tubes. Emphasis is being given to their validation and qualification. The R&D activity can be grouped under the following categories:

—Development and validation of simplified engineering methods for the analysis of stable and unstable fracture behavior of components under static loading conditions, through elaborate computations and an experimental pipe-fracture program.

—Development and validation of similar methods for dynamic loading situations such as safe shutdown earthquake (SSE).

—Development and validation of leak-rate analysis methods for fluids flowing between through-wall cracks.

—Evaluation and eventual development and qualification of means of flow-rate detection.

—Development of the necessary material data (such as J-resistance curves) for the application of analytical methods mentioned above.

The objective of this paper is to present the application of fracture-mechanics methodology used in France to demonstrate the LBB behavior of PWR components. The results presented represent a synthesis of the various studies conducted in a view of the applicability of this concept on French PWRs.

2 GENERAL APPROACH ON LBB STUDIES

Demonstrations of the applicability of LBB to large pipes are usually carried out in two phases; an approach slightly different from the United States

Nuclear Regulatory Commission (USNRC) methodology:[2]

—a main phase, to demonstrate that a leak from a through-wall crack can be detected for crack lengths smaller than the critical crack size, and
—a complementary phase, to demonstrate that the risk of developing a through-wall flow is small during the planned lifetime of the plant and that any realistic end-of-life defect would be small enough not to affect the integrity of the structure.

The main phase comprises three steps:

(i) Examine the stability of a through-wall defect under safe shutdown earthquake (SSE) + normal operating load to evaluate the critical crack lengths in the component.
(ii) Calculate the crack area and corresponding leak rates for through-wall cracks.
(iii) Compare the evaluated leak rates with the detectable leak rates, to show the available margins.

The complementary phase also consists of three steps:

(i) Postulation of an initial defect: based on experience from fabrication and in-service inspection, one can determine two types of defects (realistic and exceptional), to be postulated for each component. Usually, these hypothetical defects are considered in highly stressed areas of the component.
(ii) Fatigue-crack-growth studies: both hypothetical defects are assumed to propagate under the reactor transients which constitute the fatigue loading conditions in normal and upset operation for the reactor's lifetime.
(iii) Stability of the end-of-life defect: the stability of the end-of-life defect is analyzed under maximum (SSE + normal operating) loading conditions.

3 APPLICATION OF LBB CONCEPT ON A 4-LOOP PWR

An exercise was conducted[3] on the hot-leg of the primary piping of a French 4-loop PWR of 1300 MWe to show that:

—the simple mechanics of the cracked piping are valid in a given domain of application. This validation of engineering rules and the limits of their application are obtained by a three-dimensional finite-element computation on the real piping,

—design of a 4-loop PWR using RCC-M code[4] or ASME code[5] possess the LBB characteristics and give a first estimation of the safety margins.

Using a through-wall crack of length $3T$ (T = the thickness of the pipe) which remains stable under the maximum load (SSE + normal operating load) with a small stable crack growth of 0·2 mm, it was shown that the application of simplified methods[6–9] was justified for the analysis of the piping considered and that the concept of LBB is applicable to the piping analyzed. Since then, more confidence has been gained on the methodology used.

Another application on a 4-loop French design PWR is underway. The analysis has been conducted on:

—primary loop including surge line,
—auxiliary piping,
—secondary piping (feedwater and steam lines) inside the containment.

The results obtained to date are very encouraging and show the applicability of the LBB concept to the actual design.

4 RESEARCH IN PROGRESS

Some questions on material properties and simplified methods in their present state of application still remain to be answered:

—transposition of material data from specimen to real structures, including the problem of large crack growth,
—choice of the initial crack length for fatigue analysis,
—knowledge of real loading (in particular, thermal loading),
—validation of crack stability analysis under static and dynamic conditions, stress classification and superposition of loading, weld and base metal interface,
—validation of leak flow-rate models and crack-area determinations,
—evaluation of leak-detection systems.

To answer such questions, a large research and development program has been underway between Framatome, CEA and EDF since 1986.

The results presented here are from the International Program on Piping Integrity Research Group (IPIRG). In addition, the results from other programs, such as the NRC Degraded Piping Program (DPIII) and different Electric Power Research Institute (EPRI) and Japanese programs, are analyzed.

5 PRESENTATION OF THE FRENCH R&D PROGRAM

The French program covers: static and dynamic tests, development of specific computer models, validation and comparison of engineering methods and formalization of recommendations for LBB analysis.

5.1 Prototypical tests

These are four-point bending tests, at 300°C without pressure, on carbon and stainless steel pipes up to 700 mm external diameter and different radius/thickness ratios, with circumferential through-wall cracks and part-through cracks in base metal and in welded joints (Fig. 1). About 20 tests are scheduled. The objective is to verify conclusions reached in the Degraded Piping Program for materials and pipes used in French PWR plants. Tables 1 and 2 present the test matrix and the first results that confirm the unconservatism of the net-section collapse load for large diameter pipes.

TABLE 1
Initial Prototypical Test Program

No.	*M*	*D*	*T*	*CA*	*CL*	*CT*	*Load*	*Situation*
0	CS	406	21	90°	BM	TW	M	Completed, preliminary test
1	CS	406	21	120°	BM	TW	M	Completed
2	CS	406	21	60°	BM	TW	M	Completed
3	CS	406	21	120°	W	TW	M	Completed
4	CS	406	21	120°	BM	SC	M	
5	CS	406	21	30°	BM	TW	M	Completed, not computed
6	CS	406	21	30°	BM	TW	C	Completed, not computed
7	SS	406	40	40°	BM	TW	M	Completed
8	SS	406	40	120°	BM	TW	M	Completed
9	SS	406	40	40°	W	TW	M	Completed, not computed
10	SS	406	40	120°	BM	SC	M	
11	CS	168	11	0	—	—	M	Completed, not computed
12	CS	168	11	30°	BM	TW	M	Completed
13	CS	168	11	120°	BM	TW	M	Completed
14	CS	168	11	30°	W	TW	M	
15	CS	168	11	30°	BM	TW	C	
16	SS	168	18	40°	BM	TW	M	
17	SS	168	18	120°	BM	TW	M	
18	SS	168	18	40°	W	TW	M	
19	SS	168	18	40°	BM	TW	C	

M, Material; CS, carbon steel; SS, stainless steel; D, diameter (mm); T, thickness (mm); CA, total crack angle; CL, crack location; BM, base metal; W, weld joint; CT, crack type; TW, through-wall crack; SC, surface crack; M, monotonic load; C, cyclic load.

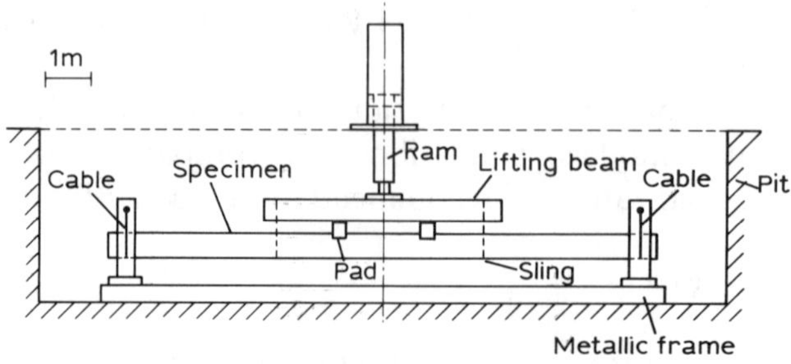

Fig. 1. EDF test facility for large diameter pipes (up to 406 mm (16″)) at 300°C without pressure.

5.2 Analytical tests

Stainless steel pipe, 100 mm in diameter, will be subjected to four-point bending, without pressure (Fig. 2). The first results are depicted in Fig. 3. They show good agreement on net-section collapse load for these diameter pipes; but a tendency to be little unconservative for small crack angles (less than 45°). This last remark has to be confirmed on a complementary set of tests that is in progress. The results will be used to verify the collapse load and the *J*-estimation directly from experiment,[10] and to validate a one-dimensional cracked beam element that can be used in a finite-element computer code.

TABLE 2
Prototypical Test Results

No.	*M*	*D*	*T*	*ICA*	M_i	M_{max}	*FCA*	M_L	M_i/M_L (%)	M_{max}/M_L (%)
0[a]	CS	406·4	21	90	—	465	113	533	—	87
1	CS	406·4	21	120	295	362	135	405	73	89
2	CS	406·4	21	60	436	628	77	669	65	94
3	CW	406·4	21	120	313	375	132	405	77	93
7	SS	406·4	40·5	40	1 159	1 293	52	1 457	79	88
8	SS	406·4	40·5	120	639	696	132	775	82	90
12	CS	168·8	11·1	30	104	106	41	100	103	105
13	CS	168·3	10·9	120	34	42	142	50	67	83

[a] Preliminary test.
M, Material; CS, carbon steel; SS, stainless steel; CW, carbon-steel weld; D, diameter (mm), T, thickness (mm); ICA, initial crack angle; FCA, final crack angle; M_L, limit moment based on SF = 0·5 ($Sy + Su$); M_i, initiation moment; M_{max}, maximum moment.

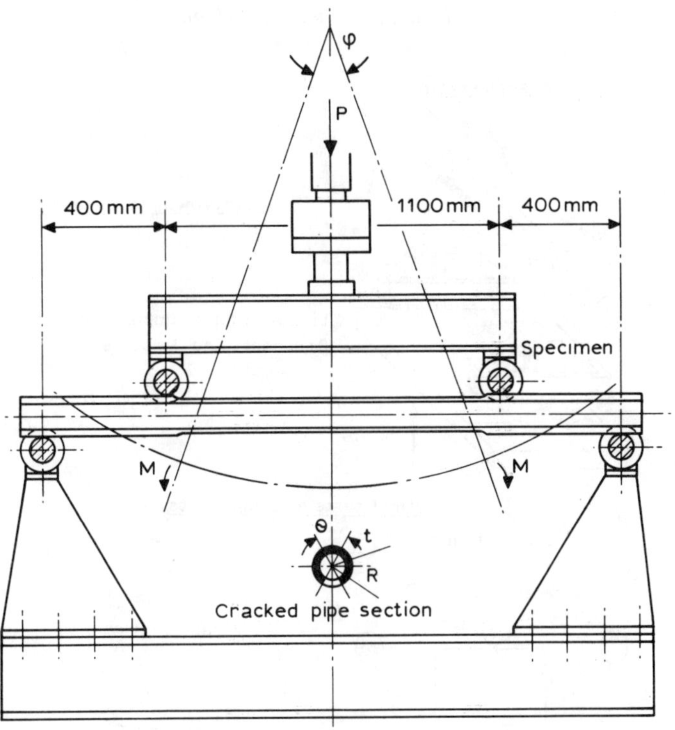

Fig. 2. CEA test facility for small diameter pipes (up to 102 mm (4″)) at room temperature without pressure.

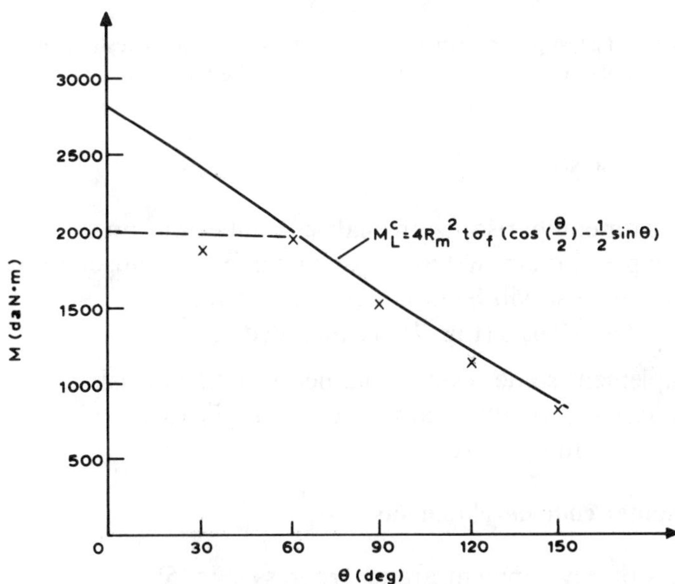

Fig. 3. Verification of net-section stress criteria at crack initiation for 102 mm (4″) stainless steel pipe tests at room temperature with different total crack angles. The dotted line is a tendency for small crack angle (under verification with a set of new tests). The solid line is based on net-section collapse load theory.[15] ×, Experimental points.

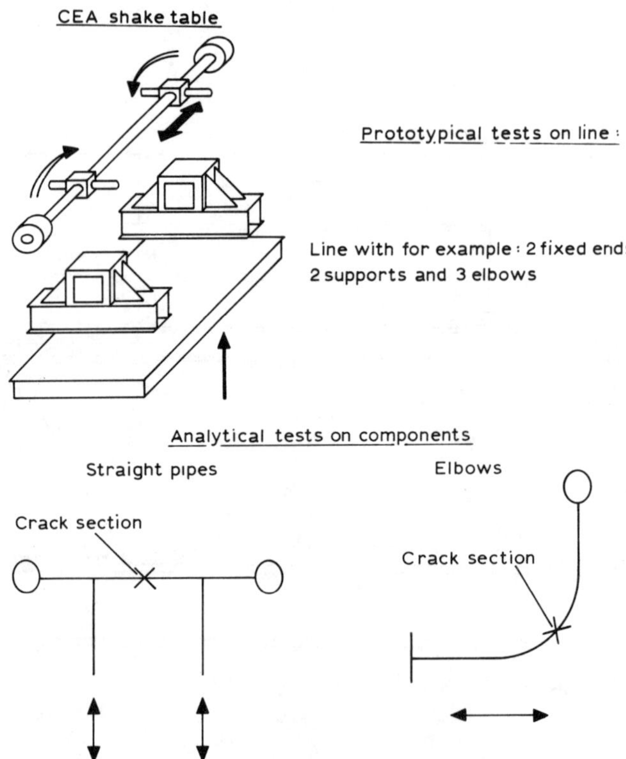

Fig. 4. Dynamic test program including tests on straight pipes, tests on elbows and tests on a simple line. These tests will be done on the CEA shake table facility.

5.3 Dynamic tests

Some tests for comparison with analysis will be done on simple components (straight pipes and elbows) under dynamic loads (sinusoidal or seismic). A more complex test will be done on a simple line.

These different tests (Fig. 4) are intended to:

—complement some results obtained in IPIRG Task 1,
—demonstrate to safety authorities the applicability of our methodology on representative tests.

5.4 Computer code developments

Two types of development are in progress (Fig. 5):

—computation of J by different methods, such as a virtual crack-extension method on a 3-D shell model, that will become, after validation, a reference method;
—development of a hinge element with constitutive equations taking into

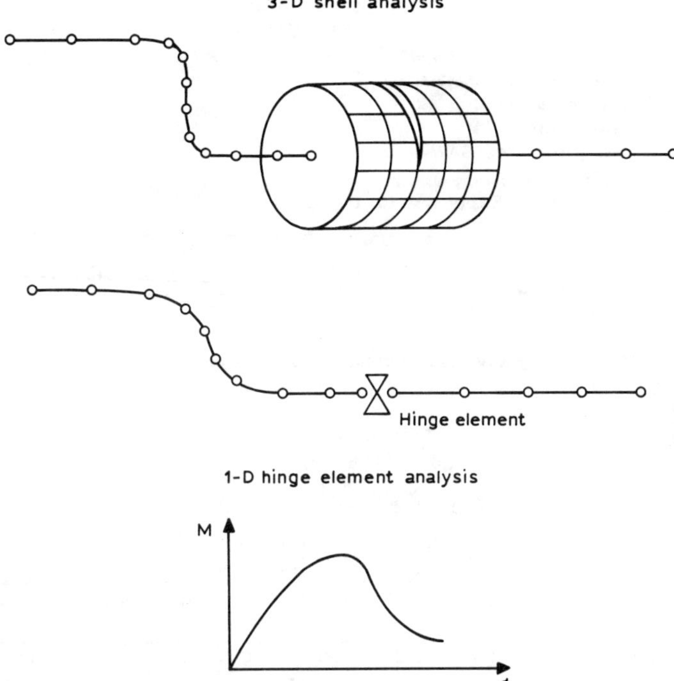

Fig. 5. Computer code models. Different developments follow the test program: one-dimensional hinge element with a complex applied moment versus rotation (M, ϕ) relationship including stable-crack growth and crack-closure effects.

account the crack, the stable crack growth, the plasticity and the different crack closure.[11,12] This is an important method for practical applications.

5.5 Engineering methods

All the available tests in France or in the literature (DPIII, IPIRG, other programs)[13–15] will be used to validate the different engineering *J*-estimation schemes (Figs 6 and 7): R6, EPRI, Paris methods, NRC pipe.[6–16]

Special treatment is planned for some difficult issues which are expected to arise in applying the simplified engineering methods. For example:

—prediction of initiation before the limit moment,
—classification of secondary and primary stresses,
—superposition of loadings,
—strain-hardening effects,
—behavior under torsion loading,
—leak-area evaluations,
—screening criterion.[17]

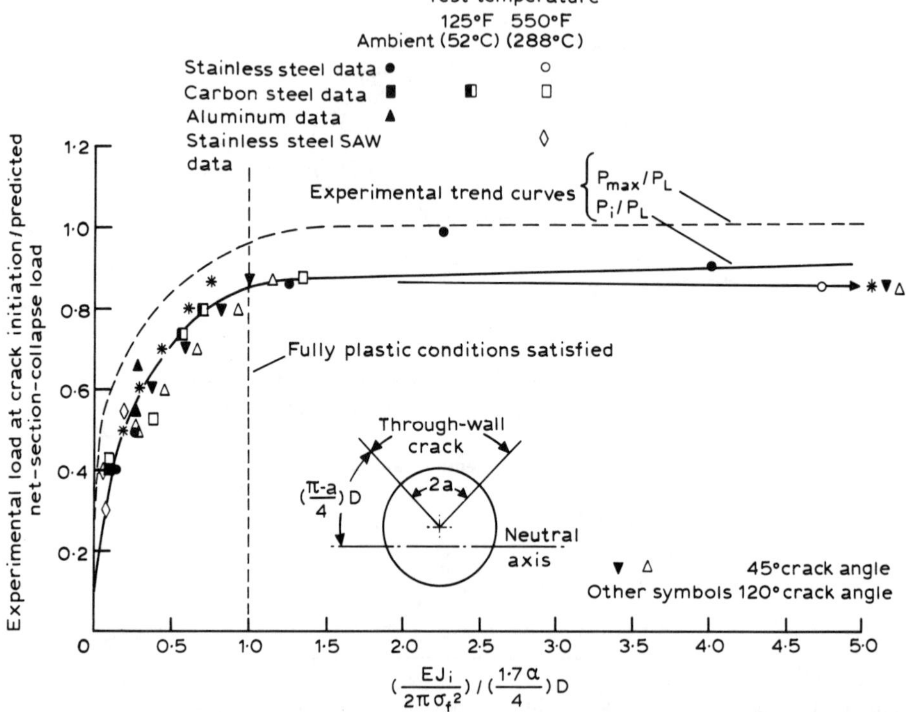

Fig. 6. New screening criteria:[17] modification of the Battelle screening criteria based on two curves that fit the Degraded Piping Program data[15] to take into account the influence of the crack angle. The proposal is a slight modification of the term on the y-axis that maintain a good fitting of the Degraded Piping Program tests for 120° crack angle and that obtain a good fitting using the same curves for other crack angles.

To complete this program, there are some other programs related to, but not specific to, LBB:

—leak-rate model and crack-area estimation,[18]
—leak-rate monitoring; global and local measurements,
—non-destructive examinations,
—technological aspect, such as limiting the number of welds or increasing the quality of some special weld joints.

6 CONCLUSIONS

At present, practical applications of the leak-before-break concept are limited in French nuclear power plants.

A feasibility study on a French 4-loop PWR demonstrates the

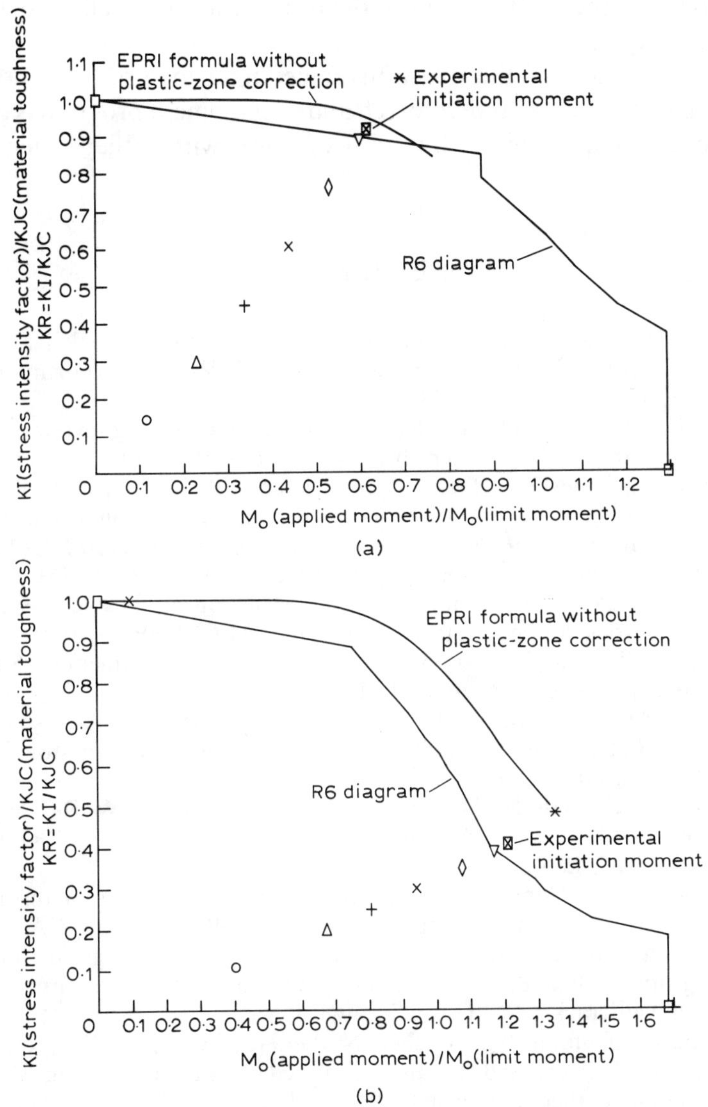

Fig. 7. Comparison between different *J*-estimation schemes for two specific tests: (a) 152 mm (6″) carbon-steel pipe (diameter 168 mm) at 300°C with 120° crack angle. Finite element computation: ○, $M = 5783$ MN; △, $M = 11\,430$ MN; +, $M = 16\,806$ MN; ×, $M = 21\,865$ MN; ◇, $M = 26\,526$ MN; ▽, $M = 30\,000$ MN; ⊗, $M = 30\,743$ MN, ∗, $M = 34\,662$ MN. (b) 102 mm (4″) stainless-steel pipe (diameter 100 mm) at room temperature with 120° crack angle. Finite element computation: ○, $M = 3000$ MN; △, $M = 5000$ MN; +, $M = 6000$ MN; ×, $M = 7000$ MN; ◇, $M = 8000$ MN; ▽, $M = 8700$ MN; ⊗, $M = 9000$ MN. The EPRI method[7] is based on Ramberg-Osgood fitting of stress–strain curve and *J*-estimation by interpolation of different finite element computations. The R6 method[6] is based on a point by point descrition of stress–strain-curve and *J*-estimation by correction of the elastic stress intensity factor.

applicability of the LBB concept to the primary loop as well as to auxiliary and secondary piping.

To answer some of the questions concerning material properties and engineering methods, a large research and development program is going on in France, complemented by an exchange with other international programs.

REFERENCES

1. Faidy, C., Jamet, Ph. & Bhandari, S., Developments in leak before break approach in France: Progress in regulatory policies and supporting research. NUREG/CP-0092-Tokyo, 1987.
2. The Pipe Break Task Group, Report of the USNRC piping review committee: Evaluation of potential for pipe breaks. NUREG 1061, Vol. 3, November 1984.
3. Charras, T., Jamet, Ph., Bhandari, S. & Taupin, Ph., Application of the leak-before-break concept on a 4-loop PWR: An exercise on primary piping. *Transactions of the 9th International Conference on Structural Mechanics in Reactor Technology*. Division G, Lausanne, August 1987, pp. 261–8.
4. RCC-M. Règles de conception et de construction applicable aux matériels mécaniques des ilots nucléaires. Annexe ZG, AFCEN, Paris, 1988.
5. ASME. Boiler and Pressure Vessel Code, Section XI. American Society of Mechanical Engineers, New York, 1988.
6. Milne, I., Ainsworth, R. A., Dowling, A. R. & Stewart, A. T., Assessment of the integrity of structures containing defects. CEGB-R/H/R6, révision 3, CERL, Leatherhead, UK, 1986.
7. Kumar, V., German, M. D., Wilkening, W. W., Andrews, W. R., De Lorenzi, H. G. & Mowbray, D. F., Advances in elastic–plastic fracture. EPRI NP-3607, Palo Alto, CA, 1984.
8. General Electric Company. Elastic–plastic fracture analysis of through-wall and and surface flaws in cylinders. EPRI NP-5596, Palo Alto, CA, 1988.
9. Paris, P. & Tada, H., The application of fracture proof design methods using tearing unstability theory to nuclear piping postulating circumferential through-wall cracks. NUREG/CR-3463, 1983.
10. Moulin, D., Touboul, F., Foucher, N., Lebey, J. & Acker, D., Experimental evaluation of *J* in cracked straight and curved pipes under bending. SMIRT 10, Division G, Anaheim, August 1989.
11. Petit, M. & Jamet, Ph., Numerical evaluation of cracked pipes under dynamic loading using a special finite element. *Transactions of the 10th International Conference on Structural Mechanics in Reactor Technology*. Division G, Anaheim, August 1989, pp. 341–6.
12. Brochard, J., Petit, M. & Millard, A., A special cracked pipe element for leak before break applications. *Transactions of the 10th Internationnal Conference on Structural Mechanics in Reactor Technology*. Division G, Anaheim, August 1989, pp. 357–62.
13. Taupin, Ph., Franco, C. & Bandhari, S., Analysis of the US degraded piping test results using two-criteria approach of the CEGB. Paper to be presented at ASME Pressure Vessel and Pipings Conference, Nashville, 1990.

14. Faidy, C., Coustillas, F., Leak-before-break and through wall cracked pipe under pure bending. *Proceedings of the International Symposium on Pressure Vessel Technology and Nuclear Codes and Standards.* Séoul, April 1989.
15. Wilkowski, G. M., *et al. Degraded Piping Program, Phase II.* NUREG/CR-4082, Vols 1–6, April 1986.
16. Gilles, P. & Brust, F. W., Approximate methods for fracture analysis of tubular members subjected to combine tensile and bending loads. *Proceedings of the OMAE Conference.* The Hague, Netherlands, March 1989.
17. Champonier, F. P. & Hilsenkopf, P. An alternative approach to the Battelle screening criterion for pipes with through-wall cracks. *Int. J. Pres. Ves. & Piping,* **38** (1989) 69–80.
18. Bandhari, S., Faidy, C. & Acker, D., Computation of leak areas of circumferential cracks in piping for application in demonstrating leak before break behaviour. *Transactions of the 10th International Conference on Structural Mechanics in Reactor Technology.* Division G, Anaheim, August 1989, pp. 333–9.

Int. J. Pres. Ves. & Piping **43** (1990) 165–179

Leak-Before-Break in Steam Generator Tubes

B. Flesch

Electricité de France, SPT, 13–27 Esplanade Charles-de-Gaulle,
La Défense, 92060 Paris, Cedex 57, France

&

B. Cochet

Framatome, Tour Fiat, La Défense, 92040 Paris, Cedex 16, France

ABSTRACT

The steam generator tubing constitutes one of the main barriers against the release of activity to the environment. The capacity of the tubing to withstand safely the loads exerted on it during normal operation and faulted conditions is therefore the most important factor in steam generator safety evaluation. Another important consideration in safety evaluation is the tendency of the tubes to leak at a significant but acceptable rate under normal operating conditions before there is a risk of rupture under accidental overpressure: the Leak-Before-Break (LBB) criterion.

This paper presents the theoretical and experimental programme undertaken in France to assess the LBB criterion for PWR steam generator tubes. Criteria for instability of different types of defect have been deduced from experimental and numerical results. Leakage models have been derived from leak tesks, as well as crack-opening measurements and calculations.

1 INTRODUCTION

Of the primary system components of a pressurized water reactor (PWR), one of the most sensitive to degradation is the steam generator tube bundle described in Fig. 1. As in almost any heat exchanger, the wall between the hot

Int. J. Pres. Ves. & Piping 0308-0161/90/$03·50

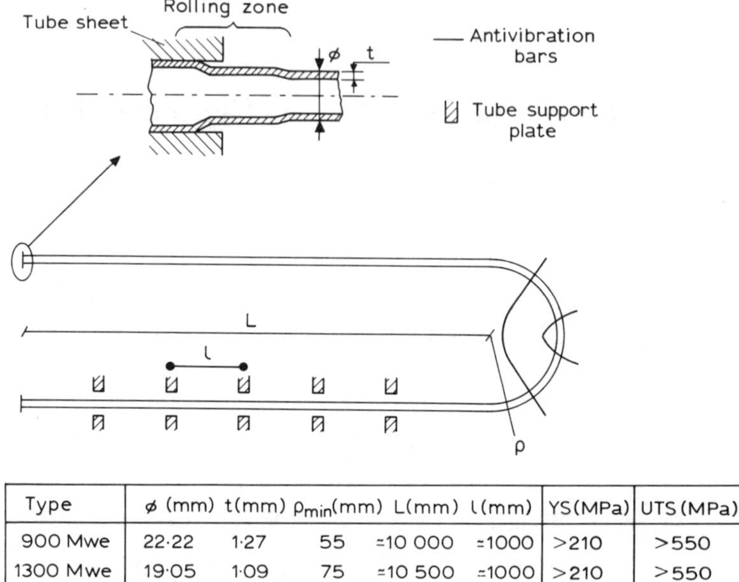

Type	ϕ (mm)	t(mm)	ρ_{min}(mm)	L(mm)	l(mm)	YS(MPa)	UTS (MPa)
900 Mwe	22·22	1·27	55	≃10 000	≃1000	>210	>550
1300 Mwe	19·05	1·09	75	≃10 500	≃1000	>210	>550

Fig. 1. Main characteristics of the tube bundle.

and cold streams is subjected to severe thermal, mechanical and environmental conditions. This is particularly true of a 1000 MW (thermal) steam generator.

The main source of degradation in Electricité de France (EDF) steam generator tubes is corrosion. This takes the form of primary side stress corrosion cracking in the roll transition zone, and in the small radius U-bends and, to a lesser extent, secondary side cracking. Mechanical damage caused by vibration and wear is also present, for example, between tubes and anti-vibration bars, between tubes in large radius U-bends of 1300 MWe unit steam generators and, in some cases, between tubes and loose parts.

As the steam generator tubes represent a considerable proportion of the main primary system (80% of the second containment barrier, 10 000 tubes and 220 000 m of tubing for 900 MWe units, and considerably more in 1300 MWe units) and must therefore satisfy safety and availability criteria, we have developed a new, and particularly stringent monitoring and maintenance policy which differs significantly from that used for other main primary system components.

Steam generator tube maintenance has two main objectives:

- With respect to safety: keeping the tube rupture probability sufficiently low. This means determining suitable methods of compensating for the increased risk associated with tube degradation.

- With respect to availability: limiting the number of shutdowns due to out-of-specification primary-to-secondary leakage. On the other hand, corresponding actions should not unacceptably shorten the steam generator service life.

Application of the LBB criterion to steam generator tubes is the principal monitoring method used by EDF in implementing this policy. This major design initiative has made it possible to demonstrate that virtually all the tube damage encountered in service complies with this criterion, under which any risk of rupture due to overpressure in accident situations is necessarily preceded by detectable primary-to-secondary system leakage during normal operation without significant consequences.

These studies, for which EDF has enlisted the assistance of Framatome and CEA, consist of two parts for each of the types of damage identified:

—establishing mechanical models for prediction of critical pressures;
—establishing leak rate models from mechanical dimensions (break area) and thermohydraulics (fluid flow laws).

These models were developed using a series of theoretical and experimental studies which, as far as possible, were compared with experience from the reactor. This paper describes the overall progress of the programme.

2 STEAM GENERATOR TUBE RUPTURE CRITERIA

Assessment of the burst risk for steam generator tubes subjected to internal pressure, thermal loading and external loading representative of mechanical rupture in accident conditions necessitates experimental validation of rupture criteria.

More than 1000 burst tests on Inconel 600 and Inconel 690 tubing with nominal diameters of 19·05 mm and 22·22 mm have been performed with various types of damage located in straight sections remote from discontinuities and supports (non-specific sections), in the roll transition zone at the top of the tubesheet, in the vicinity of the tube support plate or of an antivibration bar, and in U-bends.

The majority of the tests have been carried out in cold conditions, but some of them in hot conditions, showing that the instability limit pressure can be accurately estimated using plastic instability criteria, based on the tensile characteristics at the given temperature. Furthermore, comparative burst tests have shown that the limit pressure does not depend on the sharpness of the cracks; so the majority of the burst tests have been performed on tubes with machined cracks or defects; the boundary conditions on the sample reproducing the actual conditions. For the through-wall defects, mechanical seals have been used to maintain leak tightness; their influence has been measured and taken into account for the results.

2.1 Burst pressure of tubes without defects

The results of burst tests show that tube rupture occurs after extensive strain, as a consequence of the high ductility and toughness of tubes.

There is a perfect correlation between test results and a model predicting the burst pressure by referring to a plastic flow stress;[1] after simplification the criterion can be expressed by the relationship

$$\sigma_\theta = \sigma_f \tag{1}$$

where σ_θ is the membrane hoop stress due to the pressure, and σ_f the flow stress given by

$$\sigma_f = k(YS + UTS) \tag{2}$$

where YS = yield stress and UTS = ultimate tensile stress. A conventional and conservative value σ_{fconv} is obtained for $k = 0{\cdot}5$.

2.2 Axial single cracks or defects in non-specific zones

Burst tests show that the burst pressure for tubes with axial part-through defects (e.g. crack, wear) can be accurately estimated by using the following functional form[2]

$$P_r = f(\sigma_f, t, R, a, d) \tag{3}$$

where t is the wall thickness of the tube, R is the inside radius of the tube, a is the crack half-length and d is the depth of the crack. The tests have shown that, when the ligament rupture pressure P_r is smaller than the instability limit pressure of a through-wall crack with the same length, the crack remains stable after the rupture of the remaining ligament. Typical test results are given in Fig. 2.

For a through-wall defect, the instability limit pressure P_a can be reliably determined on the basis of an extension of eqn (1) for a sound tube

$$M\sigma_\theta^* = \sigma_f \tag{4}$$

where σ_θ^* is the equivalent membrane hoop stress (taking into account pressure loading on the tube walls and crack lips) and M is the bulging factor, which is also the K_I magnification factor. A conservative value of P_a can be obtained using σ_{fconv}, and for M the expression proposed by Krenk.[3]

Test results show that the combination of thermal and mechanical stresses

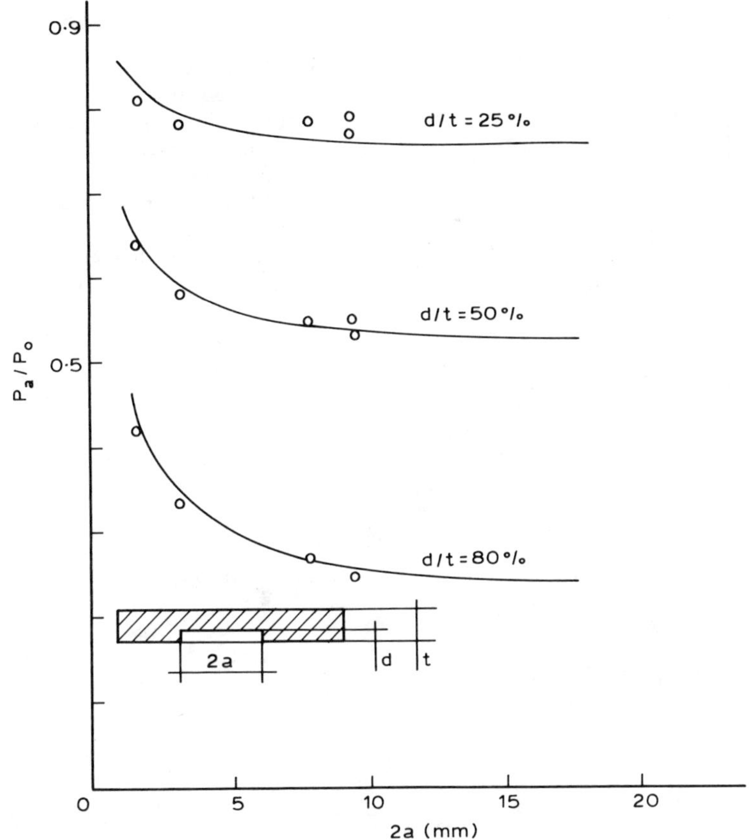

Fig. 2. Normalized burst pressure for axial defects. P_a = burst pressure for $2a$; P_0 = burst pressure for $2a = 0$ (sound tube). ——, Criterion $P_a/P_0 = 1 - \{(2ad/t)/2(a+t)\}$; ○, burst tests.

does not modify the stability limit state of a tube with an axial crack. These thermal stresses, which result from the strong thermal shock created by the injection of cold water at the tube wall on the secondary side, do not play an important role in the plastic collapse phenomena.

Equation (4) has been demonstrated in many different tests, where varying geometrical and mechanical parameters have been investigated,[2] and it has been confirmed by burst tests on tubes removed from the steam generators (see Fig. 3).

2.3 Circumferential through-wall cracks

The behaviour of tubes with circumferential through-wall defects was studied in burst tests on tubes with defect lengths between 45° and 300°. These were performed on unconstrained tubes and on tubes with a tube support plate restricting lateral displacement. The influence of the support

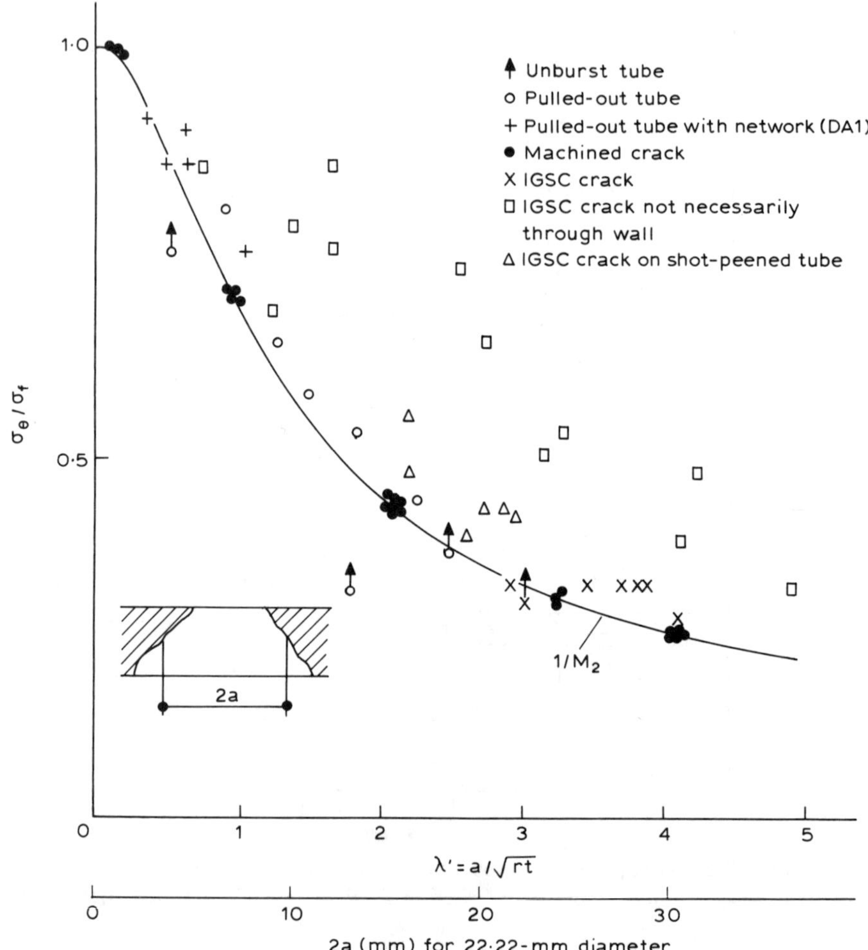

Fig. 3. Burst criterion for through-wall axial cracks in non-specific areas of tubes with diameter of 22·22 mm.

plate is shown by an increase in rupture pressure obtained for a supported tube in comparison with that obtained for an unsupported tube.

Comparison between calculation and experimental data shows a good agreement with a beam plastic hinge model in which the bending moment accounts for the interaction with the support plate.[2]

2.4 Single or multiple cracks in the kiss rolling zone

In all cases, it is experimentally confirmed that the instability limit pressure of a tube with a crack located in the rolling zone at the top of the tubesheet is higher than that of a tube with the same type of defect located in a straight

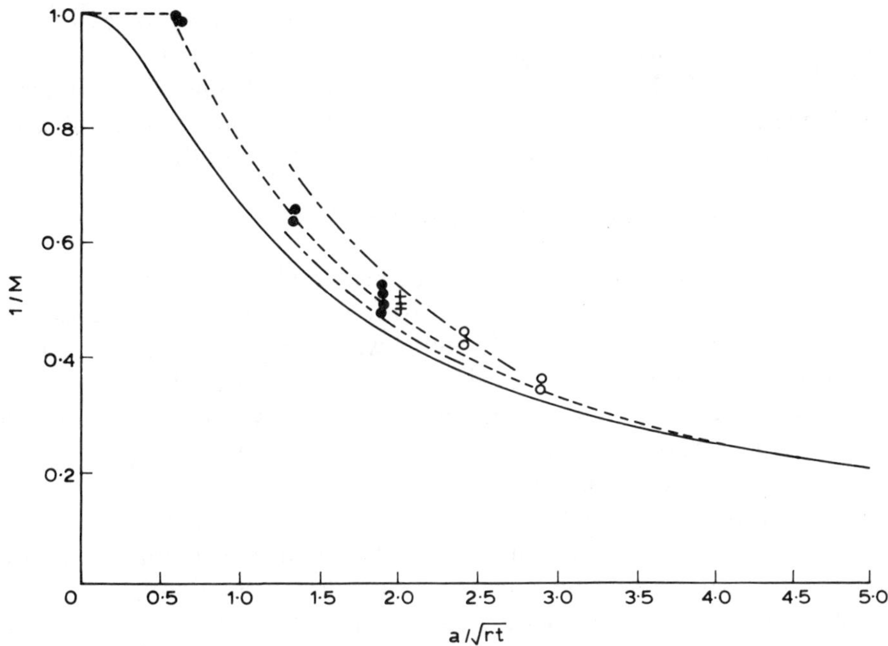

Fig. 4. Steam generator tube burst criteria for non-specific zones and kiss rolling zones. ——, Non-specific zone criterion (M2), – – –, Kiss rolling criterion (M2*); —·—, Finite element models; +, ●, 22-mm diameter tube burst; ○, 19-mm diameter tube burst.

section remote from discontinuities. The gain in critical pressure is due to two combined effects: local mechanical strengthening due to seating in the tube sheet, which limits bulging of the tube, and an increase in the flow stress due to strain hardening of the material in the kiss rolling zone during the kiss rolling operation. The gain in critical pressure is all the greater when the defect is short, becoming negligible for relatively long defects of the order of $4\sqrt{rt}$, where r is the mean radius of the tube.

In the roll transition zone, the instability criterion is an adaptation of eqn (4)

$$M^*\sigma_\theta^* = \sigma_f \tag{5}$$

where $M^* = \mu \times M$ in which $\mu = f(a/\sqrt{rt}) \leq 1$, and M relates to a non-specific zone defect (Fig. 4).

It has been demonstrated that this rupture criterion remains applicable where there are several major cracks distributed over the entire tube circumference.

2.5 Axial cracks in U-bends

Burst tests have been carried out on bends with axial cracks at different positions. In all cases, the burst pressure is greater than that corresponding to an equal crack located in a non-specific section, the gain ranging from 15 to 70%. This strengthening is ascribed to strain hardening of the material combined, in certain cases, with an effect due to the curvature of the tube, but it is only slightly affected by the degree of out-of-round (see Fig. 5).

2.6 Cracks in the vicinity of the tube support plate

Room temperature burst tests have been performed on tubes with surface and through-wall defects and stress corrosion cracks located in the vicinity of the tube support plate. Defect lengths were between 5 mm and 30 mm.

The experimental set-up was designed to simulate, where necessary, squeezing of tubes due to deformation of the support plate near the wedge supports and tube rotation due to out-of-plane bending of the support plate. The values of squeezing and rotation were determined during steam generator accident transient analysis. In all cases, with or without squeezing or rotation, the bursting pressure was higher than that obtained with the same defects in non-specific zones, the bulging deformation being locally constrained by the support plate. In many cases, bursting occurred outside the support plate intersection, in the non-specific zone at the sound tube pressure value.[2]

2.7 Double cracks

Burst tests on tubes with two aligned or offset cracks have made it possible to detect and to quantify interaction between two parallel cracks. Here again, rupture, whether it first occurs in the ligament or at opposite points on defects as if it were a single defect strengthened by the ligament, takes place by plastic instability. A simple interpretation of these tests is given by using the generalized criterion established for a single crack in a tube. If K_I is the stress intensity factor of an isolated crack located in an infinite plate and K_{i*} the stress intensity factor, under the same conditions, of the tip i of that crack interacting with the second one, the interaction factor F_i is given by: $K_\mathrm{i}^*/K_\mathrm{I} = F_\mathrm{i}$.[4]

The failure criterion is then

$$f \times F_\mathrm{i} \times M^* \times \sigma_\theta^* = \sigma_\mathrm{f} \tag{6}$$

where M^* is the bulging factor related to the isolated crack and f an empirical adjustment factor.

This formula must be written for the two tips of both cracks; the lowest value of σ_θ^* deduced from the four formulae identifies the tip of the crack where the failure will occur. The failure pressure is given in eqn (6). An interpretation of some tests by this equation is shown in Fig. 6.

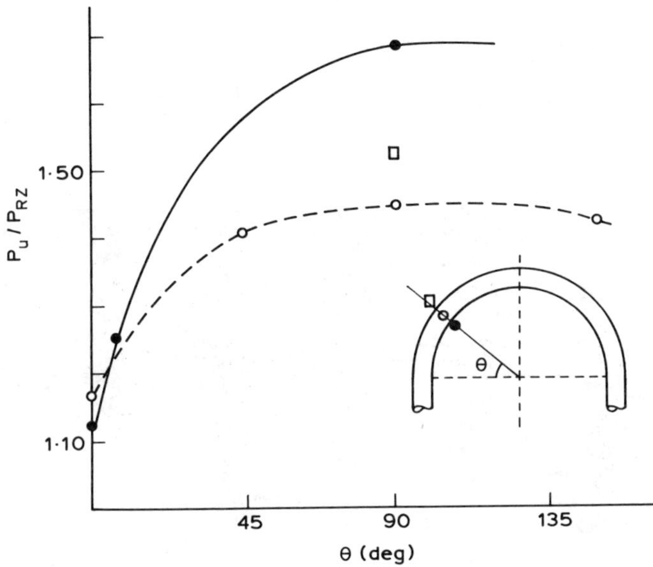

Fig. 5. Normalized burst pressure for U-bends. P_U = U-bend burst pressure; P_{RZ} = non-specific zone burst pressure. ●, inside bend; □, outside bend; ○, side of bend.

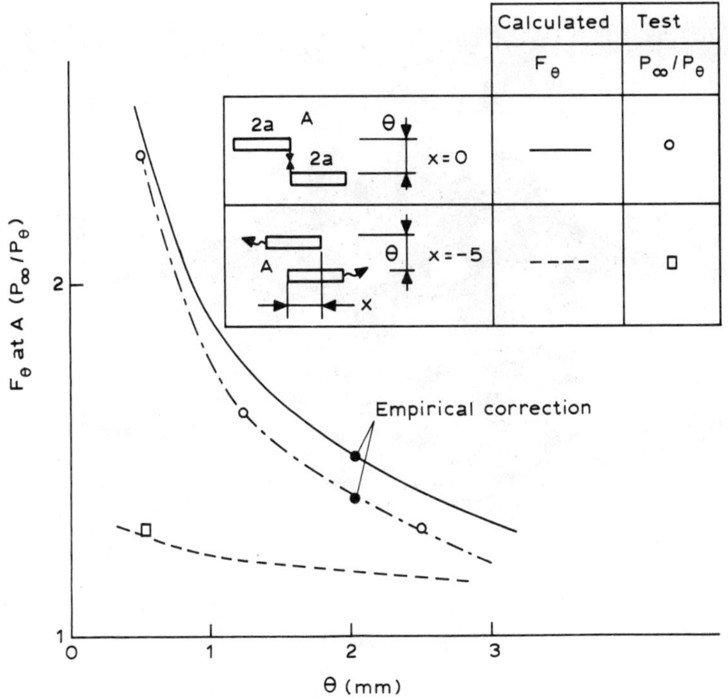

Fig. 6. Offset double defects, tests and interpretation.

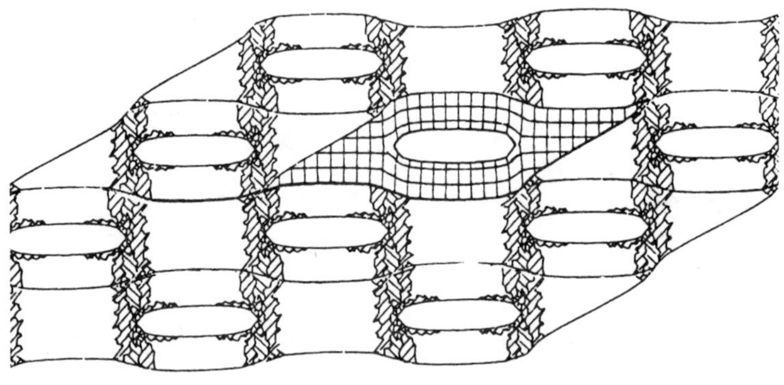

Fig. 7. Regular networks of short cracks, burst tests and finite element models.

2.8 Multiple parallel cracks

A comparison has been made between burst tests on tubes with a single defect or crack and on tubes with two or several parallel axial surface and through-wall defects of the same length with their tips at the same axial position. Study of crack behaviour after rupture of the remaining ligament of surface defects evenly distributed around the tube has shown that the bulging on one of the cracks exceeds that of the others, and this crack is the one which becomes unstable and governs the burst pressure.

The burst pressure was found to be very close to that measured with only one crack of the same length and therefore there is no significant interaction between parallel axial cracks (minimum distance between cracks 3 mm).

For other parallel defect configurations, comparative analysis of the results with a single crack reveals weakening, variable in extent, of overall tube strength in the deteriorated area. This is shown by a decrease in the instability limit pressure of the main crack which varies with the types of deterioration, the lengths and the relative positions of the various cracks.

Semi-analytical models have been developed for the case of regular networks of short cracks,[5] whether they surround a main crack or not. These models can be used to predict the critical pressure and, where applicable, the opening area of the main crack.[6] These predictions are confirmed by tests with machined cracks and are found to be conservative when applied to tubes with damage of the same type removed from steam generators (see Fig. 7).

3 CRACK OPENING AREA AND LEAK RATE

Variations in crack opening areas have also been the subject of a combined experimental and theoretical approach for some of the configurations mentioned earlier:

- cold tests, and in certain cases hot tests, have made it possible to monitor the change of the crack opening area in the vicinity of the critical pressure; displacement and photogrammetric methods have been used for these tests.
- elastoplastic calculations concerning cracked tubes have been carried out to simulate these tests.

The results of this work have made it possible to finalize analytical opening area models based on the shell theory of cracks. In the case of an axial crack, this can be expressed as follows

$$A_{\mathrm{cp}} = \Gamma_{\mathrm{cp}} A_{\mathrm{el}}$$

where A_{el} is the elastic part of the opening area and Γ_{cp} is a suitable elastoplastic correction

$$A_{\mathrm{el}} = \mu^* \frac{2\pi\sigma_\theta^* a^2}{E}, \quad \text{and} \quad \Gamma_{\mathrm{cp}} = f(YS, K_{\mathrm{I}}) \quad \text{(Dugdale correction)}$$

where μ^* is the bulging factor relating to the opening area allowing for the environment of the crack (adjacent cracks, surrounding under-thickness, nearness of tube sheet). μ^* is directly related to the factor M^*. E is Young's modulus.

Figure 8 shows comparative results for opening areas of axial cracks in typical zones: finite element elastoplastic calculations, tests and analytical models based on M^* given by Ref. 7. Leak rate measurements at room temperature and at 120°C have been made on axial stress corrosion cracks replicated in the laboratory. The results obtained, reinterpreted in the light of the preceding results concerning the opening area, have made it possible

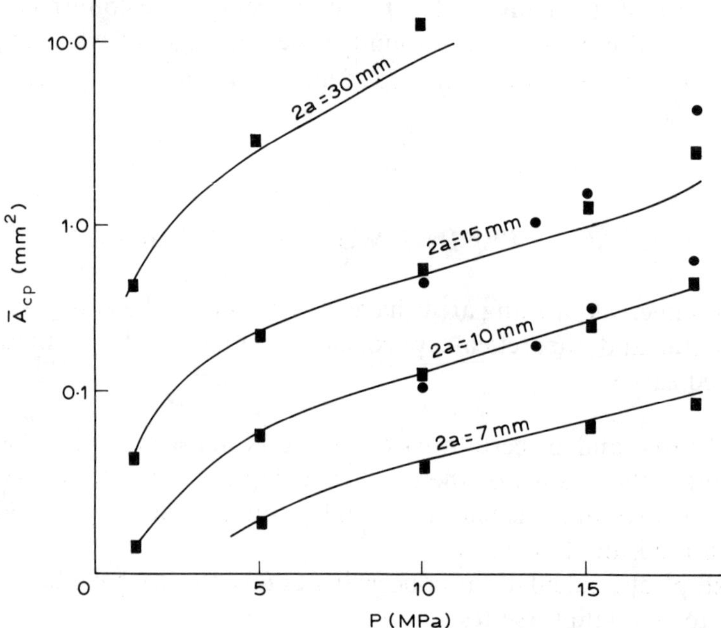

Fig. 8. Opening areas in non-specific zones, comparison between tests, calculation and analytical models. ——, Elastoplastic analytical method; ■, finite elements; ●, measurement.

to establish correlations between opening areas and leak rates, which in particular indicate a non-leakage threshold for short cracks.

The use of this correlation in a number of cracked tube configurations has made it possible to establish for each a leak rate which is allowable in operation and below which there is no rupture risk in accident overpressure situations. This allowable leak rate varies considerably with the type of damage analysed, as well as with the geometry and mechanical properties of the tube involved. An example of this is to be found in the transition zones, which are very prone to cracking (Fig. 9). This figure also gives, as a function of the hot yield strength:

- $2a_c$ the critical length in accident situations ($\Delta P = 172b$)
- $2a_{72}$ the length resulting in a leak rate of 72 litres/h under normal operating condition ($\Delta P = 100b$): for this situation the leak is limited at the value of 72 litres/h.
- $2a_o$ the length below which leakage cannot occur.

The curves given here correspond to a tube of average geometry. For a tube of reduced thickness, the margins in terms of leakage before rupture are lower.

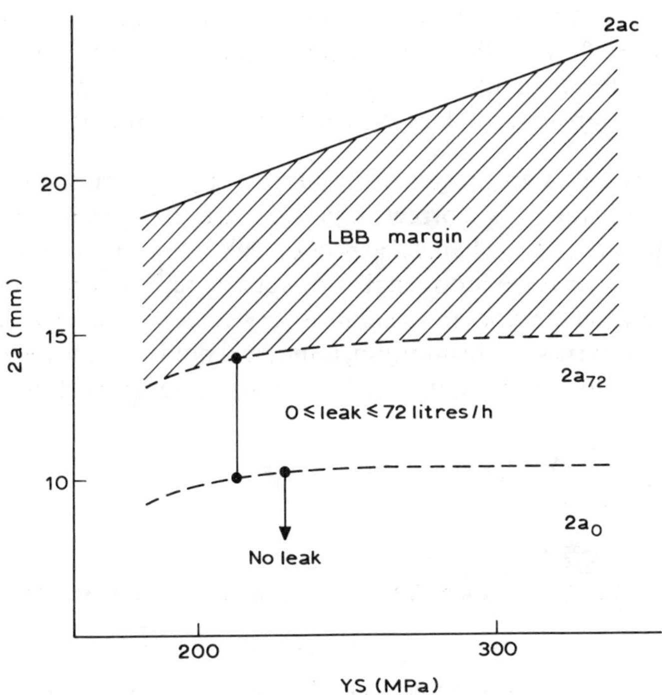

Fig. 9. Leak before break in kiss-rolling zone.

4 DISCUSSION AND FUTURE WORK

A certain number of tubes removed from steam generators have provided useful information which has been interpreted using the preceding models. Although no results call into question the Leak-Before-Break criterion, and all satisfy the critical pressure models (see Fig. 10), frequently with ample margin, some cracked tubes failed to give the expected leak rate in service, even though the leak rate observed in the laboratory, under conditions which were different from those of service, is in close agreement with the models interpreted under the laboratory conditions. A number of hypotheses have been advanced to explain the discrepancy:

(1) failure to allow for stress of a secondary nature (residual or thermal) which, although it may not affect the critical pressure, can, in certain cases, reduce the opening area;
(2) postulated boiling at the crack lips under certain thermohydraulic conditions during operation which would appear to both reduce the leak rate and contribute to the crystallization of certain boron-lithium salts, thus reducing the size of the passage open to the flow of fluid;
(3) the presence at the end of the tube, on the secondary side, of compact deposits of crud capable of blocking the crack; this hypothesis is supported by the fact that the tubes differing from the leakage model were indeed in the central part of the tube sheet where oxides resulting from secondary system corrosion are found to accumulate.

All these hypotheses are currently the subject of programmes to determine which one of them is correct. Special care has been paid to the thermohydraulic and mechanical aspect of leak rates, with the construction by EDF of experimental loops enabling accurate replication of normal and accident thermohydraulic and chemical service conditions for tubes with representative corrosion. In addition, numerical models are being developed to assess the effect of secondary stress on opening areas. Finally, in-situ

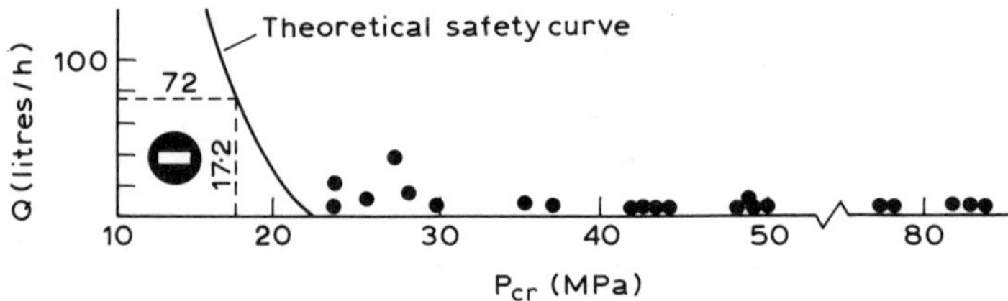

Fig. 10. Axial cracks in kiss-roll zone (pulled tubes) diagram showing normal flow rate against critical pressure (site conditions).

measurements of crud deposits, correlated with secondary system velocity studies, will make it possible to understand better the high risk areas.

Other aspects of the LBB criterion are also being studied. Models for predicting the initiation and propagation of cracks are being finalized. We require both fuller knowledge of in-service stresses (residual stress and stress due to operation) in sensitive areas as well as analysis of the mechanisms of the appearance and development of cracks in the materials used, in the operating environment.[8]

The different studies will eventually be the subject of a probabilistic review making it possible to assess the safety of steam generators based on the knowledge acquired (fabrication date and periodic inspection) and to modify the surveillance programme, where necessary, to preserve acceptable leak rates in the tubes.

REFERENCES

1. Svensson, N. L., The bursting pressure of cylindrical and spherical vessels. *J. Appl. Mech.*, **25** (March 1958) 896.
2. Cochet, B. & Flesch, B., Crack stability criteria in steam generator tubes. *9th SMIRT*, Vol. D (1988) 413–19.
3. Krenk, S., Influence of transverse shear on an axial crack in a cylindrical shell. *Int. J. of Fracture*, **14**(12) (1978) 123–44.
4. Flesch, B. & Cochet, B., Crack stability criteria in steam generator tubes. *Proc. of Int. Congress on Pressure Vessel Technology*. Beijing, September 1988.
5. Flesch, B. & De Keroulas, F., Steam generator tube cracks at Dampierre 1 EDF PWR Plant. Impact on plant safety. *Nuclear Engineering and Design*, **113** (1989) 357–68.
6. Voldoire, V., Determination by the homogenization technique of the overall behavior of steam generator tubes weakened by micro cracks arrays. *10th SMIRT*, Vol. L (1989) 103–8.
7. Kastner, W. *et al.*, Critical crack sizes in ductile piping. *Int. J. Pres. Ves. Piping*, **9** (1981) 197–219.
8. Flesch, B., Vidal, P. & Vagner, J., In service stresses in 900 MWe PWR Steam Generator U-Bends. *10th SMIRT*, Vol. D (1988) 177–82.

Int. J. Pres. Ves. & Piping **43** (1990) 181–186

A Plugging Criterion for Steam Generator Tubes based on Leak-Before-Break

A. Esteban, M. F. Bolaños & J. M. Figueras

Consejo de Seguridad Nuclear (CSN), 11 Justo Dorado, 28040 Madrid, Spain

ABSTRACT

Degradation of steam generator tubes is occurring in Spanish nuclear power plants. Degradation causes are different and give occasion for leakages.

General Criteria of Design, technical specifications and availability are concepts which must be harmonized. Historic behaviour of certain tube defects together with experiments and studies made by the Spanish Owner Group of Pressurized Water Reactors and approved by Consejo de Seguridad Nuclear have led to a new plugging criterion based on the Leak-Before-Break concept and a new limit for leakages is in operation in some plants.

INTRODUCTION

Steam generator tubes are manufactured from ductile materials which exhibit intrinsically safe Leak-Before-Break (LBB) behaviour. However, a number of issues must be addressed before attempts are made to take advantage of this behaviour by using it as a plugging criterion for defective tubes. These include: the type of defect responsible for the leak; the location of the defect; the growth rate and permissible size of the defect; the uncertainties of the eddy current tests and the leak rate associated with the defect, and finally the history of leakages which were permitted in operation in accordance with plant technical specifications.

This paper discusses the approach which has been used in Spain to justify the adoption of a plugging criterion based on LBB considerations and the requirements approved by Consejo de Seguridad Nuclear which are applied to inspection outages and operating conditions.

Several experiments and studies were carried out by the Spanish Owner Group of Pressurized Water Reactors who then submitted a new plugging criterion to Consejo de Seguridad Nuclear for their approval.

THE LBB CONCEPT AS A PLUGGING CRITERION

The steam generator tube bundle forms an important part of the primary pressure boundary in a pressurized water reactor (PWR) because a tube bundle has a large surface area and the thickness of the tubing is small (see Table 1). In practice, the steam generator tubes are more frequently affected by through-wall leakage than any other part of the primary pressure boundary. The easiest solution to these incidents is to plug the affected tubes. At present, a simple plugging criterion is most frequently used for operating steam generators: any tubes exhibiting defects in excess of a specified depth will be plugged. In Spanish steam generators, the permitted loss of wall thickness is 40%, i.e. all tubes with any form of degradation, whether uniform degradation or cracking, should be plugged when the depth of the through-wall damage exceeds 40%. This criterion is based on material strength properties and is intended to maintain the integrity of the tubes.

A second criterion, referred to as P^*, is used in Spain. This allows the continued operation of tubes containing defects which would otherwise necessitate plugging, providing that the defects are located within a distance of 38 mm below the top of the tubesheet. This criterion assumes that the tubes will be supported by the tubesheet in spite of degradation, but only applies to circumferential defects. Such tubes are re-assessed from time to time and this criterion allows many tubes to remain in service that would otherwise have to be plugged.

A new plugging criterion based on LBB concepts has also been implemented in Spain. The use of tough materials of known dimensions and mechanical properties permits estimation of the maximum size of through-wall defect that can be tolerated without the risk of tube rupture. Leakage through this defect signals its presence before it poses any threat to the integrity of the tube. This concept can be used to advantage in defining the scope for a new plugging criterion.

If LBB behaviour could be relied on under all circumstances, then a plugging criterion based on the depth of degradation would no longer be necessary since leakage could be relied on as a warning of the presence of defects. Furthermore, the radioactive nature of leakage from the primary side would enable its detection in the secondary water before the defect reached sufficient size to pose any threat to safety.

If this behaviour could be relied on for all types of defect, a new plugging

TABLE 1
Dimension of Model D-3 Steam Generator Tubes Made from Inconel 600 M.A.

Dimension	*Value*
Outside diameter (mm)	19·05
Wall thickness (mm)	1·09
Straight length (m)	6·97
Number of U-tubes/bundle	4 674

criterion could be defined which would supersede those currently in use, thus avoiding the need to plug many tubes. However, this assumption is not generally accepted and a plugging criterion based on LBB behaviour can only be used where such behaviour can be guaranteed to occur.

In summary, a plugging criterion based on LBB behaviour can be used only for particular forms of degradation with well-characterized morphologies, locations and propagation characteristics and when experience in many cases has shown that LBB has occurred.

DESIGN CONSIDERATIONS

Nuclear power plants are designed and built in accordance with accepted design criteria which define the general standards to be accomplished and which cannot be violated. In Spanish PWR power plants, following the practice in the USA, such criteria are embodied in the General Criteria of Design, Appendix A of title 10, part 50 of the Code of Federal Regulations (10CFR50).[1] Before a plugging criterion based on LBB can be implemented, it must be demonstrated that this meets the requirements of the design criteria. The general criteria mentioned above are as follows.

Criterion 14, 'Reactor Coolant Pressure Boundary', requires that there should be a low probability of abnormal leakage, rapid defect growth and gross rupture. To ensure that this criterion is not violated, it is necessary to define the abnormal leakage and to guarantee that the defects will not increase the probability of such leakage or propagate rapidly, leading to gross rupture (i.e. that LBB behaviour will take place).

Other rules, such as Regulatory Guidelines, which have lower legal status than the General Criteria, define the permissible leak rate limit for the pressure boundary which is fixed in plant technical specifications. One may define as abnormal leakages those increases beyond the permissible limits which are included in plant technical specifications. These limits are not fixed but will change as the technology improves. The increase in leak rate is high when defect growth is rapid. Clearly, the rate of leakage through cracks in tubes which are not going to be plugged must either be lower than the specified limit in the plant technical specifications or must be re-defined as normal so that the plugging criterion based on LBB does not contravene Criterion 14.

Criterion 15, 'Reactor Coolant System Design', specifies that the reactor coolant system shall be designed with sufficient margin to ensure that design conditions are not exceeded during any normal operation or anticipated operational occurrences. This requires that a plugging criterion based on LBB concepts can only be considered if design safety margins are not reduced.

Fortunately, the tubes are manufactured with adequate safety margins to tolerate the permitted wall loss of 40%, so that even if some defects are through-wall, the safety factors embodied in the design standards are not compromised. Consequently, one may define a plugging criterion based on LBB which does not contravene Criterion 15.

Both of the above General Criteria have a strong influence on the plugging criterion based on LBB. Criterion 30 is also relevant. This states that, 'Means shall be provided for detecting and, as practicable, identifying the location of the source of reactor coolant leakage'. There is clearly no conflict between this criterion and the plugging criterion based on LBB since the latter depends upon the detection and location of the defect and, in addition, may also identify its contribution to the overall leak rate.

APPLICATION

A plugging criterion based on LBB behaviour can only be applied to the specific modes of degradation with recognized morphology, location and growth characteristics for which previous experience exists in nuclear reactors. Furthermore, the approach is of particular benefit if the type of degradation is frequently encountered. The maximum permissible leakage could be reduced if the degradation mechanisms are evaluated and graded in importance. During operation the leakage by different causes is obviously not dependent on the plugging criterion.

From Spanish experience, one defect which meets this requirement is primary water stress corrosion cracking (PWSCC) in the roll transition zone.

This degradation occurs as tight, axially-oriented flaws which do not extend beyond the residual stress field of the transition zone. In Spain, the tubes with these defects have shown, statistically speaking, a crack growth rate of 4 mm between inspections. PWSCC at the transition zone develops as axial cracks. The plugging criterion is based on calculation of the greatest crack length which is stable under the worst circumstances. For steam generators, these correspond to a steam line break (SLB). The largest length of stable crack under SLB conditions, taking into account suitable safety factors, is determined and is referred to as the critical length. This value is obtained using analytical procedures and is validated by burst testing carried out in Spain.

It is arbitrary to define the critical length as the greatest crack length which is still safe or the smallest crack length that is unstable because they are coincident values at the boundary. However one prefers the first one because it seems more conservative to refer to the crack staying stable. The stresses imposed are those which are derived from loading associated with normal and accident conditions and are included in the Stress Report of the Main Supplier.[2] The stress due to rolling in the tubesheet is not taken into account.

A critical crack length of 13 mm has been obtained by means of the Folias bulging factor and using finite element calculations. Then, considering that crack growth rate is 4 mm and that eddy current test uncertainties are about 1 mm, the limit of crack length from the point of view of the plugging criterion application is 8 mm. Hence, tubes with cracks of length ≤ 8 mm may remain in service.

Burst testing has confirmed that this value of 8 mm is conservative enough. Eddy current tests in tubes which were pulled out of Spanish steam generators have shown that the crack sizing uncertainty of 1 mm is suitable. Multifrequency eddy current tests, with standard bobbin coils and rotating probes were used in the inspections.

Complementary to the above is surveillance of the leak rate operation. For this purpose, the leak rate which corresponds to the critical crack length is determined by analytical methods and leak rate measurements. Suitable safety factors are then applied to reduce this value to 5 litres/h per steam generator. This limit can be safely tolerated during operation.

A plugging criterion based on LBB has been permitted in Spain by Consejo de Seguridad Nuclear for a 3 loop 930 MW Westinghouse PWR. For this power plant, the application of this criterion has permitted the continued service of a total of 149 tubes containing cracks due to PWSCC, as follows: SG1, 29 tubes; SG2, 76 tubes and SG3, 44 tubes.

CONCLUSIONS

(1) A new plugging criterion based on LBB concepts that permits leaking tubes to remain in service subject to certain requirements has been implemented in some Spanish PWR steam generators. So far,

this has been applied to Westinghouse Model D3 units. In future, the criterion will be used where appropriate, depending on the type and extent of the degradation.

(2) Close surveillance of the primary-to-secondary leak rate is imposed during operation to ensure that safe limits are not exceeded. Hence, the original limiting conditions which were included in technical specifications have been reduced.

(3) Significant benefits have been realized since 149 tubes have remained in service which would otherwise have had to have been plugged, without adversely affecting the safety of the plant.

(4) The plugging criterion based on LBB concepts is consistent with the requirements of the General Criteria of Design. It maintains or, by closer surveillance of leak rates during operation, increases safety margins and also improves plant availability.

REFERENCES

1. Anon., *Code of Federal Regulations*, Title 10, Part 50.
2. Anon., *Stress Report of the Main Supplies.* Confidential report, Westinghouse.

Int. J. Pres. Ves. & Piping **43** (1990) 187–204

Locating a Leaking Crack by Safe Stimulation

C. E. Coleman, S. Sagat,

Reactor Materials Division, Atomic Energy of Canada Ltd,
Chalk River Nuclear Laboratories, Chalk River, Ontario, Canada K0J IJ0

G. K. Shek, D. B. Graham & M. A. Durand

Ontario Hydro, 700 University Avenue, Toronto,
Ontario, Canada M5G 1X6

ABSTRACT

A few Zr–2·5 Nb alloy pressure tubes in CANDU nuclear reactors have leaked through cracks that have grown by delayed hydride cracking (DHC). In some instances, tubes contained confirmed leaks that were leaking at a rate too low for precise identification of the leaking channel. Controlled stimulation of DHC can be used to help locate these leaks by extending the crack and increasing the leak rate without approaching crack instability. In the event of a leak being detected, a plant operator can gain time for leak location by a heating and unloading manoeuvre that will arrest crack growth and increase the critical crack length. This manoeuvre increases the safety margin against tube rupture. If required, the operator can then stimulate cracking in a controlled manner to aid in leak identification. Sequences of temperature and load manoeuvres for safe crack stimulation have been found by laboratory tests on dry specimens and the efficacy of the process has been demonstrated, partly in a power reactor, and partly in a full-scale simulation of a leaking pressure tube.

INTRODUCTION

As part of the defence against rupture of a pressure tube in a CANDU reactor, the Leak-Before-Break (LBB) criterion is used; if a crack develops

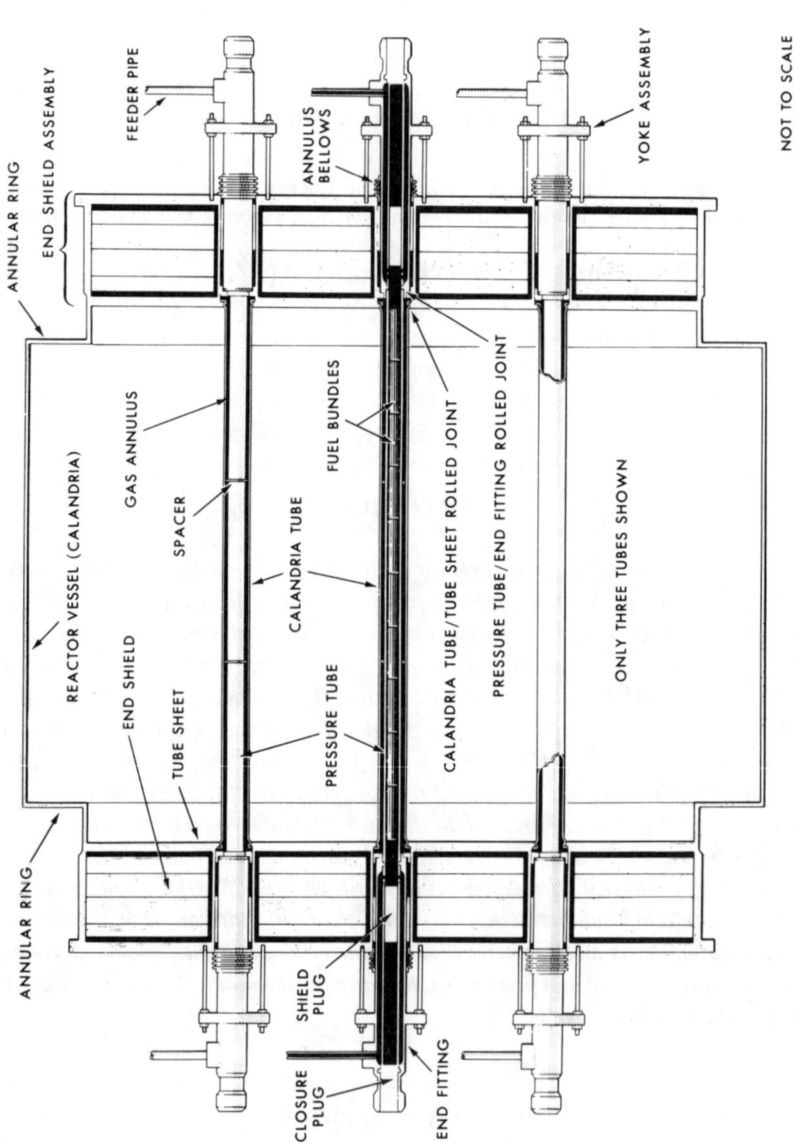

Fig. 1. Schematic diagram of a CANDU reactor.

and penetrates the tube wall, it is detected and located by coolant leakage and action is taken before the crack becomes unstable. The validity of this approach has been demonstrated on twenty-four occasions in three reactors.[1–3] Once the cracked tubes were located, the tubes were simply replaced. Occasionally the leakage was so low that location was difficult; therefore, techniques have been devised to stimulate stable cracking, and thus leaking, without increasing the chance of tube rupture. In this paper, the authors describe the properties of delayed hydride cracking (DHC) on which the principle of safe stimulation is based, its demonstration in a laboratory simulation and two experiences with leaking cracks in reactors.

THE CANDU REACTOR

The CANDU reactor consists of a large tank (calandria), containing heavy-water moderator at 70°C, that is penetrated by about 400 horizontal fuel channels arranged in a regular lattice (Fig. 1). Each channel consists of a pressure tube surrounded and insulated from the cold moderator by a calandria tube. The pressure tube contains the natural uranium fuel and heat-transport heavy water (HTS) at a pressure of about 10 MPa and temperature ranging from 250 to 315°C. The space between the pressure and calandria tubes contains dry nitrogen or carbon dioxide and is called the gas annulus. This space is maintained by garter spring spacers around the pressure tube at intervals along the fuel channel. The gas annulus system is sensitive to the presence of moisture and can be used to detect any breach of the primary pressure boundary. The pressure tubes are made from cold-worked Zr–2·5 Nb; each tube is 6 m long, has an internal diameter of 100 mm and wall thickness of 4 mm. Both ends of the pressure tubes are attached to stainless steel end fittings by a rolled joint.

The fabrication of the rolled joint produces residual stresses in the rolled portion of the pressure tube because of wall thinning and tube expansion. In early power reactors an incorrect rolling procedure produced excessive tensile stresses in the pressure tube at the limit of rolling where the tube was no longer supported by the end fitting. High tensile hoop stresses caused cracking, and subsequent leakage, by a stable crack growth mechanism called DHC. (A modified installation procedure has eliminated the high stresses in new reactors.)

DHC is the dominant mechanism of stable crack growth in Zr–2·5 Nb pressure tubes and its avoidance in pressure tubes requires vigilance during fabrication, installation and reactor operation.

DELAYED HYDRIDE CRACKING

The solubility limit of hydrogen in zirconium alloys is very low below 315°C. When the limit is exceeded, zirconium hydrides, in the form of collections of platelets, are precipitated. These hydrides are brittle. In DHC, hydrogen preferentially dissolves from hydrides in the zirconium alloy matrix and migrates up the stress gradient to the tip of a stress concentrator, such as a surface flaw, then reprecipitates. If the stress is large, at some critical condition (perhaps hydride size) the first hydride will crack, the process is then repeated and the crack grows in a series of steps. Hydrides must be present for cracking. Under steady load, and after cooling from above the solvus temperature, the crack velocity, V, is at a maximum and, above a threshold value,[4] is essentially independent of the stress intensity factor, K_I. The cracking has a temperature dependence described by:

$$V = A\exp(-B/T) \quad (1)$$

where A and B are constants and T is absolute temperature.

With varying loading and after heating the behaviour is more complicated.[4-8] If a crack is growing by DHC and K_I is raised, V is not much affected. If K_I is lowered the crack may slow or stop and an incubation time is required for cracking to continue. If the temperature is attained by heating, cracking is slower than expected from eqn (1) or negligible above a critical

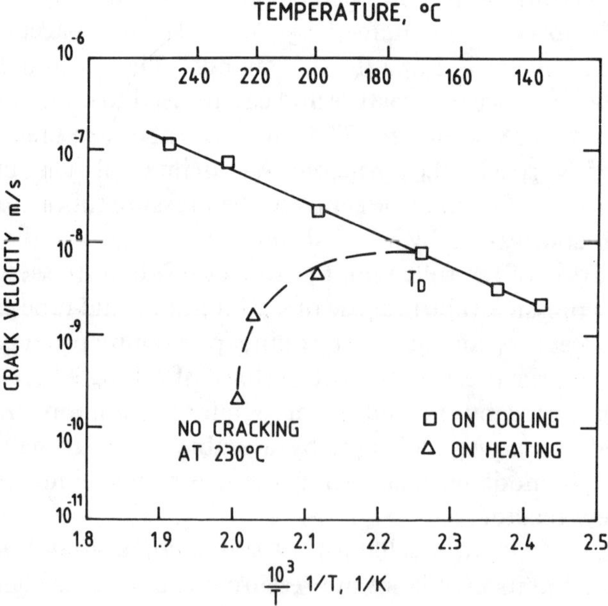

Fig. 2. Temperature dependence of velocity of DHC after heating and cooling. Results from beams of irradiated Zr–2·5 Nb.

temperature, T_D (Fig. 2). If this same temperature is now attained by cooling, the behaviour of eqn (1) is restored. Such behaviour can be used to repress DHC (avoidance manoeuvre)[4] or promote DHC (crack stimulation).

If DHC develops and the crack penetrates the tube wall, allowing leakage, the leak must be detected as soon as possible so that appropriate action can be taken before the crack reaches the critical length for instability, the critical crack length.

CRITICAL CRACK LENGTH

The critical crack length in Zr–2·5 Nb pressure tubes is estimated from measurements of fracture toughness using small compact specimens cut from the tubes[9] or directly from burst tests on full sections of tube containing through-wall axial cracks.[10] Between 200°C and 300°C the fracture toughness has little temperature dependence. Although neutron irradiation reduces flaw tolerance from its initial high value, beyond about 3×10^{25} n/m^2 the rate of decrease in toughness slows to provide an estimate of the lower bound value of critical crack length of about 50 mm at an operating pressure of 10·3 MPa. The hoop stress in the pressure tube is reduced as the HTS pressure, *P*, is lowered, and the critical crack length is increased approximately as P^{-2}.

LEAK DETECTION AND LOCATION

When HTS moisture is detected in the annulus gas system (AGS), it is assumed to be a result of a through-wall crack in a pressure tube. A successful station response to this leak involves:

—detection and verification of the leak source,
—identification of the leaking channel, and
—placing the reactor in the cold, depressurized state before the growing crack becomes unstable.

Procedures in place at the stations require regular and frequent measurement of moisture if an action limit is exceeded. Action limits include high dew point, high rate of rise of dew point, alarms based on increases of electrical conductivity across an air gap or measurement of water accumulation.

Once the leak is confirmed, the reactor is shut down to zero power but remains hot, with the required thermal and pressure manoeuvres aimed at limiting further DHC.[4,8] The gas flow in the AGS is then stopped to aid leak location.

Leak location can be accomplished by observing temperature differences. If the leak rate is sufficiently high, leaking steam condenses on the cooler calandria tube, water accumulates in the gas annulus and eventually contacts the pressure tube. The heat loss from the reactor channel to the moderator will increase, resulting in a decrease in channel outlet temperature, thereby identifying the leaking channel. The temperatures are measured with resistance thermometers placed on the outlet feeder pipes.

The leak may be at a rate too low for channel identification, for example, as low as 1 g/h. Shutting down to the cold, depressurized condition and then trying to find the leak site using non-destructive techniques, such as ultrasonic testing, in approximately 400 pressure tubes, is impractical. Analysing the frequency spectra of acoustic emission when the reactor is pressurized to less than maximum pressure at low temperatures, up to 180°C, has been used successfully to locate leaking channels in Pickering Units 3 and 4[1] and in Bruce Unit 2;[2] however, this method is not universally applicable, as described below. Thus a standard, benign scheme for leak detection is being developed. One solution is to increase leakage through the crack by subjecting the reactor to conditions that will promote DHC. This operation of stimulating DHC must be carefully controlled to avoid the possibility of exceeding the critical crack length or causing excessive growth to any non-leaking crack that may be present.

Although we need to grow a weeping crack in a controlled manner, we also need time to examine records and plan a course of action. This time can be gained by stopping or slowing the crack growth. The operations to stop or slow down the crack are unloading (actually depressurization) and either heating after a cooling manoeuvre, if the reactor is at hot shutdown, or cooling to room temperature. The unloading has the additional advantage that it increases the critical crack length for rupture and thus increases the margin against unstable fracture during subsequent manoeuvres. Once the crack is under control, the temperature of the reactor can then be cycled to encourage cracking; there may be an incubation time, but when cracking starts again, the crack length will still be far from instability. In our experiments we have measured the effects of various combinations of loading and temperature cycling so that we may exploit the behaviour of the DHC for leak location.

LABORATORY EXPERIMENTS

Small specimens

Constant K_I specimens[11] were machined from flattened strips of cold-worked Zr–2·5 Nb pressure tubing containing up to 1 at% hydrogen. The samples were fatigue pre-cracked and loaded in tension with a dead-weight. Delayed hydride cracking was monitored by dc potential drop across the crack.[8]

Notched beam specimens[6] were machined 1·3 m from the outlet (hot) end of a Zr–2·5 Nb pressure tube removed from Pickering Unit 3. This material had received a fluence of fast neutrons of about $7{\cdot}5 \times 10^{25}$ n/m^2 and contained a hydrogen isotope concentration of 0·41 at%. The specimens retained their original curvature and were loaded in pure bending with the crack growing from the inside surface of the pressure tube. Hydride cracking was detected and measured by acoustic emission.[12]

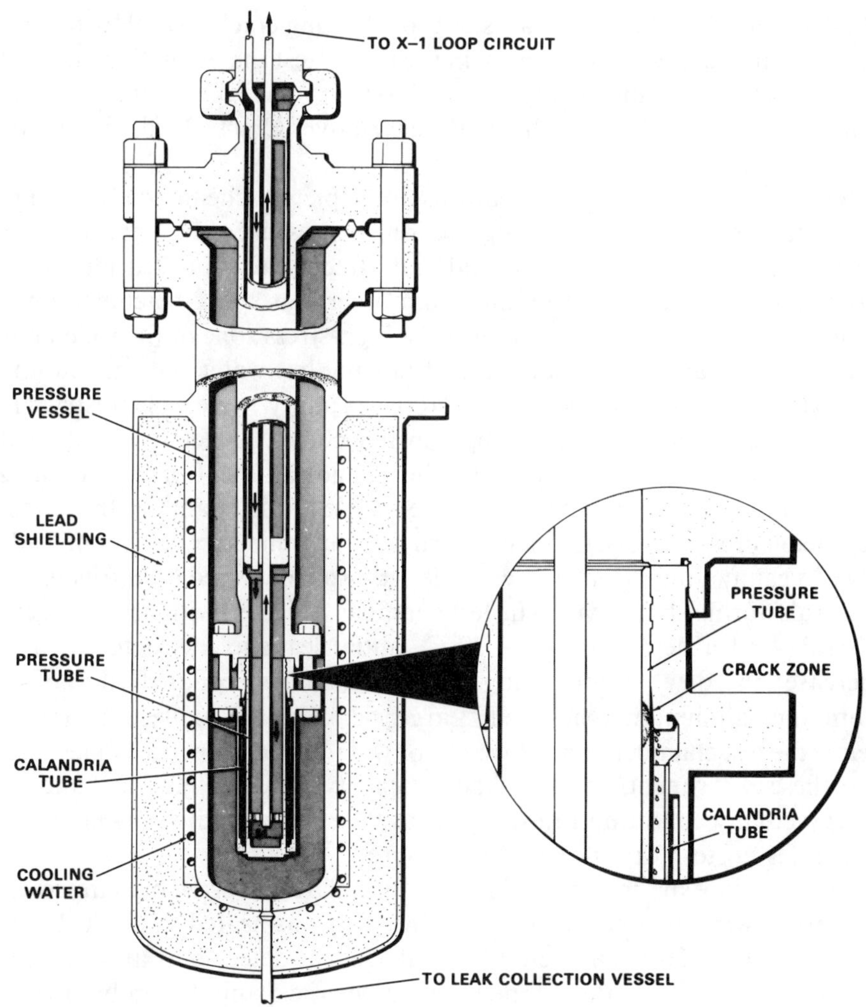

Fig. 3. Schematic diagram of CRACLE apparatus.

The specimens were cracked at a nominally constant K_I and temperature, after an initial temperature cycle to promote cracking. The cracking behaviour was then studied as a function of K_I and temperature manoeuvres. After the tests were stopped, the specimens were broken open to examine the crack to compare the actual cracking with that expected from potential drop or acoustic emission. Corrections were made using the data from the crack surface area.

Full-scale simulation

We have developed an apparatus to grow leaking cracks by DHC in actual components removed from operating reactors, test the predictions from the small specimens, and study the behaviour of leaking cracks.[13] The apparatus is called the Chalk River Active Crack Leak Evaluation (CRACLE).

The main features of the apparatus are depicted schematically in Fig. 3. The specimen is a 450-mm long section of pressure tube, removed from Pickering Unit 3, sealed at one end and attached to its rolled joint at the other end. A flaw, 1 mm deep and 3 mm long, was spark-machined on the inside surface of the pressure tube at the high-stress zone of the rolled joint. The joint was attached to a source of flowing hot water pressurized up to 10·3 MPa and maintained at CANDU primary water chemistry. The pressure tube was surrounded by a water-cooled jacket to simulate the temperature conditions imposed by the calandria tube. For protection the test section was contained in a water-cooled test chamber that was a registered pressure vessel, which in turn was surrounded by a layer of lead to absorb gamma radiation. The crack length was measured by ultrasonics at intervals during shutdowns. The test chamber was filled with dry nitrogen at 100 kPa and a dew point of −30°C. Initial leakage was detected as an increase in dew point using the change in conductivity of an aluminum–aluminum oxide cell. Larger leakage was collected in tanks below the test chamber and the weight of water was continuously measured, For these experiments the tank had a capacity of 13 kg (0·013 m^3) and leak rates (actually collection rates) as low as 0·05 kg/h could be discerned and leak rates up to 10 kg/h could be measured.

To start the experiment the flaw was stimulated to grow by temperature cycling, as with the small specimens. The loop was heated to 290°C for 1 h, then cooled to 220°C and held until leakage was detected as an increase in dew point. Tests were then done to evaluate crack stimulation by pressure and temperature cycling as a procedure for locating leaks.

RESULTS

Small specimens

Unirradiated material

All the load manoeuvre tests were performed with K_I reduction steps of either 5 or 10 MPa$\sqrt{m}$ after the specimen had been cooled from 300°C. The results showed that an active delayed hydride crack could be temporarily or permanently arrested by load reduction. The incubation periods (defined as the time between load reduction and the first sign of continuous increase of potential drop across the crack) under different load manoeuvre conditions and temperatures are summarized in Table 1. The length of incubation was found to increase with the relative size of the load drop, $\%\Delta K_I$. It also increased with decrease in final K_I and cracking temperature.

Irradiated material

Several load and temperature manoeuvre experiments were done:

(1) T_D is the temperature above which the cracking slows down if the temperature is approached by heating. This temperature was estimated from DHC tests in which the test temperature was approached by heating. The cracking started to slow down at about 180°C and completely stopped at 230°C (Fig. 2). T_D in irradiated material is similar to that in unirradiated material.[4,6]

(2) In load manoeuvre experiments, a K_I reduction of more than 20% was required to achieve an incubation time greater than 35 min at

TABLE 1
Length of the Incubation Period under Different Conditions of K_I Reductions for Unirradiated Material

ΔK_I (*MPa*$\sqrt{m}$)	% ΔK	*Temperature* (°*C*)	*Incubation time* (*h*)
30 → 25	17	250	1
25 → 20	20	250	1·5
20 → 15	25	250	2–4·4
15 → 10	33	250	no cracking
25 → 15	40	250	16 h, no cracking
25 → 20	20	210	6
20 → 15	25	210	9

TABLE 2
Length of the Incubation Period under Different conditions of K_I Reductions for Irradiated Material

ΔK_I ($MPa\sqrt{m}$)	% ΔK	*Temperature* (°C)	*Incubation time* (h)
15 → 13	13	200	0
11 → 9	18	200	0
17 → 13·5	21	200	0·6
13·5 → 10	26	200	4·2
17 → 11	35	200	2·9

200°C (Table 2). With load reduction greater than 25%, the incubation time was extended to several hours. The cracking rate immediately after the incubation time was slow but gradually increased to the value established before the load reduction. The effectiveness of load manoeuvres on DHC appears to be similar in both unirradiated and irradiated materials.

(3) Two types of temperature manoeuvre experiment were performed:

(a) A specimen containing a delayed hydride crack was cooled down from 300°C to 210°C and the crack was stimulated. During subsequent heating to 250°C, the crack growth rate slowed down but cracking never stopped. Similar cracking behaviour was

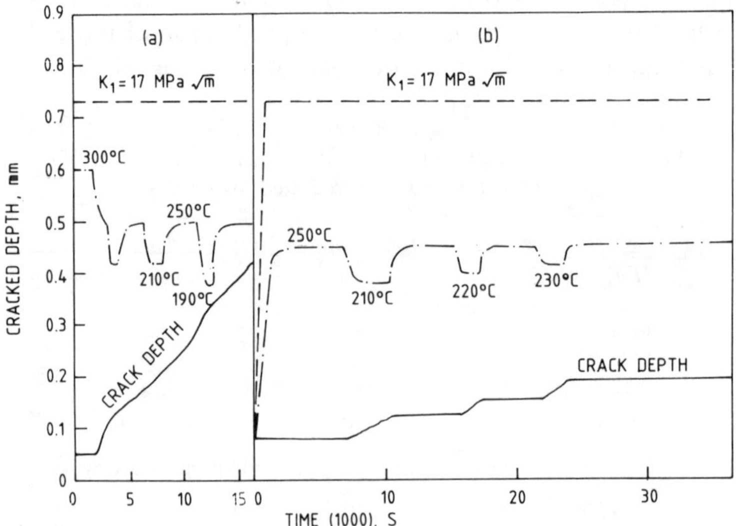

Fig. 4. Effect of cooling (a) and heating (b) on response of DHC in irradiated Zr–2·5 Nb to subsequent temperature cycles.

observed during cooling to 190°C and heating back to 250°C. The cooling seemed to stimulate cracking but heating was not effective in stopping it (Fig. 4(a)).

(b) A similar specimen was heated under stress to 250°C and then subjected to the cooling and heating cycles. No cracking occurred during the initial heating from the room temperature to 250°C. In subsequent temperature cycles, cooling stimulated cracking and heating was very effective in stopping it (Fig. 4(b)).

These tests show that the effectiveness of temperature manoeuvres to arrest cracking depends on prior history, while the effectiveness of temperature manoeuvres to stimulate cracking by cooling is not sensitive to the prior temperature history.

Full-scale simulation

The first four stages of the experiment illustrate some of the features of crack stimulation (Fig. 5). In Stage 1, the spark-machined flaw, initially 1 mm deep by 3 mm long, grew axially about 4 mm while penetrating the tube wall. Leakage was easily detected by the change in dew point about 30 h after the temperature cycle. In Stage 2, the leak rate increased rapidly from 0·1 kg/h to over 1 kg/h about 8 h after the pressure and temperature cycle. The crack lengthened 1 mm in this stage. This result appears to demonstrate the desired behaviour; the crack is made safe by pressure reduction but is stimulated to grow by the temperature reduction so that after a delay time, caused by the incubation for cracking and the time required to collect the leaked effluent, the increase in leakage is distinct.

A complication with a leaking crack is the development of a temperature gradient because the metal surrounding the crack is cooled when the hot, pressurized water flashes to steam across the crack face. Consequently, parts of the crack are growing at temperatures much less than that of the water. The temperature gradient itself also affects the behaviour of the crack.[14] With reference to the crack-tip temperature, when the crack tip is cooler than the surrounding metal, on cooling the crack velocity is moderately increased (e.g. a factor of 2 with a gradient of 20°C/mm) while on cooling T_D (Fig. 2) is shifted to higher temperatures (e.g. 260°C with a gradient of 20°C/mm). Thus on pressurizing with water heated above the T_D measured for dry cracks. a leaking crack can still grow. In Stage 3 of the experiment, the temperature and pressure were raised together and held constant at 290°C and 10·3 MPa, respectively. In 240 h the crack extended 2 mm and the leak rate slowly increased, although it appeared to peak at 0·5 kg/h after about 125 h.

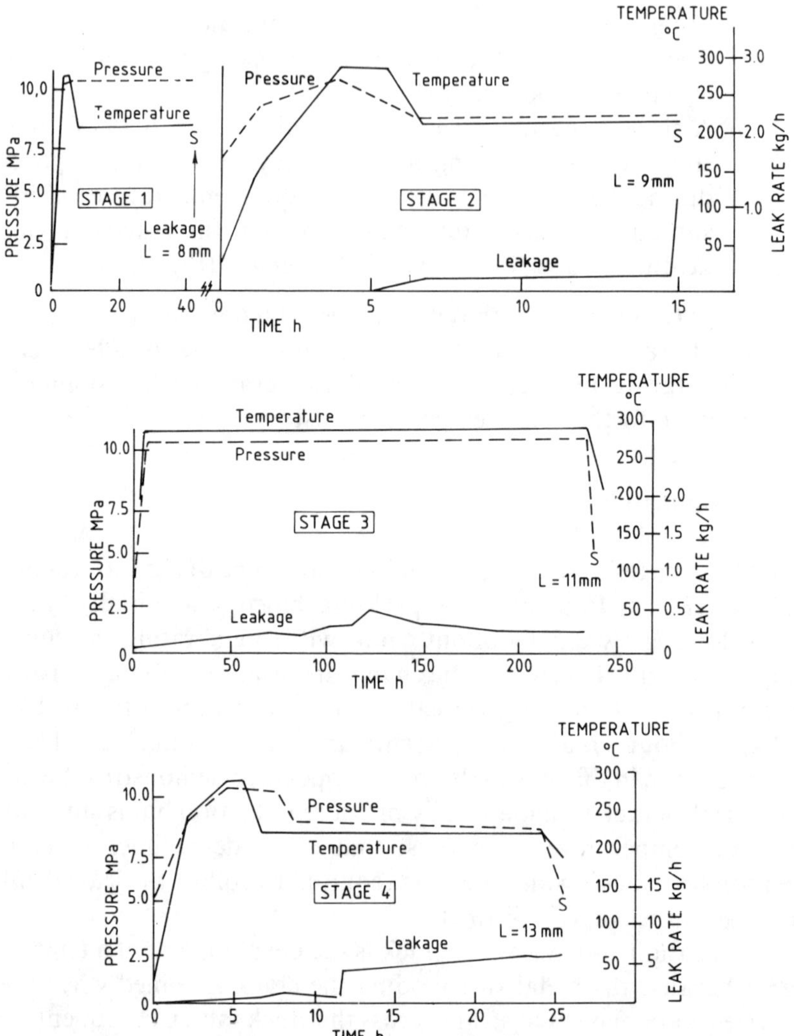

Fig. 5. Effect of temperature and pressure variations on the behaviour of a leaking crack. The experiment was shut down at *S* and the crack length, *L*, was measured by ultrasonic measurements.

When the stimulation manoeuvre was repeated, but with the pressure reduction after cooling (Stage 4) a sudden increase in leak rate was observed 3 h after the pressure reduction. The crack grew a further 3 mm to a total length of 13 mm and the final leak rate was over 5 kg/h. In subsequent tests, when the pressure was increased, the leak rate usually had a step increase, presumably because the crack was opened further.

The results of the experiments demonstrate the efficacy of cooling to stimulate DHC in both dry and leaking cracks, and show the variable behaviour of leakage through real cracks. Experience with leaking cracks in power reactors shows the need for a formal leak-location procedure for small leaks.

REACTOR EXPERIENCE WITH LEAK LOCATION

A partial leak-stimulation procedure was successfully employed in Pickering Unit 3 in 1985. During the reactor restart, following a maintenance outage, moisture was detected in the gas annulus. Analysis of the water showed that it was heavy water leaking from a fuel channel at a rate of about 1 g/h. This leak rate is too low to enable identification of a leaking channel. The records of operation revealed that while the reactor was being returned to power and being held at about 260°C at zero power, temperature control was lost and the reactor was cooled to 200°C (A in Fig. 6). No DHC avoidance manoeuvres were performed, thus the rolled joints were subjected to conditions conducive to DHC. The leakage coincided with this loss of temperature control. After reheating, the leak rate remained constant, suggesting that the crack was no longer growing.

Based on this evidence, the authors believed that a rolled joint had developed a through-wall, delayed hydride crack and that the cracking had stopped once the reactor was back at high temperature. Because the leak rate was too low for channel identification, a short DHC stimulation was performed. The reactor was cycled to 210°C for 0·5 h so that the crack would grow a small amount (B in Fig. 6); based on measurements of crack velocities, the crack should have extended 0·1–0·2 mm. Following the cycle, the reactor was returned to 260°C and the response of the gas annulus was monitored. Within about 2 h, the leaking channel was identified as F13 from

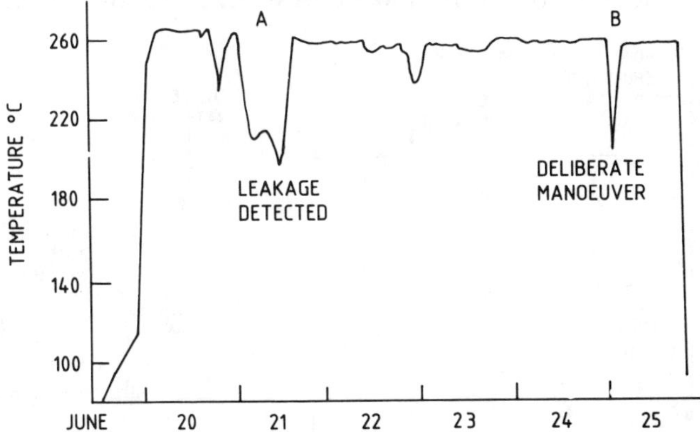

Fig. 6. Temperature history of Pickering Unit 3 when a leak was detected in fuel channel F13 in 1985. The pressure in the coolant was 8·7 MPa.

the differences in channel outlet temperatures. A through-wall crack, 27 mm long, was found by non-destructive examination of the inlet rolled joint. The pressure tube was removed from the reactor and subsequent destructive examination confirmed the DHC mechanism.

Not all leak-location exercises are as successful. In March 1986, an operational problem caused Bruce Unit 2 to poison out. After a temperature cycle, moisture was detected in the gas annulus and when the collection rate reached 7 kg/h, the reactor was cooled and depressurized. The leak was located in a string of 11 channels from their outlet temperatures. To locate the leaking channel more accurately using acoustic emission on individual channels, the reactor was pressurized cold. Unfortunately, before the leaking channel was identified in a controlled way, the pressure tube of fuel channel N06 ruptured when the water pressure reached 8·4 MPa (Fig. 7). Examination revealed that the initiating flaw was a long lamination formed during fabrication. This type of flaw is associated with tubes from ingot tops. No more laminations have been found by inspection of similar tubes. In N06, cracking initiated at several adjacent sites but the first leakage was sufficient to allow shutdown of the reactor in a controlled manner. On cold pressurization the crack was sufficiently long to exceed the critical crack length at room temperature; the crack would not have propagated unstably at elevated temperature (e.g. 250°C) at the reduced pressure.

The results of this reactor experience and the principles demonstrated by the laboratory experiments are being developed into a station procedure.

APPLICATION OF THE RESULTS TO POWER REACTORS

Following confirmation of a leak and its source, the reactor is shut down to the zero-power, hot condition and depressurized to the lowest practical

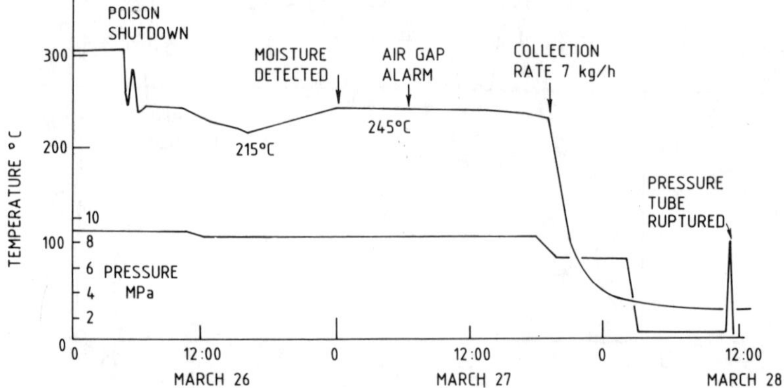

Fig. 7. Temperature and pressure history of Bruce Unit 2 when a leak was detected in 1986. The pressure tube of fuel channel N06 ruptured at room temperature.

pressure consistent with this condition; this pressure ranges from 6 to 8 MPa, depending on the reactor. The extra time, t, available for action, can be estimated from:[3]

$$t = (C - L)/2V$$

where C = critical crack length and L = length of crack at wall penetration. An illustrative evaluation, using conservative values and taking no account of incubation time, stress redistribution nor effects of leaking on crack growth, shows that at 250°C the time is increased from 18 h at 10·3 MPa, to 24 h at 8 MPa, to 33 h at 6 MPa, using $L = 16$ mm and $V = 2{\cdot}7 \times 10^{-7}$ m/s. While the crack is safe, all relevant operating records up to leakage are analysed to plan subsequent action. The necessary operating records are included in Table 3.

If the review of operating records determines that further crack stimulation is appropriate, then the procedure provided in Fig. 8, which includes activities from leak confirmation to final shutdown for channel removal, is recommended.

The approach presented for stimulating DHC has been selected to provide the most control to promote stable crack growth. The initial pressure reduction and DHC-avoidance thermal manoeuvre recommended in the procedure will stall DHC. Some indefinite incubation period will be required before cracking resumes. The reduction in operating pressure will increase the critical crack length and hence lower the risk of the leaking crack becoming unstable during the procedure.

Resumption of crack growth is stimulated by reducing the temperature by about 40°C. Conditions conducive to DHC are produced without significantly reducing the critical crack length. If crack growth does not resume or the leak is still not located after 24 h under these conditions, crack re-initiation is encouraged by a small increase in pressure, about 0·5 MPa. This pressure increase is expected to reduce the incubation period for the

TABLE 3

Records of Reactor Operation Required to Decide Further Action when Leakage is Detected

Item	*Rationale*
Detailed pressure and temperature records	Establish conditions responsible for cracking and determine the times that the reactor was exposed to these conditions.
Detailed AGS dew point records	Correlation of moisture to thermal history for an indication of cracking period to arrive at an estimate of crack size.
Operator logs	Determine times and reason for condition leading to crack growth.
Logs of channel outlet temperatures	Detailed channel temperature information.

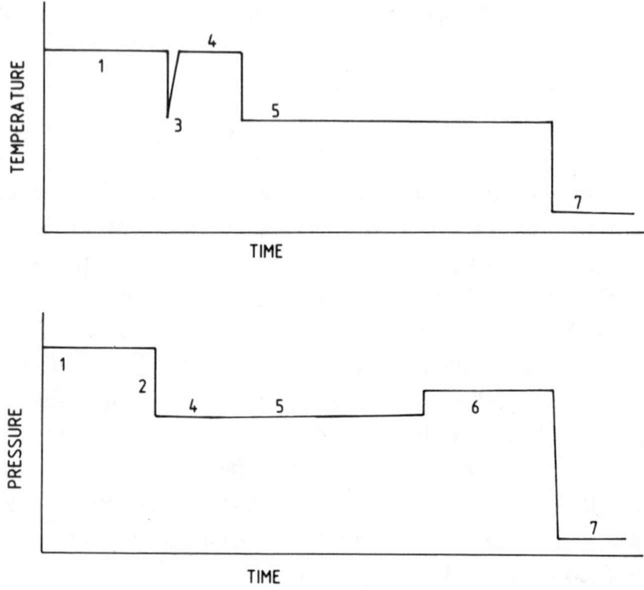

Event	*Description*	*Comment*
1	Reactor operating when leakage confirmed. Reactor is shut down but remains hot.	Start the leak identification procedure.
2	Reduce pressure from 10 MPa to 7 MPa.	Stop crack and increase the critical crack length.
3	Temperature cycled from 250°C to 210°C to 250°C.	Help control the crack by stopping DHC.
4	Reactor held at zero power hot condition with constant monitoring for moisture.	Hold condition to review records and plan subsequent action.
5	Temperature is reduced by 40°C and held for 24 h or until the channel containing the leaking pressure tube is identified.	If crack stimulation is necessary, encourage crack growth by lowering the temperature.
6	Pressure is increased by 0·5 MPa and held for 24 h or until the leak is identified.	If the leak rate has not increased, crack initiation is encouraged further by increasing the pressure by a small amount.
7	Cooled to room temperature and depressurized.	The leaking tube is identified and removed.

Fig. 8. Proposed sequence of pressure and temperature manoeuvres that safely stimulates crack growth to increase leak rate as an aid to the location of a through-wall crack in a CANDU pressure tube.

resumption of crack growth. The pressure increase will reduce the critical crack length but it will remain larger than at standard operating pressures. The procedure should result in stable DHC crack growth while limiting the risk of fast fracture.

Circumstances exist where Leak-Before-Break cannot be guaranteed and this situation is not covered by this procedure. The failure in fuel channel G16 in Pickering Unit 2 is an example.[15] The combination of high hydrogen concentration in the Zircaloy-2 material and contact between the pressure and calandria tubes (because of displaced spacers) resulted in the formation of hydride blisters at the cold spot on the outside surface of the pressure tube. With time, the length of contact between the two tubes can increase, leading to arrays of blisters. These blisters can crack during operation, leading to long cracks. Such cracks may not penetrate through the tube wall because of the thermal gradient; the hydrides necessary for DHC may not be present at the inside surface, which is hot. Hence a supercritical, non-penetrating crack can form and thus Leak-Before-Break cannot be assured. The stimulation procedure would not be used if conditions leading to cracked blisters are known to exist.

CONCLUSIONS

A combination of reactor experience and results from laboratory experiments has been used to develop a safe procedure for locating leaking cracks in Zr–2·5 Nb pressure tubes in CANDU reactors by stimulating the growth of the crack in a controlled way to increase the leak rate. Reduction of the chance of unstable rupture is achieved by pressure reduction while DHC is promoted by cooling from about 250°C down to 210°C.

ACKNOWLEDGEMENTS

The authors are grateful to H. Herrington, D. Sage, K. Weisenberg, J. J. Schankula and H. Seahra for assistance with the experiments. Some of the funding for this work was obtained from the CANDU Owners' Group under WP 690 and WP 3125.

REFERENCES

1. Perryman, E. C. W., Pickering pressure tube cracking experience. *Nucl. Energy*, **17** (1978) 95–105.

2. Dunn, J. T. & Jackman, A. H., Replacement of a cracked pressure tube in Bruce Unit 2. Atomic Energy of Canada Limited Report AECL-7537, 1982.
3. Moan, G. D., Coleman, C. E., Price, E. G., Rodgers, D. K. & Sagat, S., Leak-before-break in the pressure tubes of CANDU reactors. *Int. J. Pres. Ves. & Piping*, **43** (1990) 1–21.
4. Coleman, C. E., Cheadle, B. A., Ambler, J. F. R., Lichtenberger, P. C. & Eadie, R. L., Minimizing hydride cracking in zirconium alloys. *Can. Met. Quart.*, **24** (1985) 245–50.
5. Ambler, J. F. R. & Coleman, C. E., Acoustic emission during delayed hydrogen cracking in Zr–2·5 wt% Nb alloy. Second International Congress on Hydrogen in Metals, Paper 3C10, Paris, 1977.
6. Ambler, J. F. R., Effects of direction of approach to temperature on delayed hydride cracking behaviour of cold-worked zirconium. *ASTM STP 824*, American Society for Testing and Materials, Philadelphia, PA, 1984, pp. 653–74.
7. Puls, M. P. & Simpson, L. A., The effects of stress, temperature and hydrogen content on hydride-induced crack growth in Zr–2·5% Nb. *Met. Trans.*, **10A** (1979) 1093–105.
8. Shek, G. K. & Graham, D. B., Effects of loading and thermal manoeuvres on delayed hydride cracking in Zr–2·5 wt% Nb alloys. *ASTM STP 1023*, American Society for Testing and Materials, Philadelphia, PA, 1989, pp. 89–110.
9. Chow, C. K. & Simpson, L. A., Determination of the fracture toughness of irradiated pressure tubes using curved compact specimens. *ASTM STP 945*, American Society for Testing and Materials, Philadelphia, PA, 1988, pp. 420–38.
10. Langford, W. J. & Mooder, L. E. J., Fracture behaviour of zirconium alloy pressure tubes for Canadian nuclear power reactors. *Int. J. Pres. Ves. & Piping*, **6** (1978) 275–310.
11. Mostovoy, S., Crosley, P. B. & Ripling, E. J., Use of crack-line loaded specimens for measuring plane-strain fracture toughness. *J. of Materials*, **2** (1967) 661–81.
12. Sagat, S., Ambler, J. F. R. & Coleman, C. E., Application of acoustic emission to hydride cracking. Atomic Energy of Canada Limited Report AECL-9258, 1986.
13. Coleman, C. E. & Simpson, L. A., Evaluation of a leaking crack in an irradiated CANDU pressure tube. Atomic Energy of Canada Limited Report AECL-9733, 1988.
14. Sagat, S., Chow, C. K. & Coleman, C. E., Hydride cracking in zirconium alloys in a temperature gradient. Third International Conference on Fundamentals of Fracture, Irsee, FRG, 19–24 June 1989.
15. Field, G. J., Dunn, J. T. & Cheadle, B. A., Analysis of the pressure tube failure at Pickering NGS 'A' Unit 2. *Can. Met. Quart.*, **24** (1985) 181–8.

Int. J. Pres. Ves. & Piping **43** (1990) 205–217

Failure Probability of Nuclear Piping due to IGSCC

F. Nilsson, B. Brickstad & L. Skånberg

Royal Institute of Technology, S100 44, Stockholm, Sweden

ABSTRACT

A simple model for the estimation of the pipe break probability due to IGSCC is developed and discussed. It is partly based on analytical procedures, partly on service experiences from the Swedish BWR program. Some rough estimates of the resulting break probabilities indicate that further studies are urgently needed. A sensitivity study is performed and it is found that the uncertainties about the initial crack configuration are the most important contributors to the total uncertainty. The results of inservice inspection are studied and it is found that the inspection intervals need to be shortened if a significant reduction in the failure probabilities is to be obtained.

INTRODUCTION

In most Probabilistic Safety Assessments (PSA) the piping failure probabilities from the WASH 1400[1] study are still employed. Since these probabilities are major contributors to the total core damage probability, it is of central interest either to verify the WASH 1400 assumptions or to obtain better estimates. A direct frequency approach is not possible since to date no major pipe break has occurred in a nuclear plant. Probabilistic fracture mechanical assessments of pipe failure have been performed, but these have so far mostly considered fatigue from pre-existing defects as the damaging mechanism. In a BWR (Boiling Water Reactor) plant, however, Inter-Granular Stress Corrosion Cracking (IGSCC) in austenitic piping from an initially defect-free material is potentially the most dangerous damage mechanism (cf. Bush[2]).

Int. J. Pres. Ves. & Piping 0308-0161/90/$03·50

Some attempts to consider IGSCC in probabilistic fracture assessments have been made recently. The major difference to the previous studies is the necessity of quantifying the probability of crack initiation. Harris *et al.*[3] describe a development of the PRAISE code to account for IGSCC which results in a very complex model and no results were given for the break probability.

The present work reports on a study performed to investigate the possibilities of obtaining reasonably useful estimates of the pipe break probabilities when IGSCC can occur. Unlike the more theoretical approach of Ref. 2, this study is based as much as possible on data obtained from service records of Swedish BWRs.

DETERMINISTIC CALCULATIONS OF CRITICAL CRACK SIZE

Throughout the study a circumferential crack along the inner surface of a pipe with a depth $a(t)$ and a length $l(t)$ is considered. The inner diameter of the pipe is denoted D and the thickness h. The anticipated collapse mechanism in an austenitic pipe is ductile failure and the general concepts can be represented in a Leak-Before-Break (LBB) diagram of the type shown in Fig. 1.

Break is assumed to occur if a and l fall on the collapse line A–B. The shape of this line depends on the geometry, the loading conditions and the material properties of the pipe. Cracks to the left of A penetrate the outer surface before fracture occurs. Taking into account the crack growth Δl, it is realized that cracks with a combination of initial length and depth placed to the left of C–A will lead to LBB. Thus a critical initial length $l_{cr}^0 = l_{cr} - \Delta l$ can be defined.

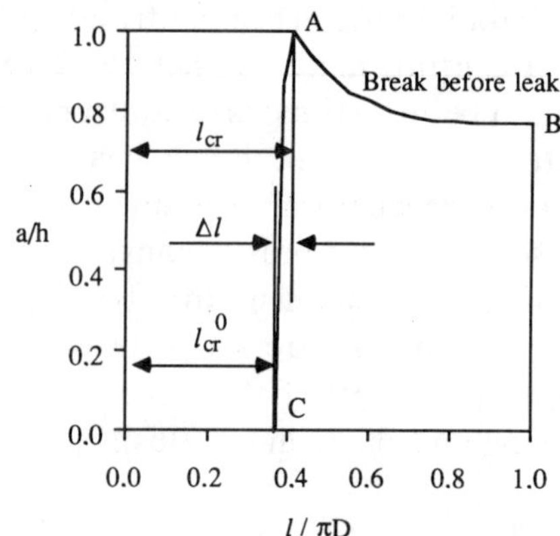

Fig. 1. Principal structure of LBB diagram.

To estimate the amount of crack growth, calculations were performed based on the following growth law taken from Hazelton.[4]

$$\dot{c} = 1{\cdot}12 \times 10^{-6} K_{\mathrm{I}}^{2{\cdot}161} \qquad (1)$$

where the local growth rate, $\dot{c}$, is measured in mm/h and K_{I} in MPa$\sqrt{\mathrm{m}}$. The possible driving stresses are primary membrane (P^{m}) and global bending (P^{b}) stresses, secondary thermal expansion (P^{e}) stresses and residual stresses. Since the global bending stresses promote depth growth more than length growth these were neglected in order to obtain conservative estimates of the length growth. The primary membrane stress is due to internal pressure (7 MPa). A profile of bending residual stress, linearly varying through the thickness with a tensile stress of 211 MPa (30 ksi) on the inside and the same compressive stress on the outside was assumed. This is in accordance with the recommendations in ASME section XI (cf. Ref. 5) for wall thicknesses less than 25·4 mm. In the present study, no pipes with a greater thickness were considered. A typical result for the growth history is shown in Fig. 2 and it is noted that the circumferential growth is of the same order as the depth growth. Thus the precise amount of circumferential growth is of minor importance since it is much smaller than the critical crack length. The growth histories, in terms of the relative crack depth, were similar for the different dimensions.

The growth rate is highly dependent on the environmental conditions and substantial variations in the residual stress profiles can be expected. Since these deviations are difficult to quantify, we shall here assume that the growth is deterministic as shown in Fig. 2. To obtain some insight into how the results are affected by variations in the growth rate, a sensitivity analysis is performed below. In the sensitivity study it will for simplicity be assumed

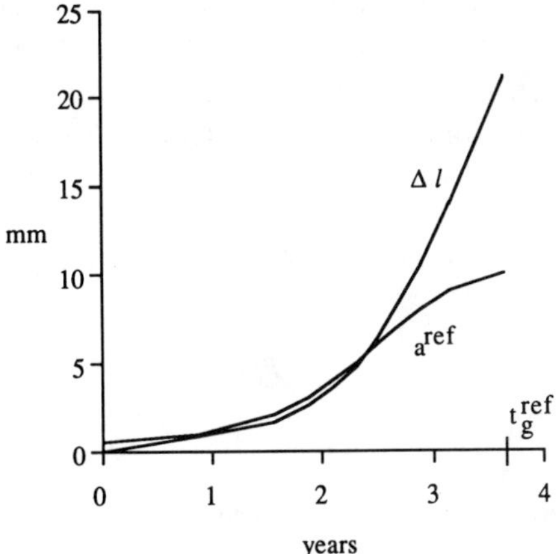

Fig. 2. Depth and circumferential growth for $h = 10$ mm.

that the effect of variations in environment, material and stress state is a simple change of the timescale, i.e.

$$a(t) = a^{\text{ref}}(t\, t_g^{\text{ref}}/t_g) \qquad (2)$$

Here, a^{ref} is the result for the actual pipe under consideration using eqn (1) and the assumed residual stress profile. The value t_g^{ref} is the time needed for the crack to grow from a length corresponding to threshold conditions and through the thickness for this case. The values $a(t)$ and t_g are the corresponding quantities for a case when the conditions differ from the reference case.

The geometry assumed for the determination of l_{cr} is that of a penetrating crack with an arc length l_{cr} along the inner surface and the arc length κl_{cr} along the outer surface. For the basic set of data, κ is assumed to be equal to 0·5. This is a compromise among the many possible critical crack configurations and has been guided by the typical appearance of IGSC cracks detected from service. For the calculation of the crack size that gives collapse, the same methods as in ASME section XI IWB-3640 (cf. Ref. 5) were adopted. When fully ductile, only the primary stresses contribute to collapse and according to Ref. 5 this can be assumed for non-flux welds, e.g. TIG (Tungsten Inert Gas) welds. For flux welds (e.g. SMAW (Shielded Metal Arc Welding) and SAW (Submerged Arc Welding) welds) instability can occur before plastic collapse. Here these cases were analysed by the same simplified method as is used in the ASME-code. An effective bending stress $P^{\text{b}}_{\text{eff}}$ is defined as

$$P^{\text{b}}_{\text{eff}} = Z(P^{\text{b}} + P^{\text{e}}/2{\cdot}77) \qquad (3)$$

where Z is a numerical factor (1·45 for SMAW and 1·56 for SAW). Once the effective stress is defined, the procedure is the same as for the non-flux case.

The stresses vary considerably between different positions in the piping system. Since it was considered impossible to treat each weld individually, a statistical evaluation of P^{b} and P^{e}, based on a sample of 316 welds in three different systems, was performed. Welds both inside and outside the containment are included in this statistical evaluation. It was found that $P^{\text{b}}_{\text{eff}}$

TABLE 1
Mean and Standard Deviation of $P^{\text{b}}_{\text{eff}}$ for Different Dimensions

Weld type	*Dimension (mm)*	*Mean P^b_{eff} MPa*	*Standard deviation P^b_{eff} MPa*
TIG	$50 \le D \le 250$	15·6	15·4
SMAW	$50 \le D \le 250$	47·8	27·1
SMAW	$D > 250$	58·0	24·4

could be approximately fitted to lognormal distributions with the means and standard deviations given in Table 1 for the considered dimensions. Since P^{m} is given by the internal pressure and by making the conservative assumption that Δl equals $2h$, for each considered dimension and material type we can obtain the critical initial crack length as a function of the stresses, $l^0_{\mathrm{cr}}(P^{\mathrm{b}}_{\mathrm{eff}}, P^{\mathrm{m}})$.

PROBABILISTIC ASSUMPTIONS AND ANALYSIS

Based on the observations discussed above, the following assumptions are made for the probabilistic analysis.

(a) The probability that a crack is initiated in the time interval $(t_{\mathrm{i}}, t_{\mathrm{i}} + \mathrm{d}t)$ in a certain weld is given by $f_{\mathrm{i}}(t_{\mathrm{i}})\,\mathrm{d}t$. Here, initiation means that a crack of observable depth (≈ 1 mm) has developed. This event is assumed to be independent of the loading. Attempts to correlate IGSCC occurrences to different factors like temperature, service stresses, and material composition have not been successful thus justifying the assumption.

(b) The length distribution of an initial crack is given by the density function $f_{\mathrm{l}}(l_0)$. This distribution is also assumed to be independent of the loading.

(c) Inservice inspections are performed at times $j\,\Delta t$ ($j = 1, 2, 3, \ldots$). The probability of not detecting a crack at an inspection is $p_{\mathrm{nd}}(a)$; assumed to be dependent only on the crack depth. This is justified since cracks near the critical size are so long that length should not be a distinguishing factor for crack detection.

(d) Leaking cracks or cracks detected at inservice inspections do not contribute to the failure probability. This implies that the methods for sizing detected cracks and the leakage detection devices must be reliable.

(e) The effective bending stress has the density function $f_{\mathrm{P}}(P^{\mathrm{b}}_{\mathrm{eff}})$ (lognormal) as discussed above.

Based on these assumptions, and assuming independence between the inspection events, it is possible to derive the probability of failure of a weld during the time interval from t to the service limit T. Consider a crack initiated at time t_{i} in a certain weld. If this crack is not detected it will cause fracture if the length is larger than the critical one. The probability p' for this to happen is given by

$$p' = \int_0^{\infty} \left[\int_{l^0_{\mathrm{cr}}(P^{\mathrm{b}}_{\mathrm{eff}}, P^{\mathrm{m}})}^{\pi D} f_{\mathrm{l}}(l_0)\,\mathrm{d}l_0 \right] f_{\mathrm{P}}(P^{\mathrm{b}}_{\mathrm{eff}})\,\mathrm{d}P^{\mathrm{b}}_{\mathrm{eff}} \qquad (4)$$

The probability that a crack initiated at time t_i remains undetected during its growth through the thickness can be expressed as eqn (5) if the random character of the initiation time is taken into account

$$p''(t,T)=\int_{t-t_g}^{T-t_g} f_i(t_i)\prod_{j=j_1}^{j_2} p_{nd}[a\{(t_j-t_i)H(t_j-t_i)\}]\,dt_i \tag{5}$$

with

$$j_1-1<t_i/\Delta t\leq j_1 \tag{6}$$

$$j_2<(t_i+t_g)/\Delta t\leq j_2+1 \tag{7}$$

In eqn (5) the dependence of p_{nd} on time through the crack depth history is shown explicitly. The value H is the Heaviside step function; j_1 denotes the first inspection event after t_i and j_2 the inspection preceding t_i+t_g. The product in eqn (5) is defined to be unity if j_2 is smaller than j_1, which happens if both t_i and t_i+t_g fall between successive inspection times. The resulting failure probability, p_f, is now

$$p_f=p'p''(t,T) \tag{8}$$

INPUT DATA

In the following, numerical calculations will firstly be performed for a set of input data that can be regarded as best estimates of the different quantities. For this set of data the reference case of crack growth is assumed. The density function for initial crack length $f_l(l_0)$ was estimated from the results of metallographic evaluation of 43 IGSCC occurrences in Swedish BWRs and the observed frequencies of crack length relative to the inner periphery are shown in Fig. 3. The observed lengths may be somewhat larger than the initial length. The crack growth is, however, small so the error of not

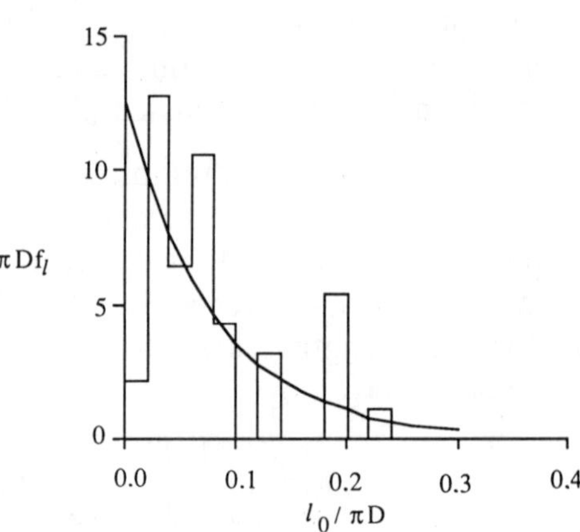

Fig. 3. Observed and fitted length distributions.

correcting for this is negligible. A truncated exponential distribution, eqn (9), was fitted to this sample. The parameter λ was chosen with $\lambda_0 = 12.6$ so that the mean values of the observed and the fitted distributions coincided.

$$f_l = \frac{\lambda}{\pi D}\exp\left(-\frac{\lambda l_0}{\pi D}\right)(1 - e^{-\lambda})^{-1}H(\pi D - l_0) \tag{9}$$

The last two factors in eqn (9) are due to the truncation. They are insignificant and can be neglected. In the following it will be assumed that eqn (9) is valid for all dimensions although the data were obtained mainly from one dimension ($D = 100$ mm). Thus it is assumed that the relative crack length has the same distribution for all dimensions.

The choice of an exponential distribution is motivated by its mathematical simplicity and because its density function is monotonously decreasing, which is a reasonable requirement. Since the critical crack length is considerably larger than the maximal observed value, the behaviour of the distribution in the range of the observed values has almost no effect on the resulting failure probabilities. It was therefore not considered worthwhile trying to find another distribution with a better fit in this interval. It is the probability that crack lengths exceed the maximum observed value that is of interest. Assuming that this probability is given by the exponential distribution, the probability $p^*(\lambda)$ that *none* of the observed values exceeds $0{\cdot}24\pi D$ can be calculated. This is given by the binomial distribution which, for 43 sample points, can be accurately approximated by a Poisson distribution. Since the probability that the crack length exceeds the maximal observed values is $\exp(-0{\cdot}24\lambda)$ from eqn (9), $p^*(\lambda)$ is given by

$$p^*(\lambda) = \exp(-43\exp(-0{\cdot}24\lambda)). \tag{10}$$

Inserting λ_0, p^* becomes 12·3%, which indicates that the assumed distribution is somewhat conservative.

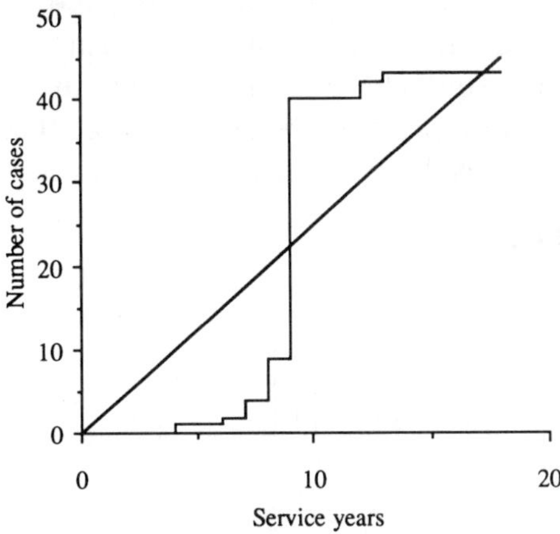

Fig. 4. Accumulated number of suspected IGSCC cases in Swedish BWRs.

In Fig. 4 the cumulative number of suspected IGSCC cases in Swedish plants up to 1988 is shown as a function of service year. To this set of data the straight line indicated in Fig. 4 was fitted corresponding to 2·5 cases of IGSCC per year. Estimating the total number of austenitic stainless steel welds in Swedish plants to be 20 000 then implies that f_i is constant and equal to $f_{i0} = 1{\cdot}25 \times 10^{-4}$. A total average of this kind is very crude and a subdivision of the failure statistics into more homogeneous groups with respect to dimension, material and environmental conditions is needed. This is planned to be performed in the continuation of this research. The value for f_{i0} is somewhat lower than the value used in a previous report from this project by Nilsson and Brickstad.[6] The reason for this is that in Ref. 6 not only IGSCC incidences in welds were included, but also IGSCC occurrences in cold-formed pipe-bends. The situation in pipe-bends is similar to that in welds but some different features also exist. In the present paper only welds are considered.

Relatively little information exists about the non-detection probability of ultrasonic inservice inspection of austenitic steel. Some data have become available recently. After an ultrasonic inservice inspection in 1982 which failed to detect IGSC cracks in the recirculation piping of the Nine Mile Point nuclear power plant, several programmes have been conducted to improve and assess the inspection effectiveness.[7–9] Based on data from some of these programmes Simonen and Woo[10] have given the following relation for p_{nd} for inspection of stainless steel with near-side access ('near-side access' is inspection that can be performed from the same side of the weld as potential cracking is situated).

$$p_{nd} = 1 - \phi[c_1 + c_2 \ln(a/h)] \tag{11}$$

Here ϕ is the normalized Gaussian distribution function. The values c_1 and c_2 are constants given in Ref. 10 and shown in Table 2 and thought to be representative for the different categories of inspection teams indicated by the labels in Table 2. A 'poor' team represents a lower bound performance among the participating teams, a 'good' team represents a better team in the round robin trials and an 'advanced' team represents the performance that may be achieved with further improved procedures and existing technology.

TABLE 2
Coefficients in the Non-detection Function

Type of inspecting team	c_1	c_2
Poor	0·240	1·485
Good	1·526	0·533
Advanced	3·630	1·106

For the basic set of input data in this study the values of c_1 and c_2 corresponding to a 'good' team of inspectors are assumed. This is motivated by preliminary results from a current Swedish programme. In general, p_{nd} for IGSC cracks also depends on crack length,[11] but this dependence is probably negligible in terms of p_{nd}. Note that p_{nd} is a measure of the ability to detect and locate IGSC cracks, only; not the ability to size them correctly.

If there is no near-side access then p_{nd} increases dramatically. Results from round robin trials indicate that the far-side detection performance is no better than pure chance.[8]

The inspection interval Δt is six years in Sweden, which is larger than t_g^{ref} as calculated for the reference cases.

RESULTS

The time-independent part of the failure probability, p', was obtained by numerical integration of eqn (4) and the resulting values are given in Table 3. The SMAW welds give about a six times larger value of p' than the TIG-welds because l_{cr}^0 is different for the two cases.

Since f_{i0} is constant the integrand in eqn (5) will be periodic with period Δt and it is thus sufficient to integrate over one period and then sum the contributions from the different periods. The quantity $p''(t, T)/(T-t)$ will oscillate slightly around the average rate $\bar{p}''$ defined from

$$\begin{aligned} \bar{p}''\frac{\Delta t}{f_{i0}} &= H(\Delta t - t_g)\left[\Delta t - t_g + \int_{\Delta t - t_g}^{\Delta t} p_{nd}[a(\Delta t - t_i)]\,dt_i\right] \\ &+ H(t_g - \Delta t)\int_0^{2\Delta t - t_g} p_{nd}[a(\Delta t - t_i)]\,dt_i \qquad (12) \\ &+ H(t_g - \Delta t)\int_{2\Delta t - t_g}^{\Delta t} p_{nd}[a(\Delta t - t_i)]p_{nd}[a(2\Delta t - t_i)]\,dt_i \qquad t_g < 2\Delta t \end{aligned}$$

A numerical integration of eqn (12) was performed and the resulting values for the different type of welds are given in Table 3 together with the

TABLE 3
Failure Probabilities

Weld type	*Dimension mm*	p''	$\bar{p}''$	$\bar{p}_f$
TIG	$50 \leq D \leq 250$	$1{\cdot}49 \times 10^{-4}$	$6{\cdot}75 \times 10^{-5}$	$1{\cdot}01 \times 10^{-8}$
SMAW	$50 \leq D \leq 250$	$8{\cdot}90 \times 10^{-4}$	$6{\cdot}75 \times 10^{-5}$	$6{\cdot}00 \times 10^{-8}$
SMAW	$D > 250$	$1{\cdot}40 \times 10^{-3}$	$7{\cdot}43 \times 10^{-5}$	$1{\cdot}04 \times 10^{-7}$

average failure rate $\bar{p}_f = p'\bar{p}''$. The results for $\bar{p}''$ are almost the same for the different dimensions due to the similarities in crack history. The results for p' show that the big difference is between the TIG welds and the SMAW welds; about a factor of six.

SENSITIVITIES AND UNCERTAINTIES

Substantial variations in the different assumed quantities may occur, and therefore the sensitivity of the failure probabilities to variations in the assumptions will be examined. One quantity at a time is varied while the others are fixed according to the basic set of data. In Fig. 5 the time-dependent part of the failure probability $\bar{p}''$ is shown as a function of the inspection interval Δt for the different sets of coefficients given in Table 2. It has been normalized with respect to f_{i0}, which is the rate that would result if no inspections at all were performed. The basic set of data shows a rather modest effect of the inservice inspections, the failure rate is decreased by 45% in comparison with the case of no inspection at all. The results for an advanced team of inspectors is not much different, the decrease being 55% for $\Delta t = 6$ years. If Δt is decreased the differences are larger. For $\Delta t = 3$ years, the team of good inspectors would have a failure rate of 18% of the uninspected value while the team of advanced inspectors would have a failure rate as low as 3·2% of the uninspected value. The poor team of inspectors differ very little from the uninspected value for all inspection intervals.

From eqns (2) and (12) it is evident that a change of t_g has the same effect as a change of $1/\Delta t$. It can thus be concluded from Fig. 5 that a decrease of t_g

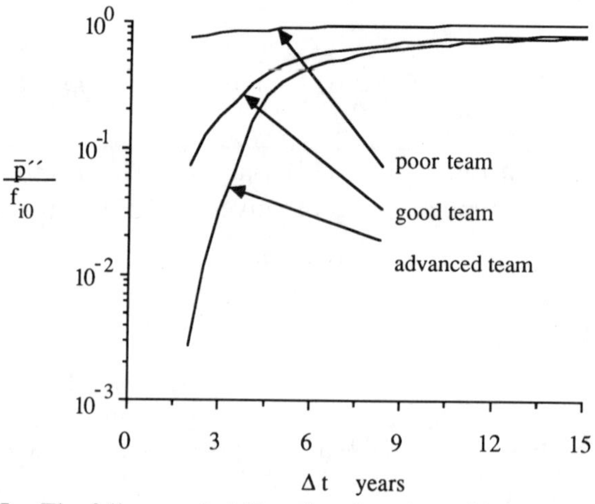

Fig. 5. The failure probability $\bar{p}''$ as function of inspection interval.

would have a rather small negative effect while an increase would have a more marked positive effect. In this respect the model is conservative, at least for the welds that are subjected to high quality inservice inspections.

The average failure probability depends linearly on the crack initiation rate, as is evident from eqn (12). The variability of f_{i0} is difficult to estimate for the above mentioned reasons, but it appears unlikely that it should differ from the assumed by more than a factor of three in either direction.

In Fig. 6 the sensitivity of the failure probability to the initial crack length distribution is visualized. The value p' for SMAW-welds in the diameter interval $50\,\text{mm} \leq D \leq 250\,\text{mm}$ is plotted as a function of the parameter $1/\lambda$. As is evident from Fig. 6, p' is very sensitive to variations in λ. It should be remembered that λ is determined from data that are of no direct interest for the failure probability. To gain some insight into the possible variation range of λ the probability p^* defined by eqn (11) is also plotted in Fig. 6. The choice $\lambda = \lambda_0$ results in a value of p^* equal to 12·3%. The value of λ that gives p^* equal to 50% corresponds to a failure probability that is 13% of the one for the basic data set. Thus the failure probability obtained from the basic data set is probably too high.

The effect of the parameter κ for the collapse geometry has also been investigated. The range of p' is about one order of magnitude in either direction as κ is varied from 0·5 to 0·1 or from 0·5 to 1·0.

In conclusion, the uncertainties of the initial crack length distribution are the most important factors controlling uncertainty of the failure probability. The analysis suggests that the choice $\lambda = \lambda_0$ gives a failure probability that might be too large but further data are needed to make more precise statements.

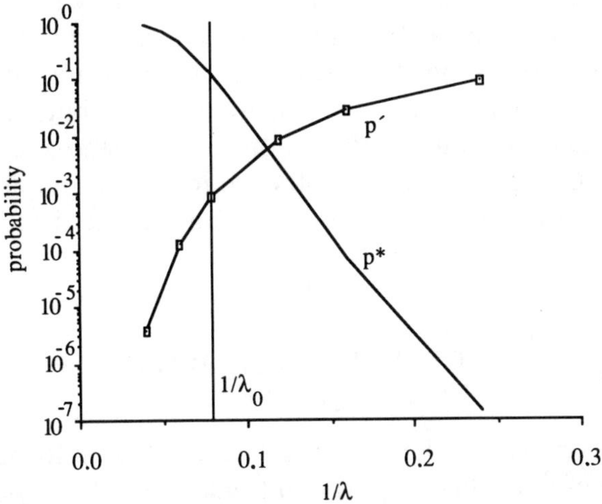

Fig. 6. The dependence of the failure probability on λ.

SYSTEM ASPECTS

To get some feeling for the relevance to our piping, an example is considered. Assume that the total number of austenitic stainless steel welds in a plant is 2220, consistent with the previous total estimate and that 50% of these are TIG welds of dimension $50\,\text{mm} \leq D \leq 250\,\text{mm}$, 45% SMAW welds of the same dimension and 5% SMAW welds with $D > 250\,\text{mm}$. Assuming that all welds are inspected with the same frequency and the same crack initiation rate prevails in all pipe groups, the resulting failure rate for a plant is estimated to $8{\cdot}25 \times 10^{-5}$ per reactor year or $7{\cdot}43 \times 10^{-4}$ for all nine plants. Thus, it is roughly one order of magnitude larger than those assumed in Ref. 1. Bearing in mind the conservatism in the initial crack length assumption, the results are not very different from the WASH 1400 assumptions. Clearly, this example is not wholly realistic: a significant part of the welds are, for example, never inspected, but it still gives an idea of the relation between the component (weld) failure probabilities and the system reliability. Note that these numbers are in qualitative agreement with estimates in Ref. 2.

CONCLUSIONS

The model gives a reasonably good representation of the different sources of pipe break probabilities and also describes the contributors in an appropriate manner. Since the assumptions are few and simple, the model is robust and can be used conveniently for sensitivity studies. The greatest source of error is the input data; the final probability seems to be most sensitive to the assumptions about the initial crack length distribution. If a more reliable estimate of the absolute probability is desired, it is of primary importance to continue the efforts to define the distribution of the initial crack lengths. One obvious continuation is to extend the base for the statistics to service cracks detected in other countries. A complementary task is to proceed with theoretical modelling along the lines in Ref. 3.

ACKNOWLEDGEMENT

This work was funded by the Swedish Nuclear Power Inspectorate. The authors are grateful for this support and to Dr P. Ståhle for helpful advice.

REFERENCES

1. WASH-1400, Reactor safety study: An assessment of accident risks in US commercial nuclear power plants. NUREG/CR-75/014, USNRC, Washington DC, 1975.
2. Bush, S. H., Statistics of pressure vessel and piping failures. *J. Pressure Vessel Technol. Trans. ASME*, **110** (1988) 225–33.
3. Harris, D. O., Dedhia, D. D., Eason, E. D. & Patterson, S. D., Probability of failure in BWR reactor coolant piping. NUREG/CR-4792, Vol. 3, USNRC, Washington DC, 1986.
4. Hazelton, W. S., Technical report on material selection and processing guidelines for BWR coolant pressure boundary piping. NUREG 0313, rev. 2, USNRC, 1986.
5. Section XI Task Group for Piping Flaw Evaluation, ASME Code, Evaluation of flaws in austenitic steel piping. *J. Pressure Vessel Techn., Trans. ASME*, **108** (1986) 352–66.
6. Nilsson, F. & Brickstad, B., An estimation of the probability of failure for BWR-piping in Sweden, in *Structural Mechanics in Reactor Technology*, Trans. SMiRT 10, Anaheim, CA, 14–18 August 1989, ed. A. H. Hadjian. The American Association for Structural Mechanics in Reactor Technology, Los Angeles, 1989, pp. 99–104.
7. Doctor, S. R., Bates, D. J., Heasler, P. G. & Spanner, J. C., Nondestructive examination (NDE) reliability for inservice inspection of light water reactors. NUREG/CR-4469, Vol. 1–6, USNRC, Washington DC, 1984–86.
8. Wheeler, W. A., Rankin, W. L., Spanner, J. C. & Badalamente, R. V., Human factors study conducted in conjunction with a mini-round-robin assessment of ultrasonic technician performance. NUREG/CR-4600, USNRC, Washington DC, 1986.
9. Taylor, T. T., Spanner, J. C., Heasler, P. G., Doctor, S. R. & Deffebaugh, J. D., An evaluation of human reliability in ultrasonic in-service inspection for intergranular stress-corrosion cracks through round-robin testing. *Materials Eval.*, **47** (1989) 338–44.
10. Simonen, F. A. & Woo, H. H., Analyses of the impact of inservice inspection using a piping reliability model. NUREG/CR-3869, USNRC, Washington DC, 1984.
11. Doctor, S. R. & Heasler, P. G., Round robin study on defect parameters, in *Proc. 9th International Conference on Nondestructive Evaluation in the Nuclear Industry*, Tokyo, 25–28 April 1988, ed. K. Iida, J. E. Doherty & X. Edelman. ASM International, 1988, pp. 515–20.

Int. J. Pres. Ves. & Piping **43** (1990) 219–227

Comments on Probabilities of Leaks and Breaks of Safety-Related Piping in PWR Plants

S. Beliczey & H. Schulz

Gesellschaft für Reaktorsicherheit (GRS)mbH,
Schwertnergasse 1, D-5000 Köln 1, Germany

ABSTRACT

Leaks or failures with a safety significance in Cl 1 or Cl 2 piping of nuclear power plants (NPP) in Germany are very rare events. This excellent record in operating experience is matched by many NPPs in other countries.

The advances achieved in the understanding of fracture behaviour, in the methods of non-destructive testing and surveillance, together with operating experiences, can be used in the re-evaluation of piping systems that have been designed and manufactured to the standards given at the time of construction.

Comments and examples are presented for determining the probability of leaks and breaks in the whole range of Cl 1 and Cl 2 piping systems.

1 INTRODUCTION

The position and licensing decisions on leak-before-break (LBB) in the Federal Republic of Germany, as well as the results achieved in the related safety research programmes, have been reported previously in international seminars on leak-before-break.[1,2] Since 1981, the 'leak-before-break' or 'break exclusion' concept has been applied successfully for large diameter high pressure piping that contains primary coolant within the containment (Class 1 piping) of pressurized water reactors (PWRs). All plants designed to this standard, as set out in the guidelines of the 'Reaktorsicherheitskommission' (RSK)[3,4] are now in full power operation.

In 1988, a decision was made to reanalyse the safety of each plant in a reasonable time interval ($\approx$ 10 years), in addition to the continuous review of

Int. J. Pres. Ves. & Piping 0308-0161/90/$03·50

plant behaviour with respect to safety related items.[5] Within these analyses, questions on the probability of leaks and breaks of safety-related piping have to be discussed for each individual plant.

Safety-related piping comprises:

—piping that contains primary coolant;
—piping that is necessary to cope with accident situations;
—piping that, when rupturing, causes damage that cannot be kept at a sufficiently low level with respect to nuclear safety;
—piping that, when leaking, can cause sequences that lead to accidents.

The assumptions that led to the results of the first important risk study, WASH-1400, do not represent the present state of knowledge and these results should not be applied. For pipe failures as initiating events for loss of coolant, an approach has been developed in the framework of the German risk study for a reference PWR plant.[6]

The conditions that have to be considered when studying the leak-before-break behaviour of a piping section are determined by the design, the properties of the materials, the manufacturing, the installation, the inspection and surveillance and the general layout. Further, the operating experience has to be evaluated on a plant-specific basis, as well as on a generic basis, in order to include potential failure mechanisms. The different aspects are discussed below and the methods used in the risk study are summarized.

2 REVIEW OF REQUIREMENTS AND ASSUMPTIONS REGARDING LEAK-BEFORE-BREAK BEHAVIOUR

The following comments are directed to

—basic requirements
—methodology and
—related operating experience

when existing systems are to be reassessed.

2.1 Design aspects

The loads used for the design of the plant have to be checked against the present design practice, especially if the operating experience has identified important, unexpected loads, e.g. resulting from striations in fluid temperature due to discontinuous operation or stoppage of the feedwater system or temperature oscillation in the vicinity of leaking valves. The malfunction of snubbers, check-valves and pumps is another source of concern.

External events also affect design. The effect of seismic-induced loads on pipe failure has been investigated in several studies.[7,8] Experimental work,[9,10] as well as on-site damage evaluations after seismic events,[11] strongly supports the view that seismic-induced loads on large and medium diameter, high pressure piping do not make significant contributions to the failure rates. For small diameter piping, external loads may be caused by repair work, material transport and failure of auxiliary shielding. The contributions of these events to failure rates are difficult to assess. Using check lists is an effective method to complement system analysis.

In the re-evaluation of piping stresses, special consideration has to be given to the surface conditions with respect to peak stresses. Modern methods of NDT can determine notch radii and therefore aid selection of stress concentration factors.

2.2 Material aspects

The documentation of material certificates has always been practised for pressure-retaining components. This documentation can be used for the re-evaluation of material properties. Past and present acceptance values have to be discussed, together with the distribution of measured properties. For most forged and wrought materials, e.g. 22 NiMoCr 37 (ASTM-A 508 Cl 2) or 20 MnMoNi 55 (ASTM-A 533 Gr. B Cl 1), the change of properties due to ageing can be neglected, but this is not the case for some cast materials (some austenitic stainless steels (chromium–nickel)). The developments in code requirements for materials are mainly concentrated on the large diameter safety-related piping. The standards on small diameter of austenitic material have not changed significantly since the first commercial PWRs started operating. Considerable progress has been achieved, however, in the application of mechanized welding equipment which has resulted in a large decrease in repair rates. For the assessment of the fracture resistance properties, complementary measurements as well as the available data are usually necessary. In cases where comparable material is not available, 'cut-outs' from relevant sections of the structure should be used.

2.3 Manufacturing and installation aspects

Technological developments and standardization in the last 20 years have led to a high quality standard for piping systems. In plants that have been built according to less stringent specifications, special attention should be given to the:

—*As-built geometry.* Important factors are peak stresses and reliability of NDT. The determination of unfavourable misalignments and weld surface conditions can be done in a step-wise approach by reviewing the documentation of radiographic films and on-site measurements at selected areas.

—*Preheat temperatures and post-weld heat-treatment applied during manufacturing.* This information is needed for evaluating residual stresses and susceptibility to corrosion.

—*Attachments and end caps.* In earlier designs, weld-on attachments were used in several instances instead of forged integrated attachments. This design may cause local residual stresses at the weld and limit the sensitivity of NDT. To limit shield gas volumes during welding, auxiliary nozzles have been used in the past in some systems. These dead ends may be subject to stagnant flow conditions and to the accumulation of impurities.

Improvements of installation techniques are derived from the evaluation of operating experience. The analysis of damage, performed within the German system of evaluation of abnormal occurrences, is able to detect causes that are due to deficiencies of the installation. A continuous information exchange on operating experience prevents the same installation deficiencies in similar NPPs.

2.4 Inspection aspects

The resolution of quantitative, non-destructive examination is continually improving. The contours of faults can be better identified; the evaluation of the shape of faults enables us to make more reliable predictions on the size of a potential leak that develops from a defect.

Improved ultrasonic inspections can be successfully applied to welds, which were originally only tested by radiography, so their integrity can be reassessed. Thus uncertainties in making predictions on leak possibilities can be reduced.

As a result, ultrasonic testing of high resolution can be used to characterize the state of piping sections that have not previously been tested to such standards.

2.5 Surveillance aspects

Instrumentation and surveillance is continuously improved and modern equipment can be installed in addition to existing safety-related instrumentation.

To complement the stress analysis:

—loadings and their frequencies can be monitored by the measurement of process variables, and local temperature distributions;
—stresses and strains can be determined by the measurement of process variables, local temperatures, strains and displacements and by calculation of fatigue at spots that are to be reassessed;
—the chemical properties of the media can be monitored by measuring pH, O_2-concentration and the electrical conductivity.

Further methods are given by the monitoring of vibrations and acoustic emissions.

3 CALCULATION OF PROBABILITY OF LEAKS AND BREAKS

3.1 Frequencies of leaks and breaks of piping that meet the requirements of break exclusion

No data exist that could be used to devise a satisfactory probabilistic statement on the frequencies of leaks from different sizes of a piping manufactured to a high standard, thus giving a quantitative statement on the leak-before-break behaviour. Calculations made so far[12,13] use flaw distributions determined from ultrasonic inspections of welds regarded as representative of the piping studied. However, flaws potentially important for failures are much larger than those seen at such inspections, and the mechanism of the development of large flaws is probably different from the one responsible for the flaws that have been experienced.

In the absence of better information, the distribution of the size and number of faults that have been considered for failure mechanisms in those calculations has been determined from a set of flaws that contains only sizes closely arranged around some average value. Inference for the values in the tails of the distribution is done by fitting tractable mathematical functions to the values that have been experienced. This treatment is not physically justified.

These studies yield frequencies of break in the range of 10^{-9}–10^{-15}/year for large size austenitic, as well as ferritic, primary coolant piping. The authors do not accept the assumptions of these studies and therefore do not use such small values.

For piping with a nominal bore DN $\geq$ 250, manufactured and operated to the standards of 'basic safety', as defined in the RSK guidelines,[3] the authors propose the statement that the frequency of breaks is less than 10^{-7}/year. The authors think this value provides conservatism to the assessments (DN 250 means, according to German standard specification, a pipe with an approximate diameter of 250 mm).

3.2 Frequencies of leaks and breaks of piping that do not meet the requirements of break exclusion

Structures have to be identified that are liable to failure. Experience shows that cracks occur in the vicinity of discontinuities, such as wall thickness changes or at the branches, junctions and turns in the piping. At such discontinuities welds are usually present.

Straight pipe sections do not fail if temperature striation can be ruled out. Using these ideas, it is concluded that the frequency of failures is not determined by the length of the pipe but rather by the number of structures liable to failure (the authors call such structures 'risk relevant structures'). Thus, operating experience, and the statistics drawn from it, are not being related to a plant or a group of plants but to the number of risk relevant structures of various sizes that are present in all plants considered.

The use of statistical data is more convincing for this kind of piping. For safety-related piping in a PWR, not only is the number of leak and break occurrences low, but also the amount of operating experience is small compared with that of other piping systems. The statistics of the frequencies of leak and break occurrences have a large uncertainty. Further, if no additional mathematical models were used, a zero occurrence for both a leaking crack and a break within the same diameter category would imply the same frequency for both kinds of failure, within the uncertainties given by such a statistic.

This situation calls for methods that use further considerations in addition to statistics. A method that uses the ratios of occurrence rates of several leak mechanisms has been devised from the phenomenology of the mechanisms involved.

The method the authors propose here will use statistical data only where they do not lead to too large an uncertainty in the evaluation of frequencies. This requirement can be best met if statistics are only applied to mechanisms that can be regarded as precursors for further and less frequent events. The frequency of a major leak, such as a severance, is then estimated from the statistics of a precursor event, e.g. of any small leak. The conditional probability of a severance, conditional on the occurrence of a leak in general, is to be multiplied by the frequency of a leak in general. It is considered that the frequency of the subset, 'a major leak', out of the set, 'a leak in general', can be neglected compared with that of 'a leak in general'. The frequency of a small leak thus is virtually the same as that of 'a leak in general'.

A representative distribution of the number and the size of flaws and statistics of the activation of crack-generating mechanisms are not known. However, a rough estimate of the ratio of the frequency of a break to that of a leak in general λ_B/λ_L is proposed.

To estimate this ratio, for piping made of the same material and designed for the same interior pressure, the following general statements can be made:

—Loadings due to vibrations, not taken into account at the construction and the design stages, are of decreasing influence with increasing diameter.

—Loadings from inertial forces originating from the liquid flow (e.g. closing actions of valves) can be predicted more accurately during design as the diameter is increased.

—The number of layers of weld beads is larger, thus the influence of faults in weld beads is smaller as the pipe diameter is increased.

—Conditions during manufacturing can be better controlled, and the quality assurance is higher, with larger pipes.

—The number of recurring inspections is greater for larger pipes.

—The reliability of early leak detection is increased, due to the larger amount of leakage, with increasing pipe diameter.

Operating experience with PWR primary circuit piping shows that for DN 25 piping (DN = nominal diameter in mm), taking $\lambda_B/\lambda_L = 0{\cdot}1$ is of the right order of magnitude.

Probabilistic fracture mechanics calculations for large diameter piping[11,12] (DN 350–800) show λ_B-values that are by about six orders of magnitude less than the corresponding λ_L-values. However, as mentioned before, the assumptions of these calculations are not completely satisfactory. Experiences from various non-nuclear piping systems suggest λ_B/λ_L-values of the order of 10^{-2}.[14]

These considerations and the intention of being conservative with the λ_B/λ_L-values has led the authors to a simple assumption for the dependency of λ_B/λ_L on the nominal diameter of the piping considered. The authors use the simple relationship

$$\lambda_B/\lambda_L = 2{\cdot}5/\mathrm{DN} \qquad (1)$$

which yields a λ_B/λ_L-value of 0·1 for DN 25 piping and of 0·01 for DN 250. In primary circuit piping in the range of DN 80–150 there have been no leaks. For piping of DN 50, three small leaks have been detected. It was desirable to use a common sample for the diameter range DN 50–150 to avoid a small sample for zero leak occurrences. Though the diameters are different, the potential leak mechanisms due to loadings and manufacturing quality are quite similar.

Leak frequency depends on piping size through:[13]

$$\lambda_{\mathrm{L,D}} = C\left(\frac{L_{\mathrm{D}} \times D}{t_D^x}\right) \tag{2}$$

for the same stress.

Where:

$\lambda_{\mathrm{L,D}}$ is the leak frequency of the piping fraction with diameter D.
L_{D} is the length of the piping with diameter D.
t_{D} is the wall thickness of piping with diameter D.
The factor C is to be determined from the overall number of occurrences, N, of a leak in the diameter range of the piping considered.

$$C = \frac{N}{\sum_{\mathrm{D}}(L_{\mathrm{D}} \times D/t_{\mathrm{D}}^{x}) \times T} \tag{3}$$

where T is the operating experience (years). The exponent x can be determined from different theories, but also from statistics on extended piping systems. In accordance with the previous statement about the risk relevance of the discontinuities rather than the length of the piping, the authors have used the number of risk relevant structures for the quantity L in formulas (2) and (3).

In the present study, the D/t ratio of the piping in the range DN 50–150 was constant at $D/t = 10$. The exponent was chosen as two, based on Ref. 14. The effect of different diameters for the determination of the leak frequency from statistics was calculated using eqn (2).

The results of calculations performed according to the ideas described here have been published in Ref. 6.

4 CONCLUSIONS

A set of material, manufacturing and operational conditions for a piping system in a PWR plant can be defined that permits us to make use of the break exclusion principle.

For piping that does not fit into this set, the use of statistics can be improved by applying some knowledge of the ratio of the frequencies of different leak mechanisms. This knowledge is drawn from considerations of the relative importance of the potential causes and modes of failures, depending on the dimensions of the piping considered.

REFERENCES

1. Schulz, H., Current position and actual licensing decisions on Leak-Before-Break in the Federal Republic of Germany. CSNI Report No. 82, NUREG/CP-0051, Monterey, 1, 2 September 1983.
2. Schulz, H., Latest development of LBB policy and future subjects in West Germany. NUREG/CP-0092, Tokyo, 14–15 May 1987.
3. Reaktor Sicherheits Kommission (RSK), *Guidelines for Pressurized Water Reactors*, Regulatory guide. Gesellschaft für Reaktorsicherheit (GRS) mbH Köln, 3rd Edition, October 1981.
4. Kerntechnischer Ausschuss (KTA), Sicherheitstechnische Regel des KTA, Komponenten des Primärkreises von Leichtwasserreaktoren, KTA 3201, March 1984, Regulatory guide.
5. Gast, K., Comment at the panel session. Proceedings of the International Nuclear Plant Ageing Symposium, NUREG/CP-0100, 30–31 August 1988 and 1 September 1988.
6. Beliczey, S. & Schulz, H., The probability of leakage in piping systems of pressurized water reactors on the basis of fracture mechanics and operating experience. *Nucl. Engn. Design*, **102** (1987) 431–8.
7. USNRC, Probability of pipe fracture in the primary coolant loop of a PWR plant. NUREG/CR-2189, Vol. 3, August 1981.
8. Industrieanlagen-Betriebsgesellschaft MBH, Probabilistische Untersuchung des Rißfortschrittsverhaltens von Reaktorkomponenten. TF-1605, SR 0125, 9 March 1984.
9. Mikatsch, D., Charalambus, B. & Haas, E., Vergleich der Ergebnisse von Rüttelversuchen mit Resultaten Unterschiedlicher Berechnungsverfahren. 11. MPA-Seminar, Stuttgart, October 1985.
10. Guzy, D., In *Proc. Piping Design Criteria And Research Symposium on Current Issues Related to Nuclear Power Plant Structure, Equipment & Piping*, Raleigh, NC, 10–12 December 1986.
11. Seismic Design Task Group and Sevenson & Associates, Inc., Report of the US NRC-piping reviews committee summary and evaluation of historical strong-motion earthquake; seismic response and damage to aboveground industrial piping. NUREG-1061, Vol. 2, Addendum, April 1985.
12. Hegemann, J., Krieger, G., Loevenich, M. P., Paul, K. D., Schmidt, Th. & Schomburg, U., Deutsche Risikostudie Kernkraftwerke Phase B. Auslösende Ereignisse und Ereignisabläufe für Kühlmittelverlustörfälle. RWTÜV Essen, January 1985.
13. Harris, D. O., Lim, E. Y. & Dedhia, D. D., Probability of pipe fracture in the primary coolant loop of a PWR-plant, Vol. 5. Lawrence Livermore Laboratory, NUREG/CR-2189, 1981.
14. Thomas, H. M., Pipe and vessel failure probability. *Rel. Engng.* **2**, (1981) 83–124.

Int. J. Pres. Ves. & Piping **43** (1990) 229–239

A Probabilistic Approach to Leak-Before-Break in CANDU Pressure Tubes

J. R. Walker*

Materials and Mechanics Branch, Atomic Energy of Canada Ltd,
Whiteshell Nuclear Research Establishment, Pinawa, Manitoba, Canada R0E 1L0

ABSTRACT

In the CANDU® *reactor, the coolant passes through the core in zirconium alloy pressure tubes. A few of these pressure tubes have leaked at cracks near the rolled joint where the pressure tube is attached to the end fitting.*

A probabilistic methodology, and associated computer code (called MARATHON*), has been developed to calculate the time from first leakage to unstable fracture in a probabilistic format. The methodology explicitly uses material property distributions, and allows the risk associated with leak-before-break to be estimated. A model of the leak detection system is included to calculate the time between leak detection and unstable fracture. The sensitivity of the risk to changing reactor conditions allows the optimization of reactor management after leak detection.*

Preliminary material property distributions show the probability of unstable fracture is very low, and that ample time is available to shut down the unit and locate the leaking tube.

1 INTRODUCTION

The CANDU nuclear reactor uses natural uranium dioxide fuel, and heavy water (D_2O) for both heat transport and moderator. The coolant is pressurized to approximately 10 MPa, and passes through the core in a number of zirconium alloy pressure tubes at temperatures between approximately 250°C (inlet end) and approximately 300°C (outlet end). In

* Present address: Fuel Engineering Branch, AECL Research, Chalk River Laboratories, Chalk River, Ontario, Canada K0J 1J0.

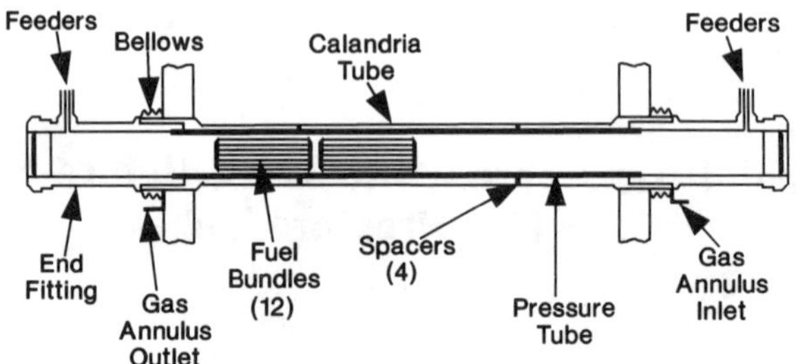

Fig. 1. Schematic diagram of a CANDU fuel channel.

commercial CANDU units, the pressure tubes are fabricated from Zr–2·5Nb, are 6 m in length, have an inside diameter of 103 mm, and a wall thickness of 4·2 mm. Depending on unit size, they number between 380 and 480 per unit. Each pressure tube is surrounded by a gas annulus and isolated from the moderator by a Zircaloy-2 calandria tube. The pressure tube–calandria tube combination, with its associated end fittings, makes up a CANDU fuel channel (Fig. 1).

The presence of large amounts of moisture in the annulus gas system indicates that a leaking crack may be present in a pressure tube. A few pressure tubes have leaked at through-wall semi-elliptical axial cracks near the rolled joint where the pressure tube is attached to the end fitting. Examination of these tubes showed that crack initiation had occurred at their inside surface and had been caused (with two exceptions) by high residual stresses resulting from incorrect over-rolling of the joints during installation. In the exceptional cases, the cracks initiated at manufacturing defects in the high residual stress region near the rolled joints. The concentration of hydrogen was sufficient to allow these cracks to grow by the delayed hydride cracking mechanism (DHC).[1] In every case, the rolled joint cracks exhibited leak-before-break (LBB), i.e. corrective measures were taken before the crack had reached the critical crack length (CCL) for catastrophic failure. Measures have been taken to reduce the residual stresses at the end fitting joints, and work is in progress to reduce the uptake of hydrogen. However, it is not possible to rule out the possibility that further cracks will grow in the future as additional service increases the irradiation fluence and the hydrogen concentration.

The annulus gas system (AGS) is a series-parallel arrangement that locates the leaking pressure tube to be amongst a particular set of pressure tubes. Identification of the actual leaking tube is achieved by additional measures, such as measurement of channel outlet temperatures (COTs). The presence of D_2O in the annulus gas system will initiate the shutdown of the reactor to prevent the possibility of fuel channel damage. To unambiguously identify the leaking pressure tube, it is desirable to keep the reactor partially pressurized at hot shutdown so that the COT measurements can be made. A premature full depressurization may cause an extended shutdown while the leaking tube is located. However, fast fracture could occur during an excessively long hot shutdown period.

A methodology has been developed to calculate the time from first leakage to unstable fracture in a probabilistic format. The methodology is based on probabilistic fracture mechanics techniques, and allows the risk associated with LBB to be estimated. Additionally, the sensitivity of the risk to changing reactor conditions allows the optimization of reactor management after leak detection.

2 PROBABILISTIC METHODOLOGY

The time between first leakage and unstable fracture is a function of the crack length at wall penetration, the critical crack length, and the DHC crack growth velocity. The values of these parameters vary from tube to tube depending upon the metallurgical properties and history of individual tubes. Thus, the values of the parameters that determine the time between leak and break are distributed over the tube population in the reactor unit.

A probabilistic methodology has been developed to explicitly use the material property distributions in the calculation of the time between leak and break (t_{LBB}). The methodology is implemented in a Fortran-77 computer code called MARATHON. This code employs Monte Carlo techniques to calculate the cumulative distribution function (CDF) of t_{LBB}. The CDF of t_{LBB} can be compared with the time required to alarm the AGS leak detectors, cool and depressurize the unit. Hence, the risk associated with unstable fracture can be determined. This paper will concentrate on giving a physical description of the MARATHON models, together with examples of the use of the code. A detailed description of the numerical techniques used is given elsewhere.[2]

2.1 The crack growth model

To produce a probabilistic estimate of t_{LBB}, it is necessary to describe a model that reasonably represents a leak near the rolled joint (RJ) of a pressure tube.

In the MARATHON model it is assumed that the local stresses are such that crack initiation can occur, i.e. it is assumed that the stress intensity factor (K_I) is above the threshold required to cause crack initiation (K_{IH}). It is also assumed that the hydrogen concentration exceeds the terminal solid solubility such that hydride precipitates are present to allow the crack to propagate by

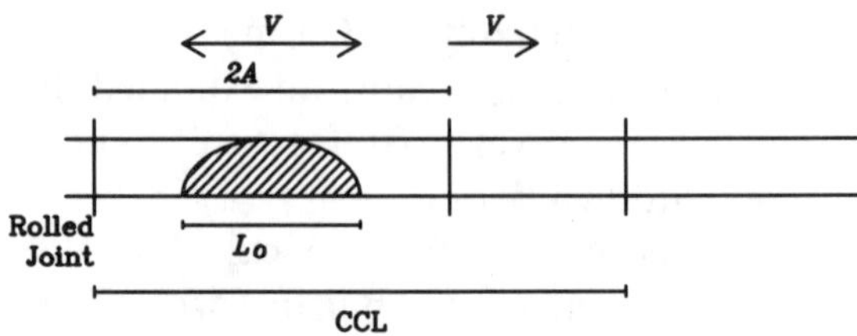

Fig. 2. The MARATHON crack growth model.

DHC. The semi-elliptical DHC crack has an initial length L_0 when it penetrates through the wall of the pressure tube and begins leaking. The centre of the crack is a distance A from the rolled joint (Fig. 2). The crack will grow at both ends until it reaches a length equal to $2A$, at which point the crack will be arrested by the compressive stresses in the rolled joint; only single-ended growth can occur from that point on. A composite micrograph of a typical through-wall DHC crack is given in Ref. 3. For the range of stress intensity factors found in pressure tube flaws, if $K_I > K_{IH}$ then the delayed hydride crack velocity (V) is essentially independent of stress intensity factor (Fig. 3).[4] Hence, the DHC velocity is not a function of the crack length.

If the delayed hydride crack velocity and the critical crack length are both constants, then t_{LBB} is given by

$$t_{LBB} = \begin{cases} \dfrac{(\mathrm{CCL} - L_0)}{2V}, & \text{if } L_0 < \mathrm{CCL} < 2A, \\ \dfrac{(\mathrm{CCL} - (L_0/2) - A)}{V}, & \text{if } \mathrm{CCL} > 2A \geq L_0. \end{cases} \qquad (1)$$

In the MARATHON model, V and the CCL are selected from probability density functions (PDFs) and a probabilistic estimate of t_{LBB} is obtained by

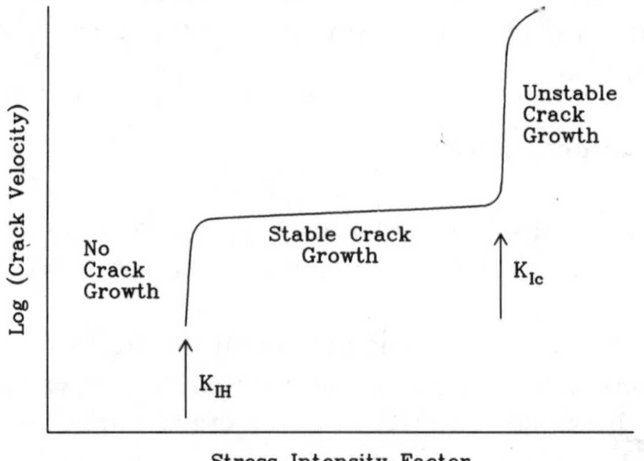

Fig. 3. Schematic diagram of the dependence of crack velocity on stress intensity factor.

using the Monte Carlo method. There is some evidence that V and the CCL may be correlated, since the same metallurgical properties control both. This is allowed for by including a provision for a correlation between the V and CCL distributions.

2.2 The leak detection model

The calculation of t_{LBB} is made using the assumption that the reactor is maintained at full power. The calculated value of t_{LBB} is then compared with the time required to alarm the AGS leak detectors and cool and depressurize the unit. In addition to the estimation of t_{LBB}, MARATHON can be used to provide a probabilistic estimate of the time from leak detection to fast fracture (t_{DTB}). After a leak has been detected, the operators will commence the reactor shutdown procedure. The values of V and CCL will change with time during a shutdown sequence, due to changes in temperature and pressure. Accordingly, provision is made to alter the distributions of CCL and V with the time from leak detection. The inclusion of the effects of various possible shutdown sequences in the calculation of the CDF of t_{DTB} allows comparison between different possible shutdown sequences.

The detection of a crack requires sufficient D_2O to leak through the crack to alarm the detectors in the annulus gas system. In the latter, CO_2 (or N_2) is recirculated through a closed loop at a pressure of, typically, 70 kPa. The system is instrumented with dewpoint indicators, 'beetles' (electronic moisture indicators), sight glasses, and cold finger traps.[5] Testing has shown that the dewpoint indicators are extremely sensitive to a small ingress of moisture, while the beetle alarms are sensitive to a total accumulation of water and are slower to react than the dewpoint indicators. For reasons of conservatism, it was decided to base the MARATHON leak detection model on the slower-acting beetle alarms.

Work is in progress to determine the leakage rate (R) of cracks in CANDU pressure tubes. It is expected that the leakage rate will increase rapidly with increasing crack length.[6,7] Until the leakage properties of pressure tubes are fully quantified, the current strategy is to conservatively model R as a linear relationship of the form

$$R = \begin{cases} \mathrm{m}L + \mathrm{d}, & \text{if } L \geq \dfrac{-\mathrm{d}}{\mathrm{m}}, \\ 0, & \text{if } L < \dfrac{-\mathrm{d}}{\mathrm{m}}, \end{cases} \qquad (2)$$

where L is the crack length, m is a positive constant, and d is a negative constant. The linear relationship was chosen for consistency with a previous deterministic assessment and because it is conservative with respect to reactor and experimental data (see Section 3.2). For a leaking, single-ended, crack,

$$R = \mathrm{m}(L_0 + Vt) + \mathrm{d} \tag{3}$$

$$= \underbrace{\mathrm{m}L_0 + \mathrm{d}}_{b} + \underbrace{\mathrm{m}Vt}_{a'} \tag{4}$$

The total amount of D_2O accumulated is given by

$$\text{Total accumulated} = \int_0^t a't + b\,\mathrm{d}t \tag{5}$$

$$= at^2 + bt \tag{6}$$

where $a = a'/2$. If $-c$ is the amount of D_2O required to alarm the leak detection system, then the time required is the solution to the equation

$$0 = at^2 + bt + c \tag{7}$$

Equation (7) is quadratic with two roots, $t_1 = c/q$, and $t_2 = q/a$, where

$$q = -\tfrac{1}{2}[b + \mathrm{sign}(b)\sqrt{b^2 - 4ac}]. \tag{8}$$

Since a and b are positive, and c is negative, the only positive root to eqn (7) is $t_1 = c/q$. Having now calculated the time from through-wall penetration to leak detection, we can calculate the length of the crack at leak detection time, from whence the time from leak detection to fast fractue (t_{DTB}) can be calculated. It is only slightly more complicated to include double-ended crack growth, and this option is included in the MARATHON model.

3 EXAMPLES

3.1 t_{LBB}

Given the value of A, and the distributions of L_0, V and CCL, MARATHON will calculate the cumulative distribution function of t_{LBB}. Experience with the rolled joint cracks suggests that the crack centre (A) is located slightly inboard of the burnish mark that represents the extent of the over-rolling. A value of $A = 17{\cdot}5$ mm is suitable for these examples. Parameter distributions representative of an outlet end crack are given in Table 1. No correlations between material property distributions are included in this example and it is assumed that the reactor is maintained at full power. The particular distribution forms and numerical values given in Table 1 are derived from the experimental data that is currently available.[3] An effort has been made to

TABLE 1
Parameter Distributions for t_{LBB} Example

Parameter	*Type*	*Mean* (μ)	*Standard deviation* (σ)	*Lower limit*
CCL (mm)	*Normal*	62	6	—
L_0 (mm)	*Truncated normal*	18	3	8
V (m s^{-1})	*Log$_{10}$ normal*	−7·045	0·263	—

ensure that the material property distributions given here are realistic, but it must be appreciated, however, that they are preliminary. An experimental programme is underway to produce a statistically-valid database for the material property distributions.

The t_{LBB} results of the MARATHON calculation, including the 95% confidence bounds on the Monte Carlo calculation, are illustrated in Fig. 4. This shows, for example, that a t_{LBB} of less than 15 h will occur with a probability of less than 10^{-3}. Similarly, Fig. 4 shows that 1% of the pressure tubes will have a t_{LBB} of less than 25 h, 10% will have a t_{LBB} of less than 48 h, and (not shown) all tubes will have failed 2726 h from first leakage. A deterministic analysis[8] has indicated that less than 6 h are necessary to alarm the leak detectors and that the reactor can be economically shut down, cooled and depressurized in about 10 h from first penetration. Taking into account that the data presented in Fig. 4 are conservative, since it is assumed that the reactor is maintained at full power, the overall result

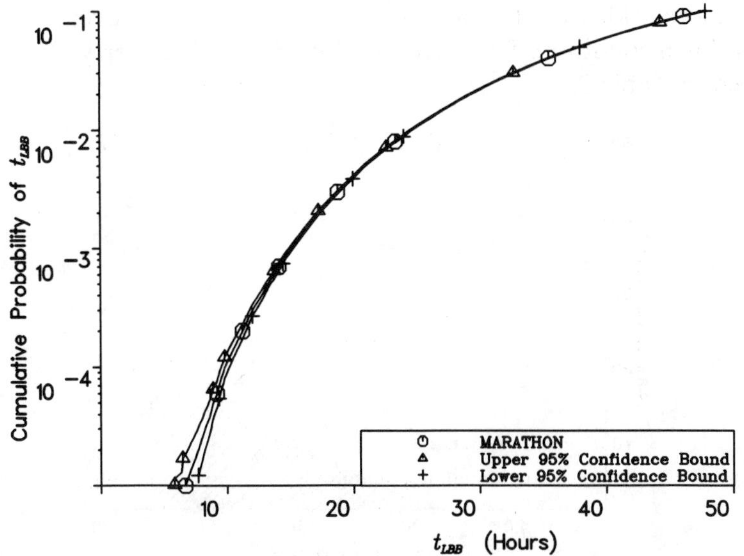

Fig. 4. Example MARATHON t_{LBB} calculation.

TABLE 2
Parameter Distributions for t_{DTB} Example

State	*Material property distribution*			
	CCL (mm)		*V (log_{10} m s^{-1})*	
	μ	σ	μ	σ
Full power	62	6	−7·045	0·263
Intermediate	70	7	−7·150	0·227
Hot shutdown	78	7	−7·265	0·191

indicates that ample margin exists to safely shut down the reactor in the event of a pressure tube crack. This conclusion is confirmed by the analysis given in the next section.

3.2 t_{DTB} and shutdown sequences

As an example of the calculation of t_{DTB}, we shall assign the parameters of the leak detection model (eqns (3) and (7)) as follows: $c = 16$ kg, m = 1·706 kg s^{-2}, and d = −4·606 kg s^{-1}. These values are conservative and represent the lower boundary for pressure tube leak rates (Fig. 5). It is assumed that the reactor is at full power until the beetle alarms, whence the reactor is moved towards hot shutdown. It is assumed that it takes 2 h[8] to proceed from full power to hot shutdown. The COT measurements, to unambiguously identify the leaking tube, are made while the reactor is in the hot shutdown condition. The corresponding material property distributions are given in Table 2.

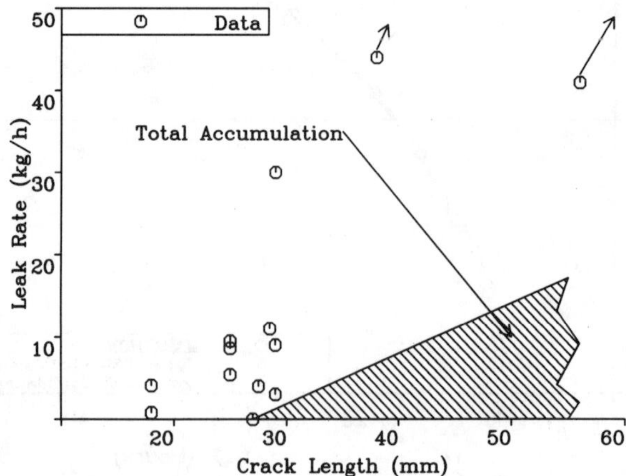

Fig. 5. The MARATHON leak rate model, showing reactor and experimental data points.

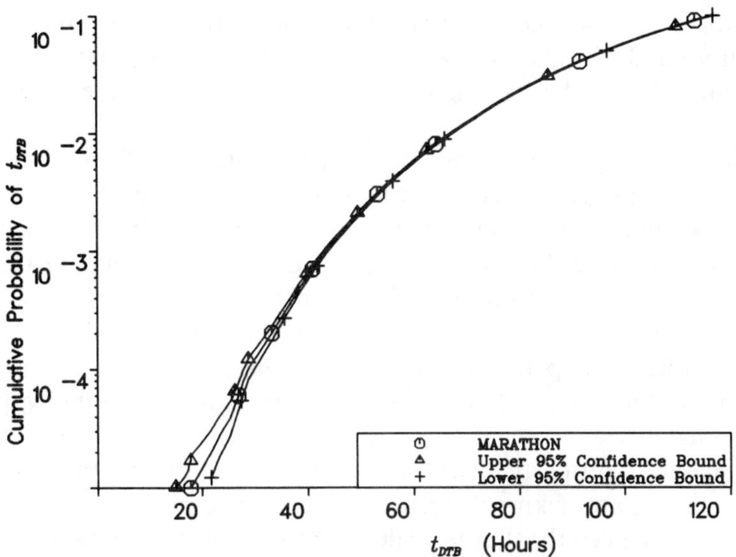

Fig. 6. Example MARATHON t_{DTB} calculation.

The results of the MARATHON simulation are given in Fig. 6, from which it can be seen that the time between leak detection and fast fracture is considerable at all reasonable probabilities. For example, at a probability of 10^{-3}, about 43 h are available to place the unit into the hot shutdown condition (taking about 2 h) and locate the leaking tube (taking at most 10 h). Even at a probability of 10^{-5}, approximately 18 h are available when at most 12 h are needed. Clearly the probability of unstable fracture is very low, and the time available to shut down the unit and locate the leaking tube is large. Different shutdown sequences (such as one designed to cause crack blunting and hence retard DHC) will imply different sets of velocity and CCL distributions and will result in different t_{DTB} probability curves and different risks. In fact, the possibility of assessing the sensitivity of the risk to changing reactor conditions will allow the optimization of reactor management after leak detection.

4 DISCUSSION

A probabilistic methodology, together with the associated computer code (MARATHON), has been developed to calculate the time from first leakage to unstable fracture in CANDU pressure tubes. The methodology explicitly uses distributions of material properties, and includes the capability to correlate these distributions. A model of the leak detection system is integrated into the code to calculate the time from leak detection to unstable fracture. The methodology allows the risk associated with LBB to be estimated. The capability to determine the sensitivity of the risk to changing reactor conditions allows the optimization of reactor management after leak detection.

Preliminary material property distribution data show that the probability of unstable fracture is very low, and that ample time is available to shut down the unit and locate the leaking tube. In the calculations presented here, no credit has been taken for the cooling effect of D_2O passing from the pressure tube to the annulus over the crack face. This will greatly reduce the DHC velocity, since V is related to temperature through an Arrhenius relationship. Hence, the time available for station response will be greater than that indicated here. Work is in progress to accurately quantify this cooling effect.

The material property distributions used as input to the MARATHON computer code are acknowledged to be preliminary. Ontario Hydro has recently commenced the 'Large Scale Fuel Channel Replacement' (LSFCR) of Pickering units 3 and 4. It is planned to cut a sample from each of the removed pressure tubes for analysis. This will provide a 'snapshot' of the material property distributions at the time of replacement, and the neutron fluence distribution will give information on the trend of irradiation-modified material properties with time. The information gained through the LSFCR exercise will be used to update the MARATHON material property distributions by the application of appropriate statistical techniques. The improved database provided by LSFCR will enable a detailed analysis of the material property distributions and allow a comprehensive sensitivity analysis of the effects of changes in material properties.

Additionally, it is intended to use the material gathered to increase our knowledge of crack initiation in CANDU pressure tubes. This information will be incorporated into the probabilistic model to determine the operating risk of CANDU reactors as a function of reactor life.

ACKNOWLEDGEMENT

The author would like to acknowledge the support provided by Ontario Hydro and the Candu Owners Group.

REFERENCES

1. Coleman, C. E. & Ambler, J. F. R., Delayed hydride cracking in Zr-2·5%Nb alloy. *Rev. Coat. Corr.*, **3** (1979) 105–57.

2. Walker, J. R., MARATHON—a computer code for the probabilistic estimation of leak-before-break time in CANDU pressure tubes. Atomic Energy of Canada Limited Report AECL-9989, 1990.
3. Moan, G. D., Coleman, C. E., Price, E. G., Rodgers, D. K. & Sagat, S., Leak-before-break in the pressure tubes of CANDU reactors. *Int. J. Pres. Ves. & Piping*, **43** (1990) 1–21.
4. Simpson, L. A. & Puls, M. P., The effects of stress, temperature and hydrogen content on hydride-induced crack growth in Zr-2·5 pct Nb. *Metall. Trans. A*, **10A** (1979) 1093–105.
5. Kenchington, J. M., Ellis, P. J. & Meranda, D. G., An overview of the development of leak detection monitoring for Ontario Hydro nuclear stations. *Proc. of 8th Annual Conference of the Canadian Nuclear Society*, Saint John, New Brunswick, 1987. Canadian Nuclear Society, Toronto, Canada, 1987.
6. Coleman, C. E. & Simpson, L. A., Evaluation of a leaking crack in an irradiated CANDU pressure tube. Atomic Energy of Canada Limited Report AECL-9733, 1988.
7. Wüthrich, C., Crack opening areas in pressure vessels and pipes. *Eng. Fract. Mech.*, **18** (1983) 1049–57.
8. Price, E. G., Moan, G. D. & Coleman, C. E., Leak before break experience in CANDU reactors. Atomic Energy of Canada Limited Report AECL-9609, 1988.

Int. J. Pres. Ves. & Piping **43** (1990) 241–253

Experiences using Three-Dimensional Finite Element Analysis for Leak-Before-Break Assessment

M. L. Vanderglas

Research Division, Ontario Hydro, 800 Kipling Avenue, Toronto, Canada M8Z 5S4

ABSTRACT

Because of practical limitations, analytical problems in fracture mechanics have often been solved using simplified geometries (e.g. plane stress/strain, shell models). We have applied the Leak-Before-Break approach extensively to the large diameter piping of a new nuclear power plant. Various piping components such as elbows, tee and branch connections with postulated cracks were analyzed. Since no credible geometric simplification was possible, fully three-dimensional (3D) analytical models were found to be essential.

The paper describes our experiences in performing 3D Finite Element (FE) analysis of these components. Included are comparisons of numerical and test results of compact specimens, material modeling considerations, handling of 3D effects, such as the variation of the J*-integral along a crack front, and especially, the effects of plasticity.*

The overall intent of the paper is not simply to present specific numerical results, but rather to give some perspective on the effort required and results attainable.

1 INTRODUCTION

Recent advances in elastic–plastic fracture mechanics have been used to perform Leak-Before-Break (LBB) assessments of the large-diameter carbon steel pipe and piping components used in CANDU reactors. Of necessity, a LBB approach is multidisciplinary, involving aspects such as leak detection, leak rate prediction, inspection and assessments of progressive degradation mechanisms such as fatigue. In the present paper, only the analytical aspect is addressed.

Int. J. Pres. Ves. & Piping 0308-0161/90/$03·50

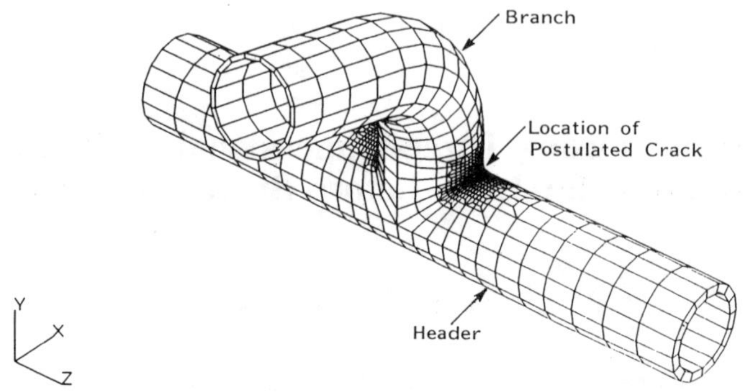

Fig. 1. Typical finite element mesh of a piping component.

The components studied (straight pipe, elbows, branch and tee connections (for example, see Fig. 1) are distinctly three-dimensional (3D). Crack geometries were idealized as through-thickness with straight fronts but 3D effects were observed which do not occur in 2D analysis and which must be considered.

Since it was not feasible to study these effects by repeated analysis of the large finite element (FE) component models (typically having 15–30 000 degrees of freedom), a number of simplified models were used to isolate individual effects.

2 AN OVERVIEW OF THE ANALYTICAL APPROACH

At the risk of oversimplification, the steps in the analysis are as follows:

(i) Obtain opening areas for a series of postulated cracks for use in leak rate predictions, then call the smallest detectable size the 'reference crack length' (RCL).

(ii) Construct models with postulated flaws equal to one times and twice the RCL. (The analysis will attempt to demonstrate that the structure can tolerate loads exceeding the maximum design loads with a detectable crack, and can tolerate a crack twice that size at maximum load.)

(iii) Obtain J-integral versus tearing modulus (J–T) 'driving force' curves for a range of load levels.

The final step in an assessment is the comparison of the 'driving force' and material J-curves. This step contains a number of contentious issues (e.g. how should material curves be extrapolated and what safety factors should be applied?) and are beyond the scope of the present paper.

3 FINITE ELEMENT AND MATERIAL MODELING

Rather than developing an in-house finite element program to perform the analysis, a commercial code (ABAQUS[1]) was employed. After a lengthy process of validation, second order solid elements[2] were chosen for a variety of reasons; among them, their ability to represent complex geometrical details, to partially emulate crack tip blunting (by 'collapsing' crack tip elements)[3] and the availability of a built-in procedure for the calculation of the J-integral for use with these elements.

The object of the analysis is to obtain the J-integral as a function of load and crack size. The J-integral[4] will be a path independent quantity and can therefore characterize the crack tip field providing that the strains are a unique function of the current state of stress (e.g. as in a deformation, total strain theory of plasticity).[5] Cases in which substantial unloadings occur are precluded.

Monotonic loadings do not guarantee such proportional stressing throughout a structure and thus the validity of a deformation theory. In very thick-walled cylinders subjected to internal pressure, for example, incremental and deformation theories give quite different results. Deformation theory predicts a continual rise in all stress components as the loading proceeds. However, incremental theory predicts that the hoop stress at the bore increases until the bore first yields, then decreases until the entire wall has yielded and afterwards increases once again.

Such behavior may be difficult to foresee in complex components. Accordingly, a number of cases were analyzed using both deformation and incremental formulations to ensure that significant but unanticipated unloadings do not occur and that both theories give sensibly identical results.

4 TENSILE PROPERTIES

The CANDU piping system is fabricated using a low carbon steel (mostly ASTM A106B with some A105 elbow forgings) for which a large amount of true-stress versus true-strain curves were available (e.g. for weld and parent metal, longitudinal and circumferential orientations, room and elevated (250°C) temperatures, and for several heats, pipe diameters and product forms).

Typical room and elevated temperature curves are shown in Figs 2 and 3. A significant difference between the two is the appearance of an initial yield plateau in the room temperature curve.

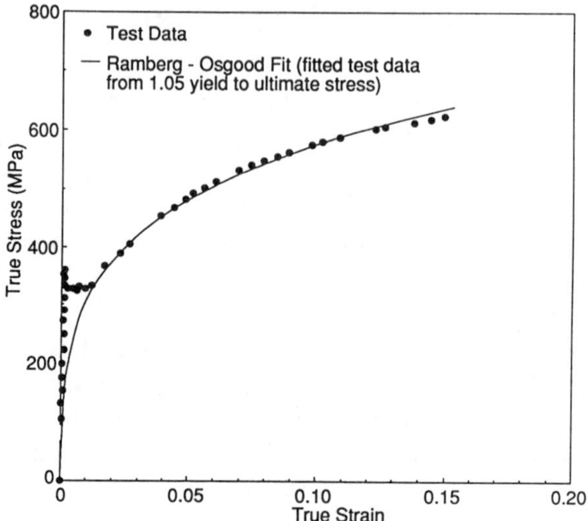

Fig. 2. Typical room temperature stress–strain curve. (A106B parent material, 20°C, strain rate is 1×10^{-3}/s.)

4.1 Numerical representation of stress–strain curves

For use in deformation plastic analyses, replicate tensile tests were curve fitted over the range 1·05 σ_0 to ultimate tensile stress using the Ramberg-Osgood[6] representation

$$\varepsilon/\varepsilon_0 = \sigma/\sigma_0 + \alpha(\sigma/\sigma_0)^n$$

where σ_0 is the initial yield stress, α and n are constants, σ and ε are the true stress and true strain. The range was selected to screen out points on the yield plateau (if there were any) to avoid biasing the fit, and points beyond maximum load since the conversion from the measured engineering stress to true stress then becomes questionable.

For tests exhibiting a substantial yield plateau, a lower bound fit was sought in the interest of conservatism. At high loads and under load control, strains should then be overpredicted with the result that values of J should also be overestimated.

For incremental plastic analyses the stress versus plastic strain curves were represented in tabular form. The data for a given case were represented by a horizontal line, with stress equal to the yield stress, until the line intercepted the Ramberg-Osgood curve fit. Thereafter, data points were generated from the fitted curve.

For 'conservative' leak rate determinations, underestimates of crack opening areas and thus leak rates were sought. Upper bound material models should then be used and for simplicity, elastic properties were used.

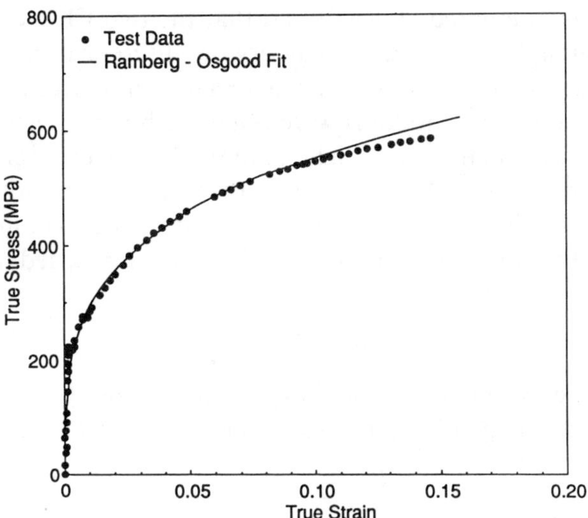

Fig. 3. Typical elevated temperature stress–strain curve. (A106B parent material, 250°C, strain rate is 1×10^{-3}/s.)

5 COMPUTATION OF THE *J*-INTEGRAL AND TEARING MODULUS

The *J*-integral was computed using the virtual crack extension technique (VCE)[7-9] which is a numerical analogue to the analytical line integral. This technique infers the *J*-integral from energy changes resulting from slight (virtual) changes in crack size.

Some confidence in results can be obtained by calculating *J* for a number of contours; each should give the same result. In applying the VCE technique to modestly complex geometries such as in Fig. 1, caution must be exercised to avoid introducing energy changes not associated with crack extension. These are often manifest as a lack of path independence.

In a typical analysis, values of *J* were obtained for five VCE contours, and for each, a value on the outer, mid-thickness and inner surface (the 3D models were discretized using a single second order element division in the thickness direction). Variations between contours of as much as 25% were obtained for a given surface, but upon integrating the results through the thickness, the resulting thickness-averaged contour variations normally decreased to 1% or less.

As an alternative check, *J* can be estimated by obtaining two separate solutions for slightly different crack lengths, then differencing the energy potentials of the two solutions and dividing by the change in crack area.

When this was attempted, it was found that the two FE meshes had to be discretized in a similar fashion, otherwise discretization errors would overwhelm the (small) energy changes associated only with crack extension.

Values of the tearing modulus were obtained by differencing values of J obtained from two analyses for slightly (about 2%) different crack lengths of otherwise identical models. In the event that values were not available for identical load levels (as might be the case if automatic load incrementation was employed) results for the two crack sizes were first fitted individually in the form

$$J(a) = \mathrm{A}(a) \times P^2 + \mathrm{B}(a) \times P^{\mathrm{n}+1}$$

where n is the Ramberg-Osgood exponent, A and B are the curve-fit constants which depend on the crack size, and P is the load or load proportionality parameter.

5.1 Variations of *J* along a crack front

In 3D, the VCE technique gives a series of J values associated with nodes along the crack front. These are inferred from energy changes obtained by perturbing the position of individual nodes along the front and may be sensitive to mesh discretization.

Figure 4 shows the nodal values of J computed along the crack front for a test case (an elastic, edge-cracked plate loaded in tension) modeled with 3D finite elements. Different distributions result as the number of element

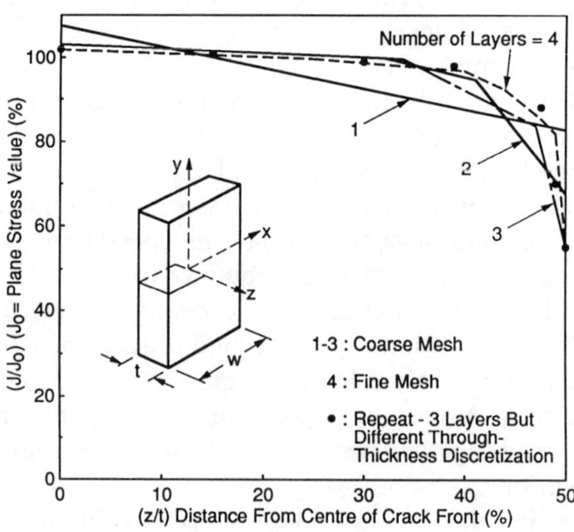

Fig. 4. Variation of J along the crack front: the effect of the number of through-thickness elements.

layers in the thickness direction is varied. At the lateral surface, computed values drop well below either the 2D plane stress or plane strain results (an even more pronounced decay occurs if plasticity is admitted).

The existence of this variation presents two difficulties:

(i) Some representative value must be extracted from the computed J distribution before analytical results can be compared with the material resistance curves.
(ii) Increasing the number of elements (through-thickness) to detect the variation will tax computer resources.

5.2 Averaging crack front *J*-distributions

The 3D analytical version of the J-integral is evaluated on a surface surrounding the crack front rather than the 2D line contour surrounding a crack tip. The 2D version can be viewed as surface integral value per unit thickness. A consistent interpretation is suggested to average the 3D numerical results.

Implicit in the VCE formulation[1] is that the values computed for the crack front nodes can be used to interpolate for some distribution, $J(s)$, using the element shape functions evaluated along the crack front, s. An average value of J can be defined as the integral of $J(s)$ along the front divided by the length of the front.

In meshes of straight-fronted through-wall cracks modelled with single second order elements through the thickness, the average J value defined in this way is

$$J_{AVG} = (J_1 + 4J_2 + J_3)/6$$

where J_1 and J_3 are the values computed for the surface nodes and J_2 is the mid-thickness value.

Table 1 shows the results obtained when the distributions shown in Fig. 4 were averaged. Although models using coarse discretizations along the crack front may give very different J-distributions, they provide excellent estimates of J-average of the more refined models.

TABLE 1

Averaged *J*-Values for 3D Edge Cracked Plates

No. of element layers	*Averaged J/J-plane stress*
1	0·9727
2	0·9732
3	0·9727
4	0·9773

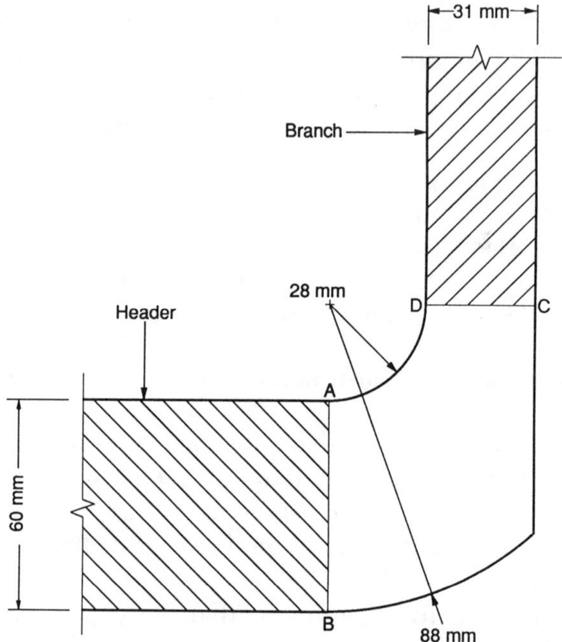

Fig. 5. Idealized through-wall crack geometry (A–B–C–D outlines the crack surface).

The suggested averaging scheme is limited to cases in which opening stresses dominate. If through-wall bending stresses dominate, for example, the scheme may be meaningless since partial crack closure is then possible.

5.3 Energy changes from sources other than crack extension

In 2D (plane stress or plane strain) applications the VCE procedure simulates self-similar extension of the crack in a flat plate of constant thickness. Figure 5 shows a cross-section of the crack postulated at the crotch of the intersection of the component in Fig. 1. The geometry is neither flat nor of constant thickness. A numerical study of a plate of non-uniform thickness was performed to simulate these geometric effects.

Referring to Fig. 6, when the VCE procedure was allowed to perturb all nodes 'straight-ahead', J became very path dependent when VCE contours crossed the change of thickness since in that case the procedure changes the volume of the component (see the insert in Fig. 6). Path independence was restored when perturbations were constrained to (approximately) preserve the original cross-section.

This is only one example in which, unless caution is exercised, the VCE procedure will contribute spurious energy changes resulting from sources other than crack extension alone. Similar effects arise when J-integral

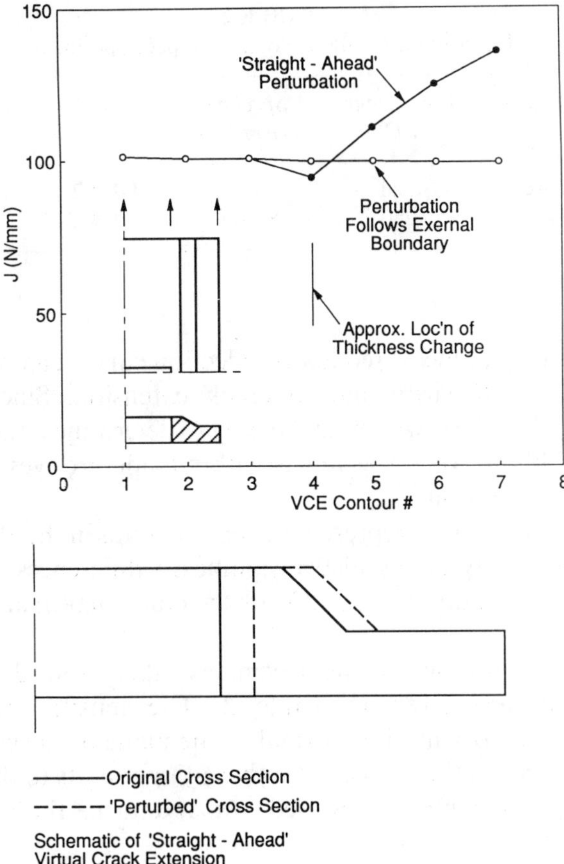

Fig. 6. Contour dependence of the J-integral showing the effect of virtual crack extension direction.

contours cross boundaries between dissimilar materials,[10] or surround holes. To determine if such effects are significant, a model can be constructed in which the crack is prevented from opening by applying suitable displacement constraints to crack surface nodes. The resulting J-integral values should then be zero. If not, there may be contributions not related to crack extension.

5.4 Comparison of analysis and compact specimen tests

To determine whether the material and finite element modeling procedures above were adequate, a series of compact specimens were tested for comparison with computed results. Compact specimens were not of interest *per se* but were chosen as an easily tested geometry which develops a non-trivial stress field.

TABLE 2
Description of Blunt Notch Compact Specimens

Specimen ID	*Material*	*Temperature (°C)*	*Specimen orientation*[a]	*Thickness*	*Description*
JB1	Parent	20	L-C	1-1/4 T	22″ NPS Sch 100[b]
JB2	Parent	250	L-C	1-1/4 T	22″ NPS Sch 100

[a] See ASTM E616.[12]
[b] 22″ = 56 cm.

The nonlinearity of measured load–deflection curves can be attributed to both material nonlinearity and to crack extension. Since comparison between test and analysis was intended only to determine if the FE modeling was adequate, blunt notch specimens without side grooves were tested to preclude crack extension.

No attempt was made to represent the notch radius in the 3D models since preliminary 2D analyses revealed insignificant differences between blunt notch (having root radii of about 2% of the crack depth) and sharp notch models.

Test conditions for two of the specimens taken from 22 in (558·8 mm) piping material are given in Table 2. The tensile properties used corresponded to an orientation normal to the plane of the crack. Analyses based on deformation theory used Ramberg-Osgood fits to the tensile data; those based on incremental theory used tabular data (i.e. the latter were more faithful to the actual tensile data).

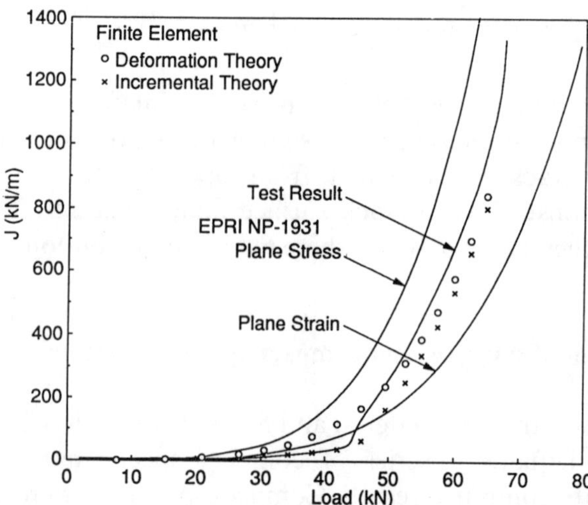

Fig. 7. Blunt notch compact specimen JB1 (room temperature): a comparison of test and calculated J versus load. (A106B parent metal, 20°C, specimen JB1.)

In addition to those obtained using the VCE method, J-versus-load curves were obtained using the ASTM E813[11] procedure based on test and calculated load deflection curves. Figures 7 and 8 compare test (solid lines) and computed results (symbols) for two specimens (JB1 and JB2). Plane stress and plane strain results (as computed using EPRI 1931 'handbook' solutions)[13] are also plotted for comparison.

The dominant influence on the predicted response is the representation of the stress–strain curve. For the room temperature test (specimen JB1), the (lower bound) Ramberg-Osgood fit is significantly at variance with the measured tensile curve, especially at low strains (see Fig. 2). However, J is overpredicted at a given load: an effect which is expected to lead to conservative estimates in applications. Incremental plastic results were somewhat closer to the measured J-load curve but in this particular comparison, tended to underestimate J at higher loads.

Tensile data for elevated temperature tests were much more amenable to curve fitting using a Ramberg-Osgood model. Calculated J-load curves obtained using both incremental and deformation theory are in much better agreement with the corresponding blunt notch compact test results (such as JB2).

The crack-front averaged J-values obtained numerically using the VCE procedure, and those obtained by applying the E813 procedure to the computed load–deflection curves were virtually identical in all cases suggesting that the proposed averaging scheme is consistent with the manner in which material test measurements are interpreted.

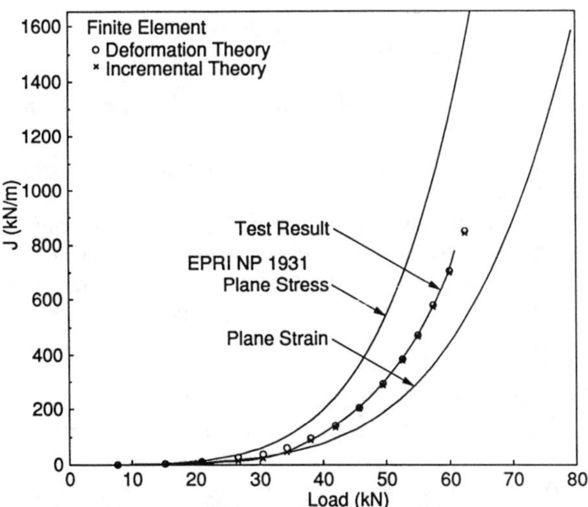

Fig. 8. Blunt notch compact specimen JB2 (elevated temperature): a comparison of test and calculated J versus load. (A106B parent metal, 250°C, specimen JB2.)

6 CONCLUSIONS

A number of features have been described which were observed in the course of a series of three-dimensional finite element analyses conducted in support of a Leak-Before-Break assessment of a CANDU primary heat transport system.

In three dimensions, the technique used to determine the J-integral (the virtual crack extension method) predicts a variation of J along the crack front. The predicted local values may be sensitive to discretization of the finite element mesh along the crack front. For stability assessments, it is suggested that it is the global response of the structure which should be determined; an averaging procedure which gives a representative value of J consistent with the value determined by current material test procedures is proposed.

The representation of the stress–strain curve has an overwhelming influence on elastic–plastic finite element results. It may not be possible to describe the curve exactly (because of material variability, lack of conformity to a particular curve-fitting procedure such as the Ramberg-Osgood representation, and because a structure consists of multiple materials). Nevertheless, a lower bound curve appears to give conservative estimates of J for structures under load control. Alternatively, to obtain conservative underestimates of crack opening areas for use in leak rate predictions, an upper bound curve is appropriate.

REFERENCES

1. Anon., ABAQUS User's Manual. Hibbert, Karlsson & Sorenson, Providence, RI, 1988.
2. Zienkiewicz, O. C., *The Finite Element Method* (3rd edn). McGraw-Hill, New York, 1971, pp. 169–71.
3. Barsoum, R. S., Triangular quarter point elements as elastic and perfectly plastic crack tip elements. *Int. J. Numer. Meth. Eng.*, **11** (1977) 85–98.
4. Rice, J. R., A path independent integral and the approximate analysis of strain concentration by notches and cracks. *J. Applied Mech.*, **35** (1968) 379–86.
5. Mendelson, A., *Plasticity: Theory and Application.* MacMillan, 1970, pp. 119–21.
6. Ramberg, W. & Osgood, W. R., Description of stress–strain curves by three parameters. National Advisory Committee for Aeronautics, Technical Note No. 902, 1943.
7. Parks, D. M., A stiffness derivative finite element technique for determination of crack tip stress intensity factors. *Int. J. Fracture*, **10** (1974) 487–502.
8. Hellen, T. K. On the method of virtual crack extensions. *Int. J. Numer. Meth. Eng.*, **9** (1975) 187–207.

9. Bakker, A., An analysis of the numerical path dependence of the J-integral. *Int. J. Pres. Ves. & Piping*, **14** (1983) 153–9.
10. Miyamoto, H. & Kikuchi, M., Evaluation of J_k integrals for a crack in two-phase materials. *Numerical Methods in Fracture Mechanics* (Proc. 2nd Int. Conf.), ed. D. R. J. Owen & A. R. Luxmore. Pineridge Press, Swansea, UK, 1980, pp. 359–70.
11. E813-81, Standard test method for J_{IC}, A measure of fracture toughness. *1986 Annual Book of ASTM Standards*, Section 3, Vol. 3.01. American Society for Testing and Materials, Philadelphia, PA, 1986, pp. 768–86.
12. E616-89, Standard terminology relating to fracture testing. *1989 Annual Book of ASTM Standards*, Section 3, Vol. 3.01. American Society for Testing and Materials, Philadelphia, PA, 1989, pp. 622–4.
13. Kumar, V., German, M. D. & Shih, C. F., An engineering approach for elastic–plastic fracture analysis. EPRI NP-1931, 1981.

Int. J. Pres. Ves. & Piping **43** (1990) 255–271

Analytical and Numerical Crack Growth Prediction for a Leak-Before-Break Assessment of a Nuclear Pressure Vessel

W. Schmitt,[a] G. Nagel,[b] A. Ockewitz,[a] L. Hodulak[a] & J. G. Blauel[a]

[a] Fraunhofer-Institut für Werkstoffmechanik, Wöhlerstrasse 11, D-7800 Freiburg, Germany

[b] Preussen Elektra AG, Treskowstrasse 5, D-3000 Hannover, Germany

ABSTRACT

*Extending the safety analysis of a nuclear reactor pressure vessel beyond the requirements of the regulations which were first laid down in the ASME-Code, the behavior of two crack sizes (1/4-*t *axial, 3/4-*t *circumferential,* a/2c = *1/6) in the upper shelf region is analyzed to demonstrate the capability of a reactor pressure vessel for Leak-Before-Break. The postulated load is a monotonic increase of pressure, ending when the crack penetrates through the wall.*

The conditions for re-initiation and ductile extension of the conservatively postulated largest possible axial and circumferential crack geometries resulting from initiation and arrest during a thermal shock are evaluated on the basis of the J*-integral concept. Three-dimensional finite element models simulating large amounts of crack growth are employed as well as analytical approximations. The results demonstrate that very high internal pressure far beyond the capacity of the plant and also far beyond the strength of the other components of the primary circuit would be required to initiate and grow the cracks, and that Leak-Before-Break behavior is confirmed.*

1 INTRODUCTION

The safety analysis of a reactor pressure vessel against non-ductile failure, as required by the regulations, contains two different criteria covering two different objectives.

Int. J. Pres. Ves. & Piping 0308-0161/90/$03·50

(1) For *normal operation and emergency conditions* the availability or re-operability after examination has to be ensured. Therefore, the exclusion of initiation and crack extension is essential.
(2) For *faulted conditions* when only safe shutdown and maintaining of the shutdown condition are required, significant propagation of the postulated flaw is acceptable; but it has to be shown that the final crack depth remains below 75% of the wall thickness.

To cover the first situation, and without consideration of non-destructive examination (NDE), the regulatory codes require us to consider an axially orientated crack with a depth of one quarter of the wall thickness. Taking into account NDE, no cracks at all have to be considered according to the codes. Since the girth weld material is leading in irradiation embrittlement and the axial extension of the weld is only a small fraction of the wall thickness, the plant-specific safety analysis according to ASME has shown that all postulated initiating axial cracks will arrest at a final size smaller than the quarter thickness crack. Therefore, this flaw size is also a very conservative assumption to analyze the second case.

Since the material of the girth weld in the core region is leading in irradiation embrittlement, circumferential flaws are of main concern for the analysis. As has been proven by the results of extended analyses of initiation and arrest, such flaws may not grow to a depth larger than the allowable 75% of the wall thickness. Therefore, in accordance with the requirement of the regulations, a circumferential flaw with a depth of three-quarters of the wall thickness is analyzed in the following.

Non-ductile failure is excluded for ferritic steels at temperatures in the upper shelf of the Charpy energy which, for the postulated flaw depth, prevail for normal operation and for a sufficient time period after crack arrest. Thus, with the exclusion of crack initiation and proven crack arrest, a third barrier against non-ductile failure is the remaining ductility of the pressure vessel wall.

The remaining load carrying capacity of the vessel containing these arrested cracks excludes ductile re-initiation, except at pressures far above the operating pressure, and Leak-Before-Break (LBB) is shown in the case of ductile re-initiation.

This paper addresses the issues of ductile re-initiation and stable growth of deep surface flaws up to leakage. A three-dimensional, elastic–plastic finite element analysis is performed to study the ductile initiation and propagation of a hypothetical semi-elliptical axial surface flaw of aspect ratio $a/2c = 1/6$ and relative depth $a/t = 1/4$. The flaw is located symmetrically about a circumferential weld. In the same way, a deep circumferential crack in the girth weld is investigated.

The material properties are assumed to be dependent on the neutron fluence in the pressure vessel wall. Initiation and propagation are assumed to be controlled by *J*-resistance curves, which are derived from correlations with Charpy energy and tensile properties developed within the Heavy-Section Steel Technology program (HSST), but using plant specific data.

While these correlations are well established for crack growth up to 2·5 mm and therefore are immediately applicable to estimate the pressure at ductile re-initiation, the large amounts of crack propagation required to address LBB call for an extension not only of the available *J*-integral resistance curves but also of the *J*-integral calculation beyond the limits currently agreed upon. Therefore, after initiation the development of the crack shape was calculated under the assumption that inevitable systematic errors—the evaluation of the *J*-integral by the virtual crack extension method for non-homogeneous material and, in particular, for a growing crack, becomes dependent on the integration regime, and the resistance curves are not known—have the same effect at any position along the crack front. Thus, the calculated development of the aspect ratio is considered to be a 'best-estimate'. The analyses also take into account recent research results that correlate ductile fracture resistance with the local state of stress ahead of the crack front.

The finite element analyses are accompanied by analytical estimates of the two-dimensional crack growth and of the stability based on the two-criteria (R6) routines. Here, different assumptions concerning material, geometry and flaw development had to be investigated. The results match well with the finite element findings.

2 MATERIAL DATA

The available data for the pressure vessel are results of the irradiation surveillance program following ASTM-E 185[1] completed by additional Charpy and fracture toughness tests. The upper shelf of the Charpy energy is well above 68 J. Fracture toughness values, conservatively converted from *J*-initiation values measured with irradiated Wedge-Open-Loaded specimens (WOL-25X), are available at temperatures up to the brittle–ductile transition. *J*-resistance curves for the irradiated material are determined using results from Charpy-V and tensile tests of the irradiation surveillance program following a procedure proposed in the HSST-program.[2] Figure 1 shows the resulting *J–R* curves for the inner and outer surface of the pressure vessel for base and weld material. Figure 2 shows the calculated resistance curves for large amounts of crack propagation modelled in the finite element study. The conventional material properties used in the finite element

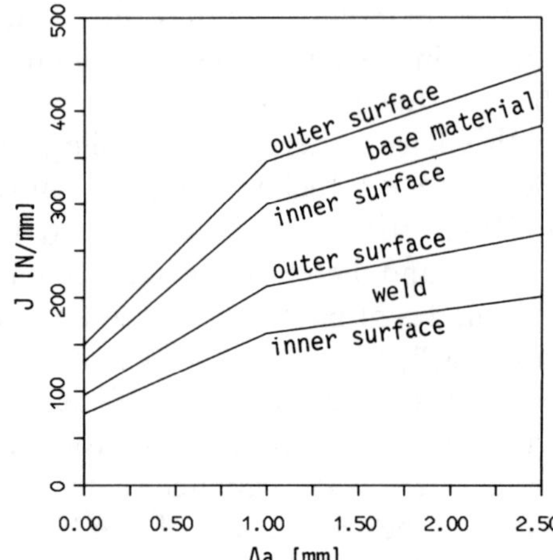

Fig. 1. *J*-resistance curves for base and weld material, inner and outer surfaces, small amounts of crack extension.

analysis are summarized in Table 1. The effect of self-attenuation of the fluence in the vessel wall is considered by adapting the material properties in a bi-linear variation through the wall thickness.

3 FINITE ELEMENT MODELS

As the analytical procedures are available only for crack depths less than half the wall thickness, numerical methods were employed to solve the problems of deep cracks and large crack propagations. The analysis makes use of recent developments in elastic–plastic fracture mechanics. Investiga-

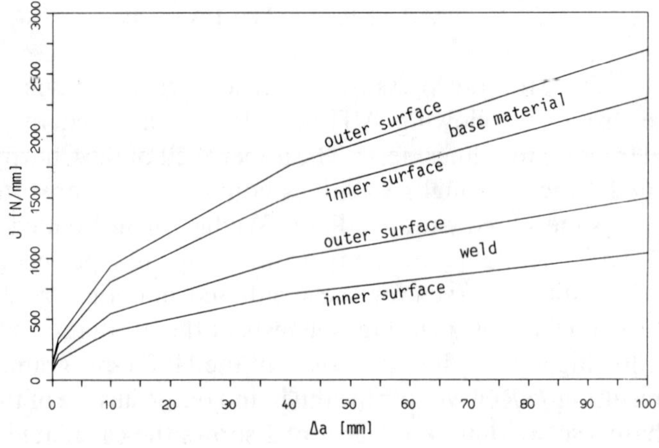

Fig. 2. *J*-resistance curves for base and weld material, inner and outer surfaces, large amounts of crack extension.

TABLE 1
Conventional Material Properties at 288°C used in the Finite Element Models

Material	*Depth from inside surface (mm)*	*Young's modulus E (MPa)*	*Tangent modulus E_t (MPa)*	$R_{p0\cdot 2}$ *(MPa)*	ν
Base	0	198 000	1 565	490	0·3
Weld	0	198 000	1 654	659	0·3
Base	120	198 000	1 504	441	0·3
Weld	120	198 000	1 532	529	0·3
Base	192	198 000	1 491	430	0·3
Weld	192	198 000	1 504	500	0·3

tions into the influence of the triaxiality of the stress (constraint) on crack resistance, discussed, for example, by Sommer,[3] have yielded results which show that with decreasing constraint the fracture resistance against stable crack growth increases. The model used here implies a linear correlation between the ratio of the hydrostatic stress over equivalent stress and the slope of the J-resistance curve. With this assumption, it was not only possible to prove these phenomena qualitatively; it was even possible to make quantitative predictions of good accuracy.

The standards for the measurement of valid fracture resistance curves restrict crack growth to a few millimeters for the commonly used specimen sizes. In this analysis, large crack propagations are allowed, to more than 100 mm, to observe the development of crack propagation. Therefore, the resistance curves were extended to large amounts of crack propagation (Fig. 2).

3.1 The finite element model

The finite element analysis was performed with the commercial program system ADINA,[4] extended by a subroutine package for fracture calculations IWM-CRACK (information through Fraunhofer-Institut für Werkstoffmechanik). In addition to the standard features of ADINA, IWM-CRACK offers the possibility to calculate the J- and C^*-integrals and to model stable crack growth both for two- and three-dimensional geometries. It also includes constraint-dependent J–R curves. In this analysis all material properties were assumed to depend on position through the wall thickness to take into account the effects of irradiation and self-attenuation. The resistance curves, as depicted in Figs 1 and 2, were modified according to the local constraint as proposed in the model described in

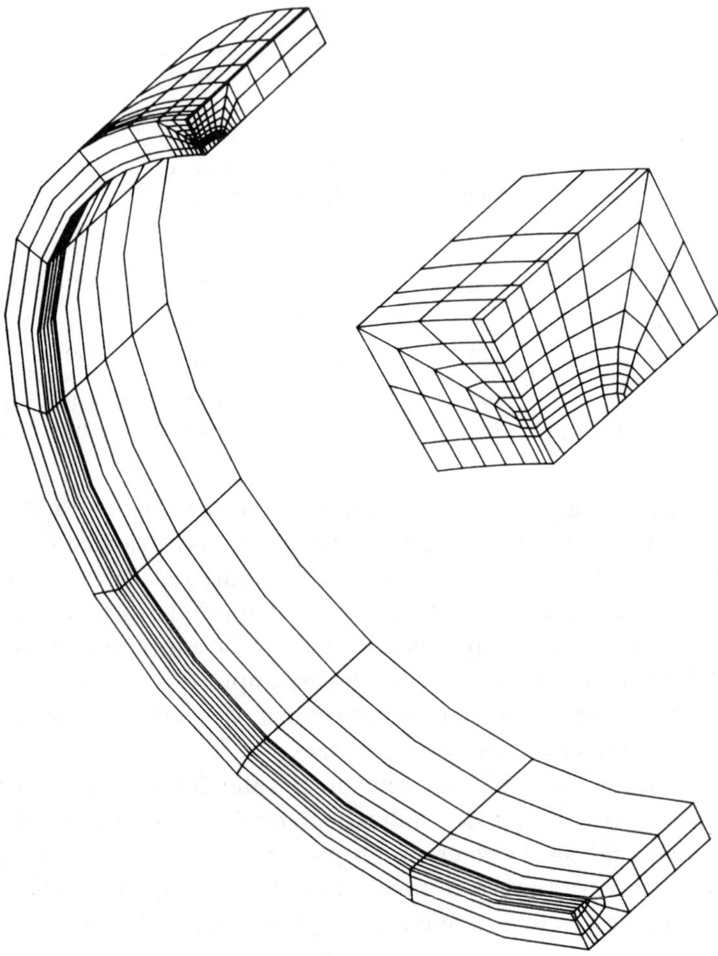

Fig. 3. Finite Element Model: total contour and cracked part.

Sommer.[3] The effect of constraint was, however, not allowed to exceed the effect of self-attenuation; the constraint-modified *J–R* curves were always within the limits given by the resistance curves for the inner surface (lower bound, highest irradiation) and for the outer surface (upper bound, lowest irradiation).

The nuclear pressure vessel under consideration has an inner radius of $R_i = 2040$ mm including an austenitic cladding of 7 mm. The wall thickness analyzed here is $t = 192$ mm (ferritic base material only). An axial flaw of a depth $a = 48$ mm is assumed to lie symmetrically about a weld; its length on the inner surface is $2c = 6a = 288$ mm. The finite element mesh consists of 485 20-noded volume elements and 2651 nodal points. For reasons of symmetry it is sufficient to generate only one half of the cylinder. The height

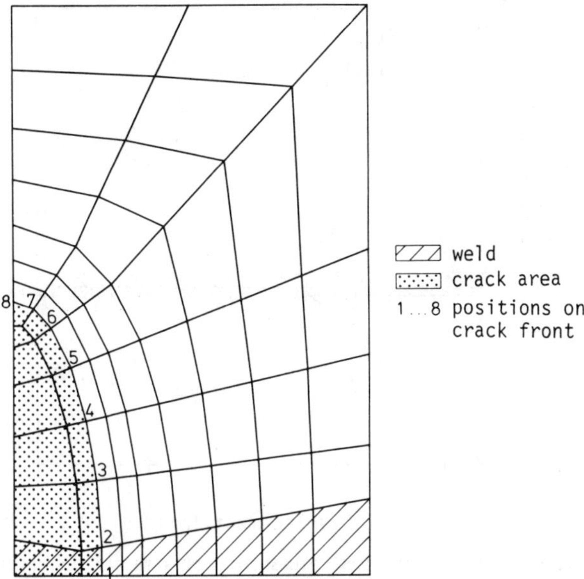

Fig. 4. Finite Element Model: details of crack and ligament area, definition of crack front positions.

modelled is 900 mm. Figure 3 gives a general idea of the finite element model. Figure 4 shows a view of the crack and ligament area with the original crack area and the weld shaded differently.

4 RESULTS

Figure 5 shows the applied internal pressure versus the radial displacement of a nodal point away from the axial crack. The calculated displacement rises linearly with internal pressure up to about 45 MPa and then turns almost horizontal to rapidly increasing displacements. It thus reflects a limit load behavior of the vessel containing a growing crack. Analytical solutions[5] for an infinitely long, homogeneous, hollow cylinder under internal pressure are also plotted in Fig. 5: the linear elastic curve (dashed line) slightly overestimates the calculated pressure at a given displacement, and pressure regimes are indicated for the start of yield (at the inner surface, points) and plastic collapse (shaded). Since the analytical solutions hold for a homogeneous vessel, the bands are defined by inserting maximum and minimum material properties into the respective equations, although the start of yield occurs at the inner surface with higher yield stress leading to the supposition that the upper bound of the pressure regime applies in this case. The comparison of the calculated curve with these estimates reveals a very satisfying agreement, taking into account the complexity of the model.

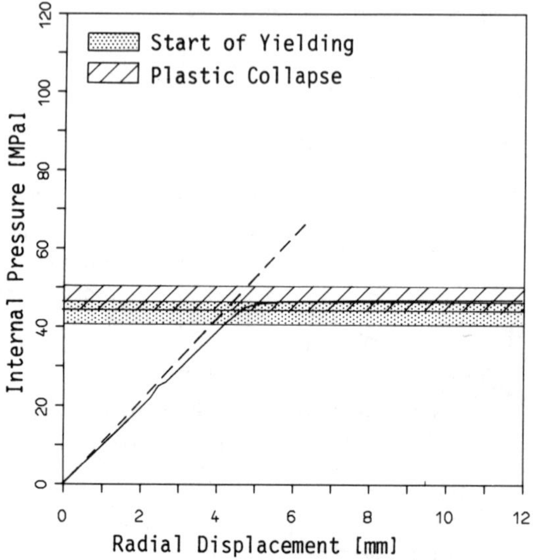

Fig. 5. Global vessel reaction: internal pressure versus radial displacement, analytical estimates shaded for start of yield and plastic collapse. ----, Linear–elastic (analytic); ——, elastic–plastic FE.

The local values of the J-integral, calculated by the virtual crack extension method over crack front length, are given in Fig. 6 for different pressures up to crack initiation. The well-known distribution of J for surface flaws is found, with a maximum in the depth direction. At 16 MPa, about operating pressure, the J-integral value calculated here agrees, to within 2%, with the elastic J obtained using analytical K-solutions following recommendations in ASME XI.

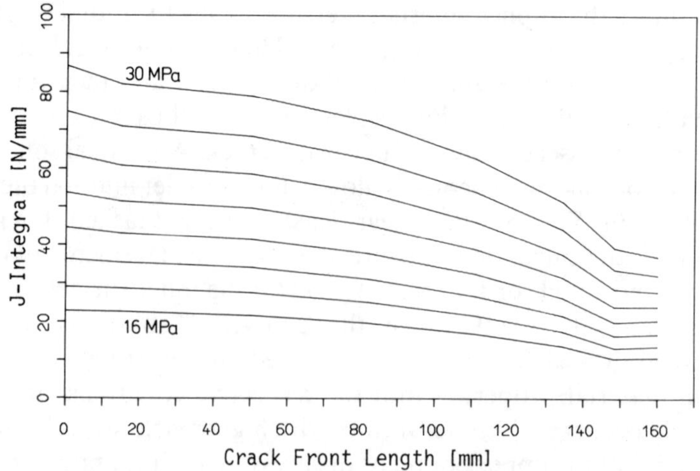

Fig. 6. Variation of the J-integral along the crack front for increasing internal pressure up to crack initiation. Lines indicate increments of 2 MPa.

With increasing load the different yield stresses of the base and the weld material lead to a further elevation of J at the deepest point in the weld. Here, at an internal pressure of $p_i = 30{\cdot}4$ MPa the local initiation toughness of $J_i = 89{\cdot}7$ N/mm (obtained from the resistance curves for weld material in Fig. 1, taking into account the appropriate fluence and constraint) is reached and the crack begins to extend according to the assumed resistance curves which depend on the local state of stress, as explained previously.

Figure 7 shows the calculated shapes of the growing crack with increasing internal pressure. At $p_i = 47$ MPa, the numerical model could no longer find a converging solution, thus determining the pressure at instability of this model. Since the numerical accuracy, especially of the J calculation, is expected to break down at large amounts of non-linearity, the crack shapes calculated are only representative for a realistic situation up to 45 MPa. Under a postulated monotonic increase in pressure, the quarter thickness crack would not initiate at pressures below about twice the operating pressure, if realistic material data are employed. A further increase of the pressure would result in crack extension about halfway through the wall at plastic collapse of the vessel, which takes place at an internal pressure about three times the operating pressure. Such pressures will not build up under any circumstances.

To investigate the conditions for LBB a further analysis was performed using exactly the same model as before, except that this time the material behavior was assumed to be elastic everywhere but in the cracked part of the structure (Fig. 3). Thus the crack propagation could be followed until the

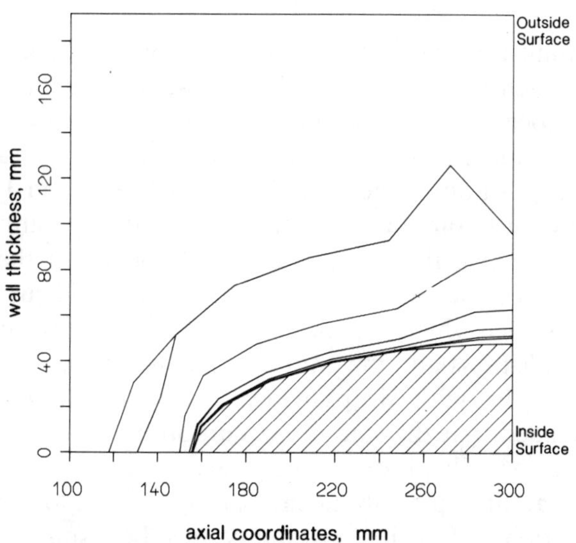

Fig. 7. Crack front development for increasing internal pressure up to plastic collapse.

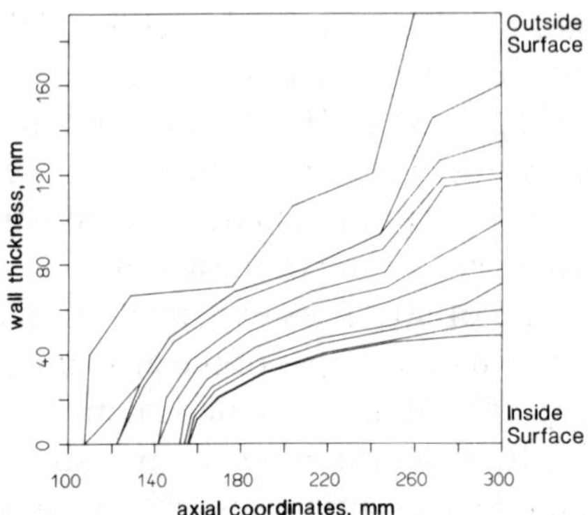

Fig. 8. Crack front development for increasing internal pressure up to wall penetration, modified material properties.

crack broke through the wall using the original crack growth data without interfering with the plastic collapse load of the vessel.

Figure 8 shows the calculated crack fronts with increasing pressure. When the crack has reached the outer wall surface half the crack length at the inside wall is almost exactly one wall thickness, t.

So far the investigations were carried out without addressing the question of crack stability. Stability requires that at any given load, a virtual increase of crack size leads to a change of the crack tip loading parameter, e.g. J, which is less than the change in the material resistance curve

$$\left.\frac{\partial J_{\text{appl}}}{\partial a}\right|_{\text{p=const}} \leq \frac{\partial J_{\text{mat}}}{\partial a}$$

This condition must be checked at every point of the crack front at every load, requiring each time a full additional load step in the finite element analysis. This procedure was not applied here. It is possible to deduce the stability statement, if the development of the local J at increasing loads is plotted for each position of the crack front. Figures 9 and 10 show these curves for the deepest point of the crack and the surface point, respectively. The dashed curves give the lower and upper bounds of the J-resistance curves, depending on the amount of irradiation; the solid curves reflect the actual local resistance which depends on the position of the crack tip with respect to wall thickness and the amount of constraint with respect to the average constraint. From the drops in local J, despite the applied increase in load, it is concluded that the stability condition holds at least up to these loads. Therefore, the fully plastic analysis carried out with realistic material data indicates stability up to about the plastic collapse load for the depth and length directions. After that, instability must be assumed at least in the depth direction. For the second model which excludes the plastic collapse,

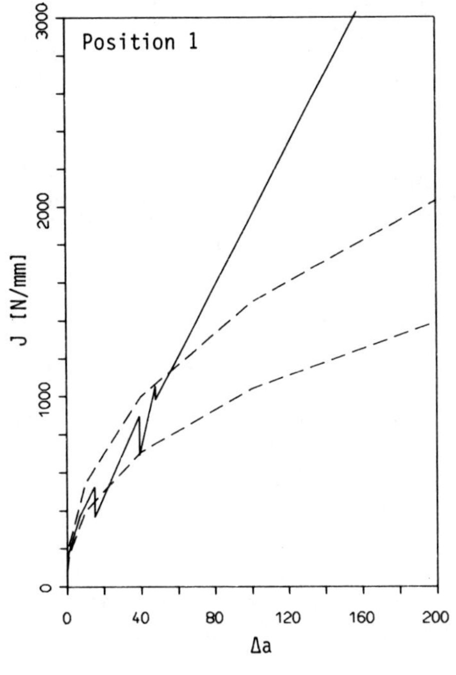

Fig. 9. Crack stability: drop of the applied J-integral despite increasing internal pressure, depth direction, position 1, Fig. 4.

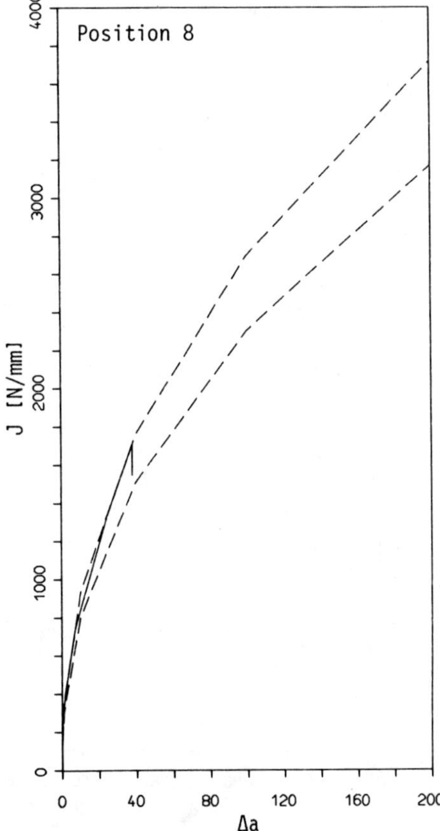

Fig. 10. Crack stability: drop of the applied J-integral despite increasing internal pressure, length direction, position 8, Fig. 4.

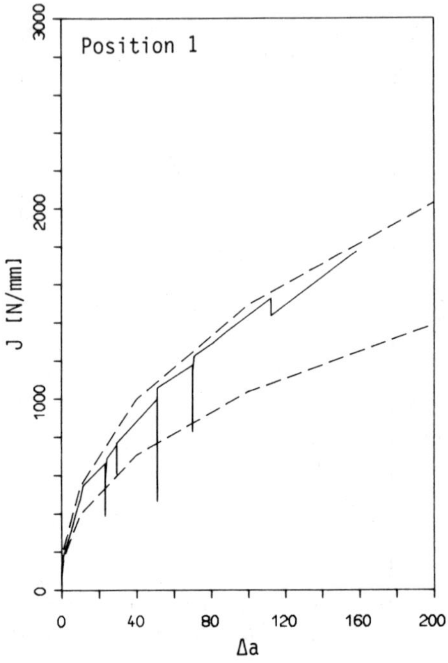

Fig. 11. Crack stability: drop of the applied J-integral despite increasing internal pressure, depth direction, position 1, Fig. 4, modified material properties.

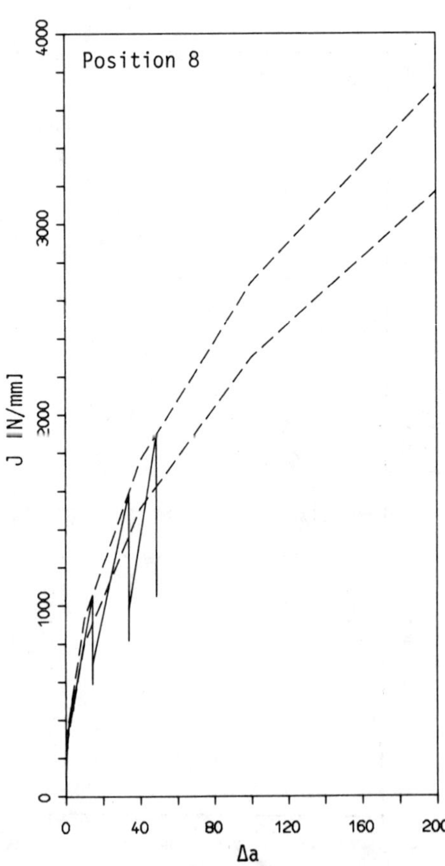

Fig. 12. Crack stability: drop of the applied J-integral despite increasing internal pressure, length direction, position 8, Fig. 4, modified material properties.

stability is determined until the end of the crack propagation (Figs 11 and 12).

In analogy to the axial flaw, a circumferential flaw of initial depth of three-quarters of the wall thickness was analyzed using exactly the same material and fracture resistance data as for the axial flaw. Ductile crack initiation occurs at $p_i = 28{\cdot}9$ MPa. The crack propagates primarily in the depth direction (Fig. 13) and penetrates the wall at a hypothetical pressure of 46·7 MPa. At this point the crack extension in the length direction is only about 5 mm on either side. The stability of this crack in the length (i.e. circumferential) direction, and therefore LBB is evident since the critical size of a through crack obtained for a higher pressure with the analytical solutions is larger than the total length extension of the part-through crack at break through.

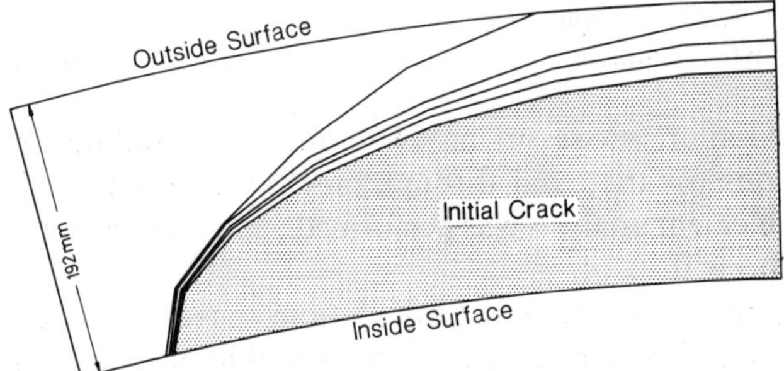

Fig. 13. Circumferential crack: development of the crack shape from initiation to wall penetration.

5 ANALYTICAL SOLUTIONS

Parallel to the costly finite element models, fast and inexpensive approximate analytical methods of elastic–plastic fracture mechanics were applied to calculate initiation, stable growth and eventual instability of the examined defects. The analytical methods were validated by a comparison with the results of the finite-element analysis and could subsequently be used as a tool for sensitivity analyses.

The calculations were done with a special version of the personal computer program system IWM-VERB (information through Fraunhofer-Institut für Werkstoffmechanik). This version involves a routine for the simulation of *J*-controlled stepwise extension of semi-elliptical surface cracks in the depth and length directions and provides checks of crack stability. For each direction, different material properties can be used. The approximate calculation of the *J*-integral is based on the

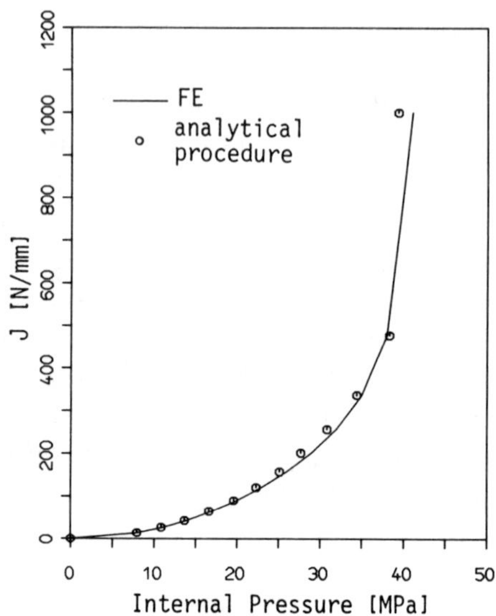

Fig. 14. Finite element analysis of an axial through crack: average J compared with analytical estimates.

two-criteria Failure Assessment Diagram. To validate the J-integral approximations for the axial through crack, a three-dimensional elastic–plastic finite element analysis was performed yielding excellent agreement of the two methods (Fig. 14).

The assessment line used in the calculations corresponds to 'Option 1' of the R6 procedure.[6] Stress intensity factors and limit loads were calculated according to Raju and Newman[7] and Kiefner *et al.*,[8] respectively.

With material data given in Table 2 the following results have been obtained for the axial crack:

pressure at the onset of stable crack growth: $p_i = 24{\cdot}6$ MPa
pressure at instability: $p_{inst} = 38{\cdot}3$ MPa
crack growth from p_i to p_{inst}:
  in depth direction from 48 mm to 95·7 mm
  in (half) length direction from 144 mm to 146·3 mm

TABLE 2
Material Properties used for the Analytical Solutions

	Yield stress (MPa)	*Ultimate stress (MPa)*	*J-R curve (N/mm, mm)*	J_i *(N/mm)*
Surface crack				
apex (weld)	430	583	$158{\cdot}1\,\Delta a^{0,424}$	79·9
surface (base)	493	642	$275{\cdot}4\,\Delta a^{0,461}$	131·1
Through crack	430	583	$275{\cdot}4\,\Delta a^{0,461}$	131·1

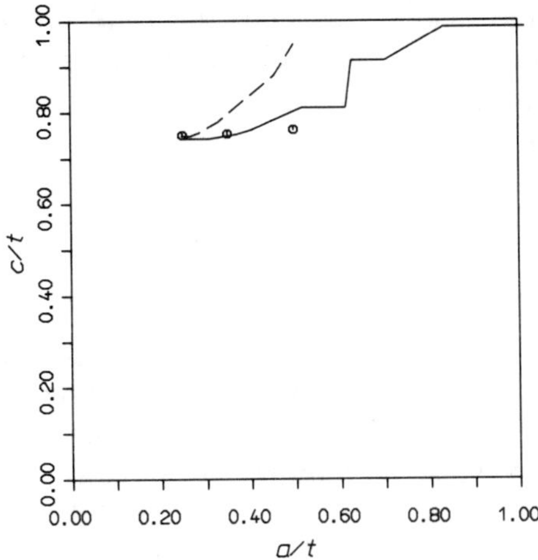

Fig. 15. Crack front development: crack growth in length direction c/t versus crack growth in depth direction a/t, analytical estimates (○), finite element with realistic material properties (----), finite element with modified material properties (——).

The critical pressures obtained in this estimate are about 20% lower than those obtained by the finite element method ($p_i = 30{\cdot}4$ and $p_{inst} = 47$ MPa). Considering the simplifications, especially the differences in the material input, the analytical method gives sufficiently accurate conservative results. Some stages of the stable crack growth in terms of normalized crack length versus normalized crack depth are shown in Fig. 15. The crack shape development obtained by all three methods are in good agreement.

In the assessment of the circumferential surface crack ($a/t = 3/4$, $a/2c = 1/6$) the following results are obtained:

pressure at the onset of stable crack growth: $p_i = 27{\cdot}7$ MPa
pressure at instability: $p_{inst} = 57{\cdot}0$ MPa
crack growth from p_i to p_{inst}:
in depth direction from 144 mm to 161 mm
in (half) length direction from 432 mm to 441 mm

Here, the calculated pressure for crack initiation agrees remarkably well with the finite element result, because, unlike for the axial flaw, the crack front is completely in the weld material. The pressure at instability is, however, significantly overestimated since only the axial stress component perpendicular to the crack area is considered in the procedure. Therefore, the plastic collapse load is not detected by this analysis.

For the instability pressure $p_{inst} = 57$ MPa the critical length $2c_{crit}$ of a through crack was established; $2c_{crit} = 918$ mm. It is the original length of a circumferential through crack which becomes unstable after several hundred millimeters of stable crack extension when loaded to 57 MPa. This size is larger than the total crack length of the surface flaw at p_{inst} of the analytical method ($2c_{inst} = 882\,\text{mm}$) and also larger than the total crack length from the finite element model related to a much smaller pressure. Therefore, LBB behavior of the circumferential crack is proven, if stable crack growth mechanisms other than ductile tearing are not taken into account.

6 CONCLUSIONS

Three-dimensional elastic–plastic finite element models including large amounts of stable crack extension have been applied to investigate the load carrying capacity of a reactor pressure vessel containing deep axial and circumferential surface flaws corresponding to crack sizes given in the code requirements and conservatively covering the results of initiation-arrest analyses of worst-case thermal shock events. Analytical procedures based on the two-criteria approach could be adapted and validated.

The results show that unrealistically high internal pressures of roughly twice the operating pressure had to be applied to initiate and to grow the cracks. The crack shape development revealed a distinct tendency to grow in the depth direction with only limited crack extension in the axial or circumferential direction.

At the plastic collapse load of the vessel, about three times the operating pressure, the axial flaw had grown about halfway through the vessel wall if realistic toughness properties were considered. Continuation of the analysis, with a modified finite element model disregarding the plastic collapse behavior to achieve more crack propagation, gave a total crack length of twice the wall thickness when the crack breaks through the wall.

The pressures necessary to grow the cracks through the wall cannot be produced by the plant. If they were produced, it would not be possible to maintain them in the primary system because the strength of the other primary components would be exceeded.

Below the plastic collapse load the crack extension, especially in the length (axial or circumferential) direction, is stable and LBB behavior is established.

The analyses presented here are part of a complete safety concept excluding crack initiation, proving crack arrest and demonstrating the high load carrying capacity of the vessel as a final barrier against non-ductile failure. The latter could be achieved by utilizing advanced but validated assessment methods, thus extending the analysis up to the physical limits far beyond the requirements of the regulations which use only linear elastic methods.

REFERENCES

1. ASTM E 185-82, Standard Practice for Conducting Surveillance Tests for Light-Water Cooled Nuclear Power Reactor Vessels, E 706 (IF), Annual Book of ASTM Standards, Vol. 03.01. American Society for Testing and Materials, Philadelphia, PA.
2. Merkle, J. G. & Johnson, R. E., Example Calculations Illustrating Methods for Analyzing Ductile Flaw Stability in Nuclear Pressure Vessels, Report NUREG-0939.
3. Sommer, E., Progress in the Assessment of Complex Components. *Proc. of the 7th International Conference on Fracture (ICF7)*, Vol. 3. Pergamon Press, Oxford, 1989, pp. 1999–2020.
4. Bathe, K. H., ADINA, a Finite Element Program for Automatic Dynamic Incremental Nonlinear Analysis. Report 82 448-1, MIT, Cambridge, MA, 1980.
5. Phillips, A., *Introduction to Plasticity*. The Ronald Press Company, New York, 1956.
6. Milne, I., Ainsworth, R. A., Dowling, A. R. & Steward, A. T. Assessment of the Integrity of Structures Containing Defects. *Int. J. Pres. Ves. & Piping*, **32** (1988) 3–104.
7. Newman, J. C. & Raju, I. S., An Empirical Stress-Intensity Factor Equation for the Surface Crack. *Engineering Fracture Mechanics*, **15** (1981) 185–92.
8. Kiefner, J. F., Maxey, W. A., Eiber, R. J. & Duffy, A. R., Failure Stress Levels of Flaws in Pressurized Cylinders *ASTM STP 536, Proc. 1972 Nat. Symp. on Fracture Mechanics*, American Society for Testing and Materials, Philadelphia, PA, 1973, pp. 461–81.

Int. J. Pres. Ves. & Piping **43** (1990) 273–284

Fracture Toughness of Weld Metals in Steel Piping for Nuclear Power Plants

K. Yoshida

Ishikawajima-Harima Heavy Industries Co. Ltd,
Yokohama Nuclear and Chemical Components Works,
1 Shir-nakanara-cho, Isogo-ku, Yokohama 235, Japan

M. Kojima, M. Iida & I. Takahashi

Ishikawajima-Harima Heavy Industries Co. Ltd,
3-1-15, Toyosu, Koto-ku, Tokyo 135, Japan

ABSTRACT

To determine toughness behaviour of dissimilar welds in steel piping and obtain data to evaluate Leak-Before-Break for these welds, an experimental study on fracture toughness was carried out. This paper provides Charpy impact properties and fracture toughness data of base and weld metals of dissimilar welds in nuclear piping.

1 INTRODUCTION

The fracture characteristics of weld joints must be determined as part of a Leak-Before-Break (LBB) evaluation. The fracture toughness of the individual materials, including welding materials, has been investigated to some extent. Landes and McCabe[1] studied the fracture toughness of stainless-steel welds, and compared the toughness of base metal, weld heat-affected zone (HAZ) and the weld metal. The results indicated that the fracture toughness of the base metal was higher than submerged arc weld (SAW) and shielded metal arc weld (SMAW) metal by a factor of 3–10, while HAZ and gas tungsten arc weld (GTAW) metal had nearly the same toughness as base metal. In the nuclear power plants, there are many dissimilar welds composed of low alloy steel, carbon steel and stainless steel. There are not enough fracture-toughness data of dissimilar welds to perform a LBB evaluation. The present paper discusses the results of fracture–toughness experiments on dissimilar welds of three material combinations. The J–R curve and J_{IC} method[2–4] are used to determine the fracture toughness in these ductile materials.

2 EXPERIMENTAL

2.1 Materials

The materials tested were JIS STS49 (which is equivalent to ASTM A106 Grade C carbon steel), ASTM A508 Class 2 low-alloy steel, Type 304

TABLE 1
Chemical Composition of the Material Investigated (Weight %)

	C	*Si*	*Mn*	*P*	*S*	*Ni*	*Cr*	*Mo*	*V*	*Cu*	*Fe*	*Co*	*Ti*
A508 Class 2	0·19	0·21	0·88	0·005	0·003	0·81	0·40	0·59	<0·010	—	—	—	—
STS49	0·21	0·20	1·42	0·014	0·001	—	—	—	—	—	—	—	—
304	0·04	0·65	1·26	0·026	0·004	8·86	18·53	—	—	—	—	—	—
182	0·030	0·43	6·62	0·011	0·004	66·88	14·90	0·05	—	0·14	8·88	0·07	0·42
308L (A508 + 304)	0·028	0·27	1·68	0·029	0·007	10·13	20·77	0·01	—	0·03	—	—	—

TABLE 2
Room Temperature Mechanical Properties of the Material Investigated

	YS (*MPa*)	*UTS* (*MPa*)	*EL* (%)	*RA* (%)
A508 Class 2	465	618	28·9	71·7
STS49	343	529	38·0	—
304	248	575	66·4	75·3
182	366	712	44·9	—
308L (A508 + 304)	—	533	30·0	—

TABLE 3
Welding Condition

Base materials	*Weld material*	*Process*	*Heat input* (*kJ/cm*)	*Interpass temperature* (°*C*)
A508 Class 2 + 304	308L	SMAW	12·4	Max. 53
A508 Class 2 + 304	182	SMAW	10·3	Max. 53
STS49	308L	GMAW	16·7	Max. 200

stainless steel, Type 308 weld metal and Alloy 182 weld metal. These materials and weld metals are used in dissimilar welds of reactor pressure-vessel-nozzle safe ends and reactor recirculation piping, residual-heat-removal piping, and reactor recirculation piping. The chemical composition and mechanical properties of these materials are shown in Tables 1 and 2, respectively. Three welds were prepared for this experiment (Table 3). A508 Class 2 low-alloy steel and STS49 carbon steel were post-weld-heat-treated (PWHT) at 615°C after weld build-up, and these were butt-welded to stainless-steel pipes by conventional welding procedures.

2.2 Test methods

2.2.1 Tensile test and Charpy impact test
The tensile test and Charpy impact test specimens were taken from a pipe at the dissimilar weld, centred on the weld. The tensile test specimen was a round bar 6·0 mm in diameter and 31·5 mm in gauge length. The specimen axis was parallel to the pipe axis. The tensile tests were carried out at room temperature and 288°C, in accordance with JIS G 0567, which is equivalent to ASTM A370.

The design of the Charpy impact test specimen was JIS Z 2202 No. 4 with a 2-mm V-notch, which is equivalent to ASTM A370. The specimen axis was parallel to the pipe axis and the notch direction was parallel to the pipe radial direction. The ductile–brittle transition curve was obtained in accordance with JIS Z 2242, which is equivalent to ASTM A375.

2.2.2 Fracture-toughness test
For the fracture-toughness tests compact specimens were used. Most specimens were 2TCT-type specimens, where the thickness was 20 mm. In STS49 carbon steel and Type 308 weld metal welded to STS49, 1TCT specimens were used. The notch axis of a specimen was parallel to the pipe circumferential direction. The fraction-toughness tests were carried out at room temperature and 288°C, in accordance with ASTM E813. In the fracture-toughness test, the specimens were loaded to pre-specified displacements, then unloaded. Specimens were then broken by monotonic loading in liquid nitrogen or fatigue loading at room temperature. Pre-crack length and ductile crack growth were estimated by averaging the measured length through the thickness on the fracture surface. The data analyses were performed using a digitizer and a personal computer connected to the load cell and clip gauge. J values were calculated by using the equation below, based on ASTM E813:

$$J = \frac{A}{Bb} f(a/W)$$

where

A = area under the load–displacement curve,
B = specimen thickness,
a = pre-crack length,
W = specimen width,
$b = W - a$,
$f(a/W) = 2(1 + c)/(1 + c^2)$,
$c = (2a/b)^2 + 2(2a/b) + 2^{1/2} - (2a/b) + 1$.

$J = 4\sigma_y a$ was applied to a blunting line, so that J_{IC} was defined as a 0·2 mm offset J value of the J–R curve, where σ_y was the material yield strength at the temperature.

3 RESULT AND DISCUSSION

3.1 Tensile test and Charpy impact test

The tensile strength of the welds was over the specified value of the materials; the elongation was also similar to the generally reported values. The tensile test results are summarized in Table 4. The Charpy impact test results at 0°C are summarized in Table 5. The absorbed energy of A508 Class 2 HAZ specimens was scattered. The scatter was thought to be due to the positioning of the notch. The HAZ of the A508 Class 2 had a grain size

TABLE 4
Tensile Test Result

	Temperature (°C)	*YS (MPa)*	*UTS (MPa)*	*EL (%)*	*RA (%)*
A508 Class 2	RT	468	608	25·7	74·4
	288	406	565	22·5	73·7
STS49	RT	298	501	38·5	75·5
	288	201	483	36·2	70·8
304	RT	238	632	79·9	79·9
	288	155	417	48·3	82·1
182	RT	431	648	35·0	49·0
	288	385	623	32·5	42·7
308L (A508 + 304)	RT	409	544	46·7	61·8
	288	328	410	25·0	61·1
308L (STS49)	RT	368	477	35·9	61·5
	288	228	404	23·0	49·4

TABLE 5
Charpy Comparisons for the Material Investigated at 0°C

	Absorbed energy (*J*)
A508BM	222
A508HAZ	268
STS49BM	177
STS49HAZ	165
304BM	291
182	136
308L (A508 + 304)	71
308L (STS49)	95

varying from coarse grains with low toughness to fine grains with high toughness and the toughness measured in the Charpy test depended in which zone the notch tip was located. The STS49 base metal had been normalized and the HAZ had a reasonably uniform fine-grained structure and uniform toughness. The fracture surface of austenitic materials was almost ductile in the test temperature range (−80 to 20°C), but the absorbed energy of the weld metal (70–100 J at 0°C) was low compared with that of the base material. The difference in the absorbed energies depended on the ferrite content and the structure. The absorbed energy of Alloy 182 was about 140 J at 0°C; this value was lower than that for Type 304 stainless steel. It was clear that the absorbed energy of the austenitic material strongly depended on the material structure. In comparing the absorbed energies of tested materials at 0°C, the following order was obtained:

$$\text{308-WM} < \text{182-WM} < \text{CS-HAZ} = \text{CS-BM} < \text{LAS-BM} < \text{LAS-HAZ} < \text{304-BM}$$

where

308 = Type 308 stainless steel,
182 = Alloy 182,
LAS = A508 Class 2 low-alloy steel,
BM = base material,
304 = Type 304 stainless steel,
CS = STS49 carbon steel,
WM = weld metal,
HAZ = weld heat-affected zone.

3.2 Fracture-toughness test

J–R curves obtained by the fracture-toughness tests are shown in Figs 1–8. The *J–R* curves for tested materials are summarized in Fig. 9, and the ductile

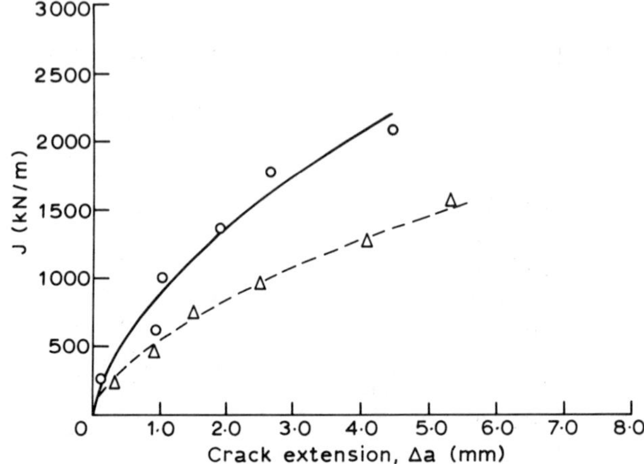

Fig. 1. *J* versus ΔaR curve for A508 low-alloy steel base metal at 20°C and 288°C. ○, A508BM (at 20°C); △, A508BM (at 288°C).

fracture-toughness (J_{IC}) values defined from each *J–R* curve are summarized in Table 6. The J_{IC} for the materials at room temperature were in the order:

182-WM < 308-WM < CS-HAZ < LAS-HAZ = LAS-BM
< CS-BM < 304-BM

The J_{IC} for materials tested at 288°C were in the following order:

CS-HAZ < LAS-BM < CS-BM

The fracture-toughness order at 288°C is similar to that at room

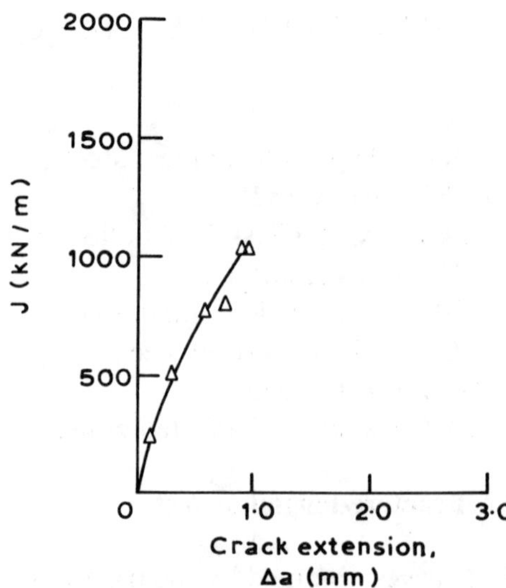

Fig. 2. *J* versus ΔaR curve for A508HAZ at 20°C. △, A508HAZ.

TABLE 6
J_{IC} Comparisons for the Material Investigated

	Temperature (°*C*)	J_{IC} (*kN/m*)
A508BM	20	637
	288	343
A508HAZ	20	637
STS49BM	20	735
	288	794
STS49HAZ	20	617
	288	216
304BM	20	1 098
182	20	108
308L (A508 + 304)	20	225
308L (STS49)	20	372

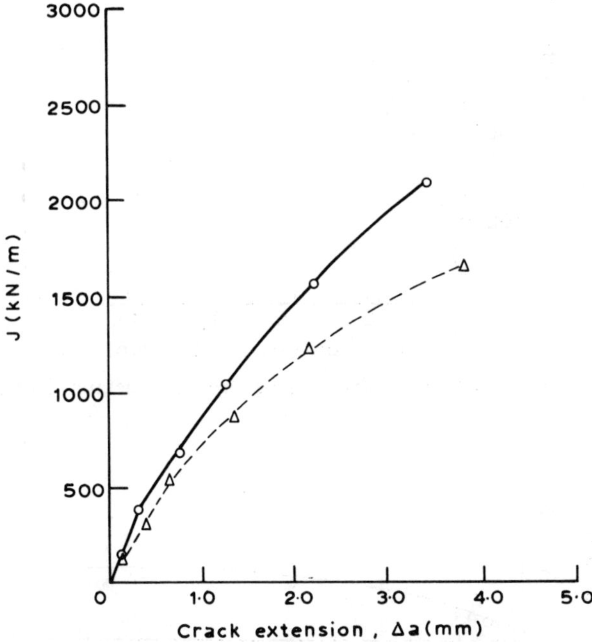

Fig. 3. *J* versus Δ*aR* curve for STS49 carbon-steel base metal at 20°C and 288°C. ○, STS49BM (at 20°C); △, STS49BM (at 288°C).

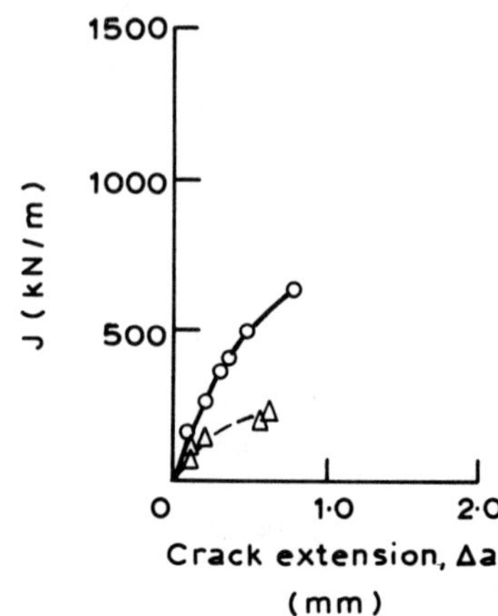

Fig. 4. *J* versus Δ*aR* curve for STS49 carbon-steel HAZ at 20°C and 288°C. ○, STS49HAZ (at 20°C); △, STS49HAZ (at 288°C).

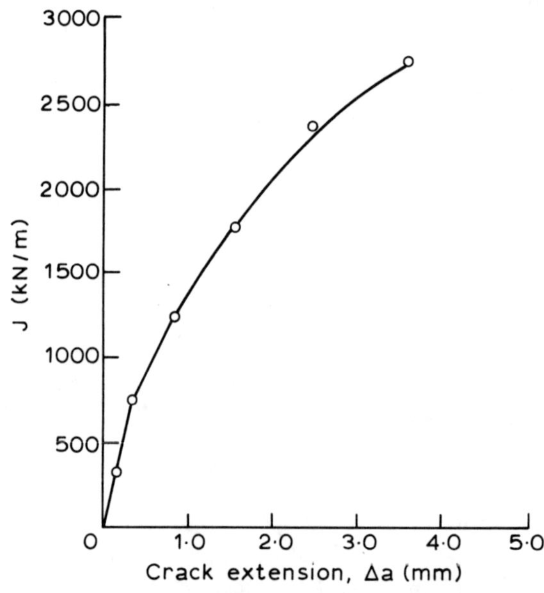

Fig. 5. *J* versus Δ*aR* curve for 304 stainless-steel base metal at 20°C. ○, 304BM.

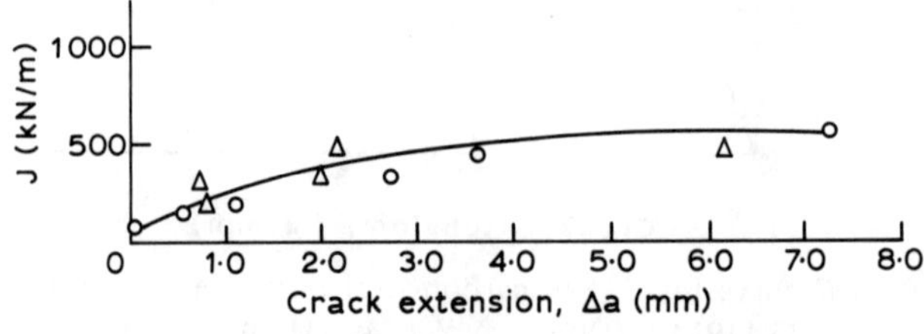

Fig. 6. *J* versus Δ*aR* curve for alloy 182 SMAW weld metal at 20°C. ○, 182WM (outside); △, 182WM (inside).

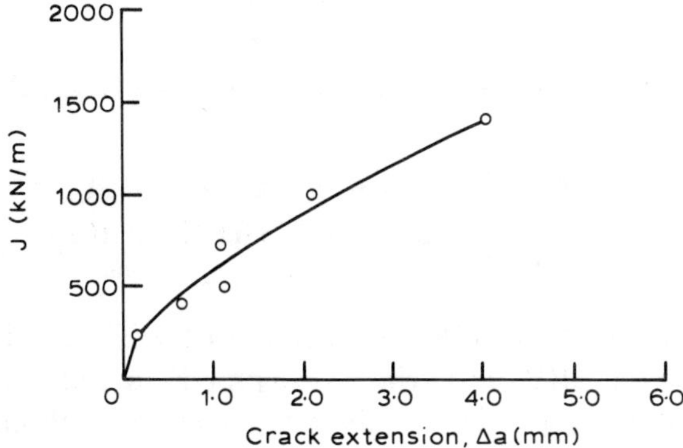

Fig. 7. *J* versus Δ*aR* curve for 308 stainless-steel GMAW weld metal at 20°C. ○, 308WM.

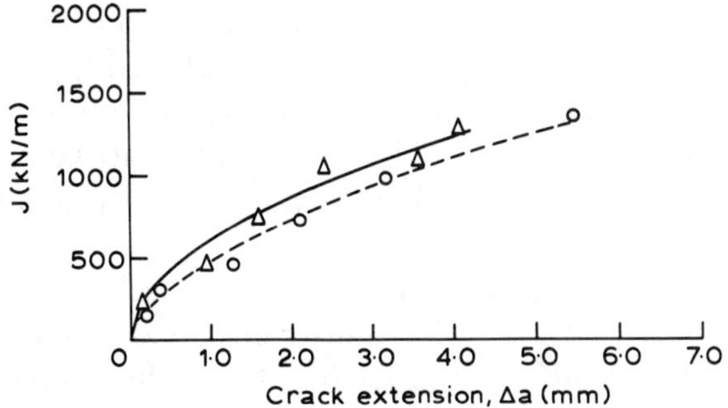

Fig. 8. *J* versus Δ*aR* curve for 308 stainless-steel SMAW weld metal at 20°C. ○, 308WM (outside); △, 308WM (inside).

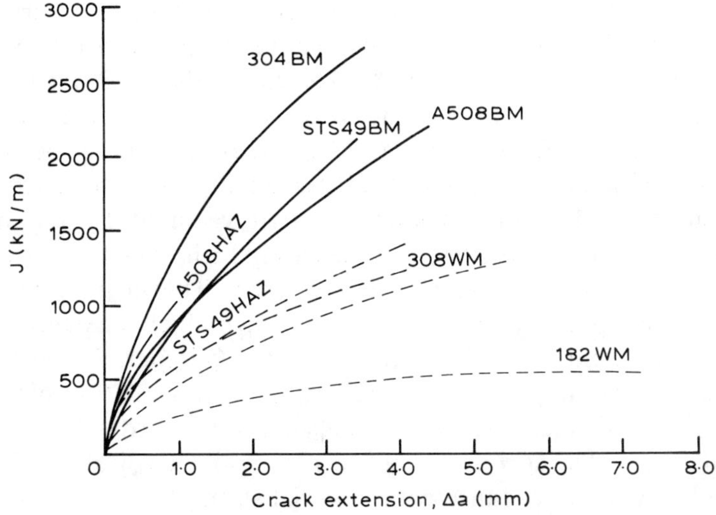

Fig. 9. *J* versus Δ*aR* curve comparing results for A508BM, A508HAZ, STS49BM, STS49HAZ, 304BM, 308WM, 182WM at 20°C.

temperature, but the material-to-material variation in J_{IC} at 288°C and room temperature is as follows:

$$\text{LAS-BM at } 288^\circ\text{C} = \text{LAS-BM at RT}$$
$$\text{CS-HAZ at } 288^\circ\text{C} < \text{CS-HAZ at RT}$$
$$\text{CS-BM at RT} < \text{CS-BM at } 288^\circ\text{C}$$

The low J_{IC} at 288°C in STS49 carbon steel HAZ should be considered when evaluating the LBB of a carbon-steel pipe weld, and these data should be gathered and analysed. The dJ/da values at 288°C obtained from J–R curves are similarly about two-thirds of that at room temperature for the three tested materials. This finding shows that we need more data on the effect of temperature on ductile crack growth in the case of STS49 base metal, which has the same J_{IC} at room temperature and 288°C. A508 Class 2 low-alloy steel, generally a high-toughness steel, had a low J_{IC} compared with STS49 carbon steel. It was not clear whether the low J_{IC} of A508 Class 2 low-alloy steel was related to the PWHT or not, but the effect of thermal aging on J_{IC} of low-alloy steel should be clarified to help evaluate plant aging.

It was reasonable that J_{IC} of Type 304 stainless steel was larger than for all other tested materials, but J_{IC} of the austenitic weld metals, Alloy 182 and Type 308, were lower than that of ferritic steels. The ductile crack front in weld-metal specimens was non-uniform on the fracture surface, and it was considered that the data scatter in the J–R curve of austenitic weld metals depended on this type of crack growth. This non-uniformity was considered to have been caused by the dendritic structure of the weld metal. However, the ductile crack growth behaviour and fracture surface observation of many austenitic and ferritic weld metals should be performed to clarify the effect of the dendritic structure on J_{IC}. The J_{IC} of Alloy 182 was lower than for Type 308. This accorded with the data scatter in the J–R curve; namely, the data scatter of Alloy 182 was larger than that of Type 308. The segregation of impurities along dendrites might affect the ductile crack growth, because the hot cracking sensitivity, caused by the impurity segregation in a nickel-based alloy, is generally high as compared with that of stainless steel, which does not have delta-ferrite. The ductile crack front of the weld-metal specimen showed a wave shape which was considered to depend on the weld-metal dendritic structure. J_{IC} generally relates to the upper shelf energy in a Charpy impact test, but this relationship was not found in this experiment. The order of J_{IC} did not accord with the order of absorbed energy. The correlation between J_{IC} and Charpy absorbed energy is shown below in accordance with material strength:[5]

$$J_{IC} = q(\text{CVN}/\text{YS})$$

where

q = coefficient which varies by strength level,
CVN = Charpy absorbed energy,
YS = yield strength.

The coefficient, q, of high-strength steel is generally smaller than that of low-strength steel. Therefore, the partial reversal of order between the Charpy absorbed energy and the J_{IC} for the tested materials was considered to be reasonable.

4 CONCLUSIONS

As a result of Charpy impact tests and fracture-toughness tests of the materials in nuclear dissimilar welds, the following determinations were made:

(1) Charpy impact energy at 0°C of the tested materials was aligned in the order shown below:

308-WM < 182-WM < CS-HZAZ
= CS-BM < LAS-BM < LAS-HAZ < 304-BM

(2) J_{IC} in fracture-toughness tests at room temperature of the tested material was aligned in the order shown below:

182-WM < 308-WM < CS-HAZ < LAS-HAZ
= LAS-BM < CS-BM < 304-BM

(3) At high temperatures, the relatively low J_{IC} for HAZs should be considered on evaluating the LBB of a carbon-steel pipe weld.

(4) Ductile crack growth of weld metals was observed to be non-uniform. This was considered to be caused by the dendritic structure of the weld metal.

REFERENCES

1. Landes, J. D. & McCabe, D. F., Toughness of austenitic stainless steel pipe welds. EPRI NP-4768, 1986.
2. Begley, J. A. & Landes, J. D., Fracture toughness part II. *ASTM STP 514*, American Society for Testing and Materials, Philadelphia, PA, 1972, pp. 1–20.
3. Landes, J. D. & Begley, J. A., Developments in fracture mechanics test methods standardization. *ASTM STP 623,* American Society for Testing and Materials, Philadelphia, PA, 1977, pp. 57–81.

4. Standard Test Method for J_{IC}. A measure of fracture toughness. ASTM E813-87, *ASTM Annual Book of Standards*, Vol. 03.01. American Society for Testing and Materials, Philadelphia, PA, 1987.
5. Fuji, A., Yamaya, I., Kitagawa, M. & Ohtomo, A., Correlation between elastic–plastic fracture toughness and Charpy impact energy for steels. *IHI Engineering Review*, **18**(1) (1965) 1–8.

Int. J. Pres. Ves. & Piping **43** (1990) 285–300

Ductile Fracture Properties for Assessing Leak-Before-Break Issues in Ferritic Weldments

O. E. Lepik & B. Mukherjee

Fracture Mechanics Unit, Nondestructive and Fracture Evaluation Section, Metallurgical Research Department, Ontario Hydro, 800 Kipling Avenue, Toronto, Ontario, Canada M8Z 5S4

ABSTRACT

A Leak-Before-Break (LBB) approach is being used by Ontario Hydro's Darlington nuclear generating station as a design alternative to pipe rupture restraint hardware on the large diameter piping of the primary heat transport system. The J*-resistance curves of four different ferritic weldments, fabricated by either the submerged arc weld (SAW) or shielded metal arc weld (SMAW) process, were determined as part of this program.*

Results indicated that the as-welded and post-weld heat treated (PWHT) welds were susceptible to varying degrees of static or dynamic strain aging at 200 and 250°C. Dynamic strain aging effects were most significant for as-welded welds, as evidenced by sudden load drops on the load-displacement curves and ductile crack jumping. The effect of loading displacement rate and PWHT on toughness was assessed and related to the weld's tensile properties and susceptibility to dynamic strain aging. Implications of strain aging to LBB assessments are discussed.

1 INTRODUCTION

The Leak-Before-Break (LBB) approach is being used by Ontario Hydro's Darlington nuclear generating station (NGS) as a design alternative to pipe rupture restraint hardware on the large diameter piping of the primary heat transport system. Ontario Hydro's Darlington LBB approach is based on the elastic–plastic fracture mechanics (EPFM) method. In support of this program, a large material test program was completed. Crack growth resistance curves and tensile stress–strain curves were obtained from every

Int. J. Pres. Ves. & Piping 0308-0161/90/$03·50

heat of 22 in (558·8 mm) and 24 in (609·6 mm) NPS Sch 100 SA106B pipe used in the heat transport piping in the first two units of Darlington NGS. These results showed that crack plane orientation (circumferential versus axial) and test temperature (20°C versus 250°C) significantly affect the *J*-resistance curves.[1] In comparison, the effect of product sizes was secondary and non-systematic. All parent material tests showed ductile fracture behavior with rising *J*-resistance curves and high crack initiation toughness J_{Ic}.

Weld material properties were also evaluated for four Ontario Hydro weld procedures used for the different pipe sizes in the Darlington LBB program. The pipes considered were 22 in (558·8 mm) and 24 in (609·6 mm) NPS Schedule 100 pipes having a nominal wall thickness of 1·375 in (34·9 mm) and 1·531 in (38·9 mm), respectively. The welding procedures (based on pipe size, wall thickness, weld process, shop or field weld, class of pipe and type of material) were developed to satisfy the requirements of the ASME Boiler and Pressure Vessel Code, Section III, NB 2300. Two welding procedures, identified as PN-107 and PN-229, were used for girth-butt welds in the 22 in (558.8 mm) NPS Sch 100 pipe. Two other welding procedures, PN-108 and PN-232, were used for the 24 in (609·6 mm) NPS Sch 100 pipe. Welding procedures PN-107 and PN-108 use the shielded metal arc weld (SMAW) process, while weld procedures PN-229 and PN-232 use the single-wire submerged arc weld (SAW) process for weld fill-up. Also, weld procedures PN-108 and PN-232 employ a post-weld heat treatment (PWHT) in accordance with the requirements of the ASME Code, which requires a PWHT of welds in pipes having a nominal wall thickness greater than 1·5 in (38·1 mm).

J-tests on as-welded PN-107 and PN-229 weld specimens at 250°C showed that the load-displacement curves contained sudden drops in load, possibly due to strain aging. *J*-resistance curves and J_{Ic}-values of the PWHT PN-108 and PN-232 welds were significantly higher than those of the as-welded welds.[2] It was also observed that the toughness of the as-welded welds increased significantly after a suitable PWHT.[2] Based on these results, all non-PWHT welds within the scope of the Darlington LBB program were post-weld heat treated.

Qualification of ferritic welds in pipes which are less than 2·5 in (63·5 mm) thick is, according to the ASME Code, based on measurements of lateral expansion in Charpy specimens. The lateral expansions for non-PWHT welds were more than twice the minimum lateral expansion required by the ASME Code. From the consideration of lateral expansion and Charpy energy, PWHT welds were marginally tougher than the non-PWHT welds. Therefore, the issue at hand was whether it is at all necessary to stress relieve welds in 22 in (558·8 mm) NPS Sch 100 pipes in the balance of plant.

To address this issue, additional tests were conducted on both as-welded and PWHT welds. In particular, the effect on toughness of the material's susceptibility to strain aging and loading rate were examined in this study.

2 EXPERIMENTAL TEST METHOD

The welding procedures detailed in Table 1 were used for the girth-butt welds tested in this program. In addition, weld coupons welded by weld procedure PN-229 were also stress relieved at 620°C for 2 h to assess the effect of PWHT.

Compact tension C (T) specimens of three different sizes (19·0, 25·4 and 31·7 mm thick), conforming to the geometry recommendations of ASTM test procedure E813,[3] were used. Specimen size was determined by the

TABLE 1
Ontario Hydro Nuclear Weld Procedures

Ontario Hydro weld procedure	*Nominal pipe size in (mm)*	*Weld process*
PN-107	12 (304·8) 22 (558·8)	—GTAW[a] root pass (ER70S-2) —SMAW[b] fill up (E7018) —for field and shop welding —no PWHT
PN-108	24 (609·6)	—GTAW root pass (ER70S-2) —SMAW fill up (E7018) —for field and shop welding —PWHT[c]
PN-229	12 (304·8) 22 (558·8)	—GTAW root pass (ER70S-2) —SMAW 2nd and 3rd pass —SAW[d] fill up (Linde 29 wire and Linde 80 flux) —for shop welding only —no PWHT
PN-232	24 (609·6)	—GTAW root pass (ER70S-2) —SMAW 2nd and 3rd pass —SAW fill up (Linde 29 wire and Linde 80 flux) —for shop welding only —PWHT[c]

[a] Gas Tungsten Arc Weld.
[b] Shielded Metal Arc Weld.
[c] Post-Weld Heat Treatment: 620°C, hold 1 h/inch; 1 h minimum.
[d] Submerged Arc Weld.

largest C (T) specimens that could be machined from the welds in the 22 in (558·8 mm) and 24 in (609·6 mm) NPS Schedule 100 pipes. The specimens were machined from the weldments in the L-C orientation, i.e. the crack plane in the specimen was in the radial-circumferential plane with crack growth in the circumferential direction. Specimens were centred along the weld toe. The specimens were pre-cracked and sidegrooved after pre-cracking to promote flat fracture and a straight crack front. The average thickness reduction from sidegrooving was approximately 20%.

A single specimen technique[4] was used to determine the J-resistance (J–R) curves. A servohydraulic test machine, interfaced to a personal computer for test control and data acquisition, was used for testing. Specimens were tested at 200 and 250°C inside a box furnace, controlled to ±2°C. Selection of these test temperatures was based on earlier tensile tests[5,6] on these weld metals, which showed that the tensile ductility, as reflected by the per cent reduction

TABLE 2
Test Matrix

Test no.	*Weld procedure*	*Weld coupon*	*Test temperature (°C)*	*Loading displacement rate (mm/s)*
J23, J24, J25	PN-107	C	250	0·04
J89, J90, J91				
J48, J49, J50	PN-107	F	250	0·04
J4, J5, J6	PN-108	A	250	0·04
J40, J41, J42	PN-108	E	250	0·04
J33, J34, J35	PN-229	D	250	0·04
J84, J85	PN-229 (PWHT)	D	250	0·04
20	PN-229	2A	200	0·000 12
19	PN-229	2A	200	0·000 25
12	PN-229	1A	200	0·002 5
14	PN-229	1B	200	0·002 5
17	PN-229	2A	200	0·002 5
13	PN-229	1B	200	0·04
11	PN-229	1A	200	0·4
15	PN-229	1B	200	0·4
18	PN-229	2A	200	0·4
7	PN-229 (PWHT)	1B	200	0·002 5
8	PN-229 (PWHT)	1B	200	0·002 5
21	PN-229 (PWHT)	2A	200	0·002 5
5	PN-229 (PWHT)	1B	200	0·04
6	PN-229 (PWHT)	1B	200	0·4
22	PN-229 (PWHT)	2A	200	0·4
16	PN-229	1B	250	0·002 5
J9, J10, J11	PN-232	B	250	0·04

of area, was a minimum in this temperature range. Displacement was monitored by a displacement gauge mounted on razor blades at the load line of the specimen. A constant loading displacement rate, ranging from 0·00012 to 0·4 mm/s and from 0·008 to 0·017 mm/s, was used for the tests at 200 and 250°C, respectively. The entire test matrix is shown in Table 2.

Crack length was estimated by using either the unloading compliance method or the DC potential drop (DCPD) method. After testing, the specimens were heat tinted to mark the extent of crack extension. Most of the specimens were then chilled in liquid nitrogen prior to being broken into two to expose the crack surfaces. The initial crack length after pre-cracking and the final length were measured on the fracture surface using a nine point average method.[4] The fracture surfaces of selected specimens were then examined by scanning electron microscopy to determine the fracture mode in local regions.

In this program J–R curves have been developed which exceed the current ASTM crack extension and plane strain J_{max} limitation. Although the requirements of J-controlled growth formulated by Hutchinson and Paris[7] are violated, J–R curves associated with larger crack extension may be necessary for application to a given structural analysis. However, since sidegrooved and approximately full thickness specimens were used, these data will still be applicable for piping analysis beyond the plane strain limitation on the J-integral. The J–R curve was established by a power law regression fit of experimental data points from J_{Ic} to test completion, and was expressed by

$$J = C_1(\Delta a)^{C_2}$$

where Δa is the crack extension and C_1 and C_2 are constants. This equation has been used extensively to establish J–R curves for ferritic steels.[8] The value J_{Ic} was calculated using the procedure described in ASTM E813-81.

The chemical composition of the PN-229 weld metal was determined by emission spectroscopy. The total and soluble nitrogen concentrations of four PN-229 weld coupons in the as-welded and PWHT condition were also determined at the mid-thickness location of the welds. Total nitrogen concentration was analyzed by the LECO inert-gas fusion process, and soluble nitrogen concentration was determined by a hot extraction technique[9] in a hydrogen atmosphere.

3 RESULTS

3.1 Chemical analysis

The chemical composition of the PN-229 weld metal is reported in Table 3.

TABLE 3
Average Chemical Composition of PN-229 Weld Metal (wt %)

C	*Mn*	*Si*	*S*	*P*	*Al*	*V*	*Ti*	*Nb*
0·066	1·02	0·58	0·015	0·013	0·014	0·003	<0·001	0·021

Results of the nitrogen analyses in Table 4 indicate that PWHT reduced the soluble nitrogen content in the PN-229 welds.

3.2 Fracture toughness tests

The load–displacement records obtained from each *J*-test were examined for features that may be attributed to static or dynamic strain aging. Investigations elsewhere[10,11] have shown the presence of 'reload peaks' and load drops, which have been linked to static and dynamic strain aging, respectively. Under certain conditions, when the unloading compliance method is used to estimate crack length, the load–displacement curve can exhibit peaks in load after each unloading increment, as illustrated in the load–displacement trace of Specimen 7 in Fig. 1. The reload peak is characteristic of a material that has undergone strain aging during the time interval taken for an unloading compliance measurement. Under conditions which promote dynamic strain aging, the load–displacement record can display sudden drops in load (Fig. 1) similar to the serrations present on the uniform strain hardening part of the material's tensile flow curve.

All weld metals tested at 200 and 250°C showed varying degrees of static

TABLE 4
Total and Soluble Nitrogen Concentration in PN-229 Weld Metal

Coupon no.	*Average nitrogen concentration (ppm)*	
	Total	*Soluble*
AW-1A	108	16
PWHT-1A	105	10
AW-1B	82	8·5
PWHT-1B	72	0
AW-2A	103	15
PWHT-2A	102	11
AW-D	95	7
PWHT-D	109	6

AW: As-welded.

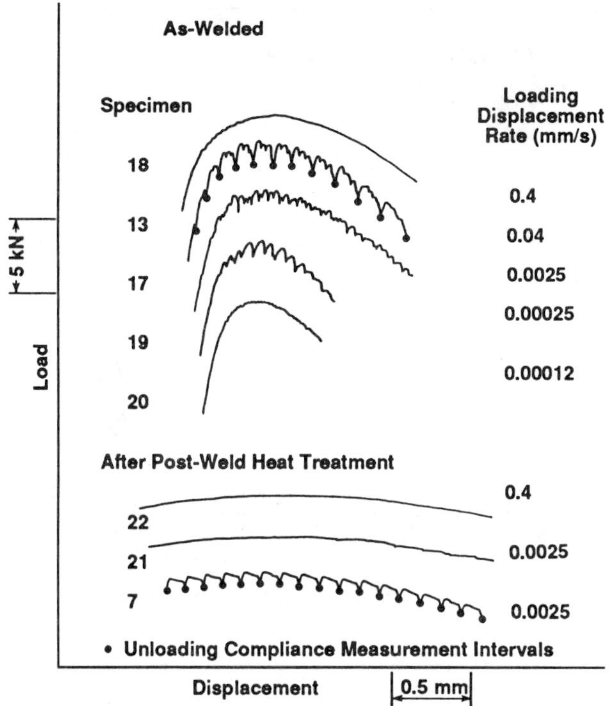

Fig. 1. Load–displacement curves in the vicinity of maximum load of PN-229 weld specimens tested at 200°C.

or dynamic strain aging. The load–displacement curves of all of the as-welded weld specimens showed numerous distinct load drops, while a small number of PWHT welds showed few small load drops. The load–displacement curves of all of the PWHT welds, for which the crack length was estimated by unloading compliance, exhibited reload peaks.

A series of J-tests on as-welded PN-229 welds showed that the loading displacement rate affected the size and number of load drops on the load–displacement curve (Fig. 1). In these tests, small serrations and load drops started prior to the specimen reaching maximum load. At the highest loading rate considered, 0·4 mm/s, small serrations occurred and disappeared soon after maximum load. As the loading rate was reduced, the number of load drops decreased and the size of the load drops increased, reaching a maximum size at 0·00025 mm/s. A further reduction in the loading rate to 0·00012 mm/s markedly decreased the size of the load drops.

Figure 2 shows the J_{Ic} initiation toughness values of the welds tested at 250°C. J_{Ic} appeared to be strongly affected by the weld metal's strain aging behavior. Welds exhibiting load drops had the lowest J_{Ic} values. Welds having only reload peaks had correspondingly higher J_{Ic} values and welds showing no discernible load drops or reload peaks had the highest J_{Ic} values.

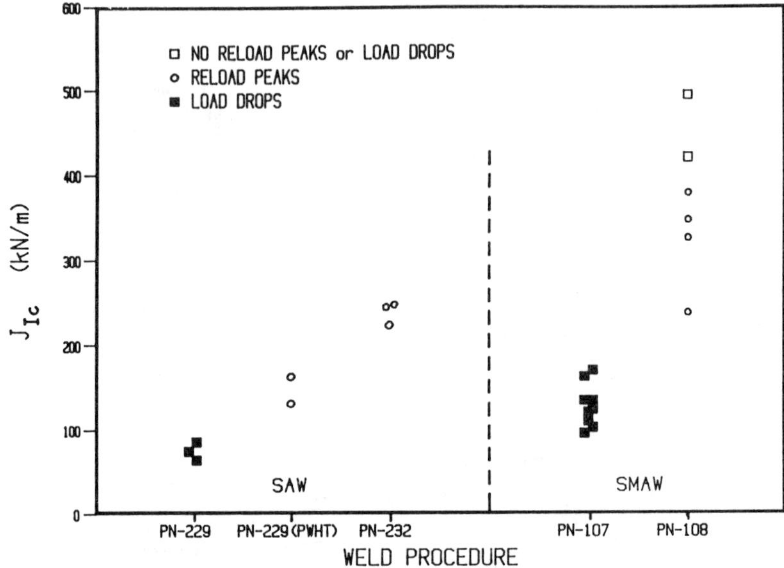

Fig. 2. Comparisons of J_{IC} for welds tested at 250°C. Features of the load–displacement records are also shown.

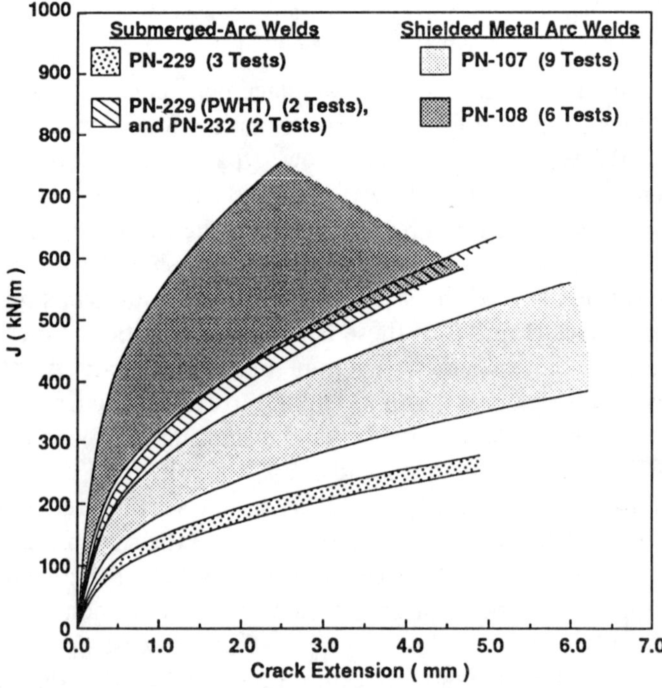

Fig. 3. Comparison of J-resistance curves of welds tested at 250°C.

The effect of PWHT on the J_{Ic} and J-resistance curves of welds tested at 250°C are shown in Figs 2 and 3, respectively. Welds made by the SAW and SMAW processes showed similar trends in fracture behavior. The as-welded welds (PN-107 and PN-229) had lower J_{Ic} values and J–R curves than similar welds that had received a PWHT (PN-108, PN-229 (PWHT) and PN-232). In the case of the PN-229 welds, J_{Ic} was increased two-fold after PWHT.

The J-resistance curves of the PN-229 welds tested at 200°C are shown in Fig. 4. The J–R curves were increased by PWHT and were dependent on the loading displacement rate. For the range of loading displacement rates considered, increasing the loading rate increased the J–R curves of the as-welded and PWHT welds. The loading rate effect on J–R curve toughness was smaller than the effect of PWHT.

3.3 Fractographic features

Fracture surfaces of the as-welded weld specimens, which were monotonically loaded in tension, were characterized by a series of macroscopic ridges

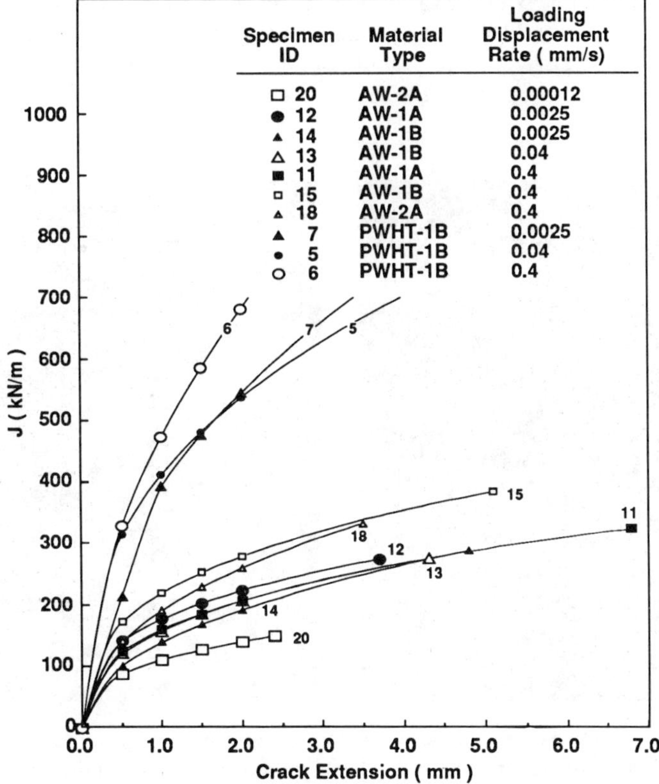

Fig. 4. J-resistance curves of PN-229 welds tested at 200°C in the as-welded condition and after PWHT. (Symbols do not represent experimental points.)

Fig. 5. Fracture surfaces of PN-229 weld specimens tested at 200°C. (a) As-welded condition; (b) after post-weld heat treatment. Arrow indicates direction of macroscopic crack propagation.

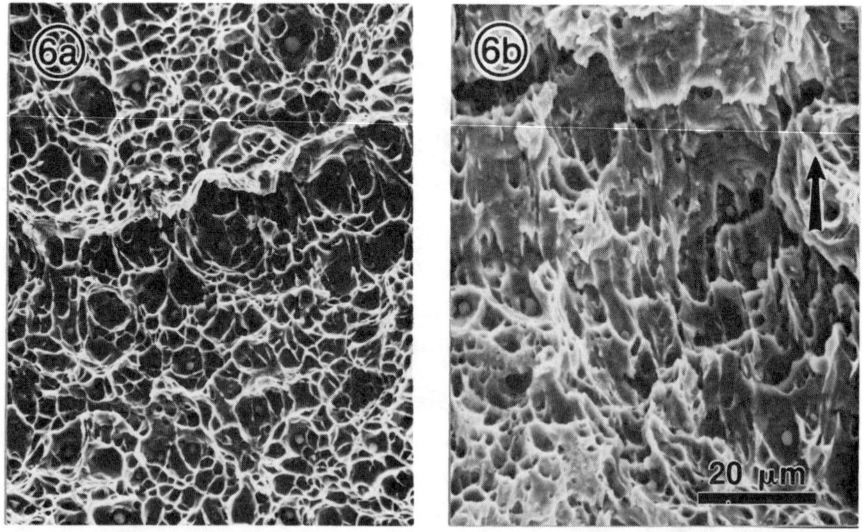

Fig. 6. Representative dimple features of PN-229 weld specimens tested at 200°C. (a) As-welded condition; (b) after post-weld heat treatment. Arrow indicates direction of macroscopic crack propagation.

that were oriented perpendicular to the direction of crack propagation (Fig. 5(a)). The ridges were essentially parallel to the initial fatigue crack front and nearly continuous from one sidegroove to the other. The different microstructural constituents of individual weld passes did not appear to affect the shape or continuity of the ridges. The PWHT weld specimens generally contained large rough regions interspersed with large secondary cracks that extended into the primary crack surface (Fig. 5(b)). Ridges of the kind present on the as-welded weld specimen fracture surfaces were not present on the fracture surfaces of the PWHT weld specimens.

Microscopically, fracture surfaces of the as-welded and PWHT weld specimens exhibited a topography typical of dimple rupture. Differences existed between the fracture surface features of the as-welded and PWHT weld specimens. In the case of the PN-229 weld specimens, the fracture surfaces consisted primarily of equi-axed and tear dimples with small regions of elongated dimples and stretched areas (Fig. 6(a)). After PWHT, the regions or bands of elongated dimples were more numerous and wider (i.e. in the direction of crack growth) (Fig. 6(b)). In addition, a larger number of secondary cracks were present on the PWHT weld specimen fracture surfaces.

4 DISCUSSION

4.1 Influence of strain aging on fracture properties

The C–Mn welds tested in this program at test temperatures approaching the operating temperature of the primary heat transport system were susceptible to strain aging. In the case of the PN-229 welds, tensile tests[6] revealed that, under comparable applied strain rates, welds in the as-welded condition were more susceptible to jerky or serrated flow than after PWHT. The serrated flow is a manifestation of dynamic strain aging and occurs when the rate of straining and the temperature are such that the solute atmospheres of nitrogen and carbon can diffuse to pin the mobile dislocations.[12] Repeated breakaway and repinning of mobile dislocations by the interstitials is known to cause the serrations in the stress–strain curve.

The different susceptibilities of the welds to static and dynamic strain aging was related to the concentration of interstitial solutes present in the as-welded and PWHT welds. With average soluble nitrogen concentrations ranging from 7 to 16 ppm, the as-welded PN-229 welds were very susceptible to dynamic strain aging at 200 and 250°C. Welds in the as-welded condition can be expected to be more susceptible to dynamic strain aging because the high cooling rates during welding promote non-equilibrium conditions, which lower the effectiveness of carbide and nitride formers in removing soluble nitrogen and carbon. Furthermore, this susceptibility can be even greater in welds that have not been adequately shielded from the atmosphere or that do not have sufficient nitride- or carbide-forming elements. The welds in the PWHT condition (PN-108 and PN-232) were only slightly susceptible to dynamic strain aging because the heat treatment reduced the concentration of soluble nitrogen, as shown in the case of the post-weld heat treated PN-229 weld metals. PWHT at temperatures of 620°C is sufficient to precipitate nitrides and carbonitrides by combining with Al, Ti, V and Si, thus lowering the free interstitial concentration.[13,14] The PWHT welds were, however, susceptible to static strain aging because the PWHT was not effective in completely removing the detrimental interstitial solutes from solid solution.

In the *J*-tests the load–displacement curves of all the as-welded welds exhibited numerous sudden drops in load while those of the PWHT welds showed few or no load fluctuations. Load drops of this nature have been reported by Miglin *et al.*[10] during unloading compliance *J*-tests on a pressure vessel steel tested at 121 and 204°C. The behavior was attributed to the material's susceptibility to dynamic strain aging.

The load drops on the load–displacement curves were related to the mode of tearing in the as-welded welds. For those specimens exhibiting large distinct load drops, a close correspondence was noted between the number of load drops and the number of ridges present on the specimen's fracture surface. In fact, tests conducted by Marschall *et al.*[15] suggest a connection between materials susceptible to dynamic strain aging tested at light water reactor temperatures, and crack jumps. Load drops corresponding to crack jumps of 6–100 mm in length were observed in the testing of a 711-mm diameter through-wall circumferentially-cracked pipe at 288°C. In this case the piping material, which was made to ASTM Specification A155 from ASTM A516 Gr 70 ferritic steel, was known to be susceptible to dynamic strain aging. With the stiff load-train used in the present test program, it is believed that the loading conditions were favorable for the formation of small crack jumps, which were manifest as small closely-spaced ridges on the specimen fracture surface.

The mode of fracture in the as-welded welds at temperatures of 200 and 250°C, consisting of small crack jumps, appeared to be directly related to the material's susceptibility to dynamic strain aging. This view is supported by the fact that the same weld material tested at 20°C did not exhibit load drops, and similar PWHT welds tested at elevated temperature showed few or no load drops. The ductile crack jumping appeared to be related to the effect of dynamic strain aging on the material's slip characteristics. The main effect of dynamic strain aging on the ductile fracture process has been considered by Brindley[16] to reduce the necking strain in mild steel by promoting the linking up of voids. Evidence of flow localization was seen on the fracture surfaces of the as-welded welds; when compared with similar PWHT welds, the as-welded welds contained narrower bands of elongated shear dimples, and fewer, smaller stretched regions.

Post-weld heat treatment markedly increased the J_{Ic} and crack growth resistance of the as-welded welds tested at 200 and 250°C in this program. This improvement in fracture toughness was attributed to the reduced tensile strength and increased ductility of the PWHT welds, as shown by the tensile tests, and to the reduced susceptibility of the welds to dynamic strain aging. In reducing the susceptibility to dynamic strain aging, PWHT also minimized the crack jumping behavior prevalent in the as-welded welds.

The susceptibility of the as-welded and PWHT welds to dynamic strain aging was also manifest by the loading displacement rate dependence of the material's toughness. J_{Ic} and J-resistance curve toughness decreased with a decrease in loading displacement rate, i.e. strain rate, for both as-welded and PWHT PN-229 welds tested at 200°C. This behavior was consistent with the tensile tests conducted on as-welded and PWHT PN-229 welds at 200°C, which showed that the ultimate tensile strength increased and material ductility (% reduction in area) decreased with decreasing strain rate.

4.2 Implication to LBB assessment

In a LBB assessment of a structure, load bearing capacity and stability of a structure are determined in the presence of a hypothetical flaw. Charpy tests, which are conducted at high loading rates using uncracked samples, do not reflect material response of cracked samples. Therefore, it is entirely appropriate that decisions on post-weld heat treatment of welded structures, where protection against failure is needed from real or hypothetical cracks, should be based on toughness obtained from cracked specimens.

Test results showed that J-resistance curves and J_{Ic} increase significantly with PWHT. This increase in toughness is accompanied by a decrease in the yield and ultimate strength and a general lowering of the stress–strain curve. A recent closed form calculation[17] on a circumferentially cracked pipe has shown that an increase in J-resistance curves does not alter instability loads significantly. The crack driving force of J applied also increases due to the material's reduced tensile properties and balances out the increase in J material.

PWHT appears to minimize dynamic strain aging in welds that were examined in this work. Load drops and sudden crack jumps have been observed in welds and parent materials when they are susceptible to dynamic strain aging. Conversely, load drops and sudden crack jumps have not been observed in specimens and pipes which have a low susceptibility to dynamic strain aging. PWHT may not alter safety factors based on instability loads, but it will change the crack extension behavior.

The dependence of *J–R* curve toughness on the loading displacement rate emphasizes the point that when a strain-rate sensitive mechanism is operative, such as dynamic strain aging over the temperature range 100–300°C in ferritic steels, the material should be tested under loading rate conditions that are representative of those imposed or postulated for the structure evaluated by an elastic–plastic fracture mechanics methodology. Alternatively, if the structural loading conditions are not known with certainty then loading conditions that yield lower bound material properties should be used for material evaluation purposes. In fact, the test conducted at the lowest loading displacement rate in this test program yielded a *J–R* curve that was lower than that obtained using loading displacement rates recommended by ASTM standard test specification E813 or E1152, and indicates clearly that toughness differences may exist between laboratory test and service conditions.

5 CONCLUSIONS

(1) At temperatures approaching primary heat transport system temperatures, as-welded C–Mn ferritic steel submerged arc and shielded metal arc weldments tested in the Darlington Leak-Before-Break program were susceptible to strain aging. Under temperature and loading displacement rate conditions where dynamic strain aging is operative, the materials exhibited small crack jumps and a marked reduction in toughness (J_{Ic} and *J*-resistance curve).

(2) Post-weld heat treatment increased the toughness and nearly eliminated the crack jumping behavior of the as-welded weldments that were susceptible to dynamic strain aging. The improved toughness (J_{Ic} and *J*-resistance curve) at 200 and 250°C was caused partly by a reduction in dynamic strain aging susceptibility from a lowering of the soluble nitrogen content in the weld metal.

(3) Toughness of the as-welded and PWHT weld metal depends on the applied strain rate. Decreasing the loading displacement rate reduced J_{Ic} and the *J*-resistance curves of the submerged arc (PN-229) welds at 200°C. The increased tensile strength at lower strain rates from dynamic strain aging was responsible for the reduced toughness.

ACKNOWLEDGEMENTS

The authors would like to thank all the people who contributed to the Darlington LBB material test program.

REFERENCES

1. Mukherjee, B., The *J*-resistance curve Leak-Before-Break test program on material for the Darlington Nuclear Generating Station. *Int. J. Pres. Ves. & Piping*, **31** (1988) 363–85.
2. Mukherjee, B., Observations on the effect of post-weld heat treatment on *J*-resistance curves of SA-106B seamless piping welds. *Nuclear Engineering and Design*, **111** (1989) 63–75.
3. Standard test method for J_{Ic}, a measure of fracture toughness. ASTM Designation E813-81, Annual Book of ASTM Standards, Vol. 03.01, 1985.
4. Standard test method for determining *J–R* curves. ASTM Designation E1152-87, Annual Book of ASTM Standards, Vol. 03.01. American Society for Testing and Materials, Philadelphia, PA, 1988.
5. Ho, E. T. C. & McGraw, M. J., Tensile properties of ASME SA106 Grade B and SA105 Darlington primary heat transport piping material at 293 and 523 K. Ontario Hydro Research Division Report No. 86-11-K, July 1986.
6. Ho, E. T. C. & McGraw, M. J., Tensile properties of AW and PWHT PN-229 weld materials in the temperature range 20–300°C under constant applied strain rate conditions. Ontario Hydro Research Division Report No. 87-319-K, January 1988.
7. Hutchinson, J. W. & Paris, P. C., Stability analysis of *J*-controlled crack growth. In *Elastic–Plastic Fracture, ASTM STP 668*, ed. J. D. Landes, J. A. Begley & G. A. Clarke, American Society for Testing and Materials, Philadelphia, PA, 1979, pp. 37–64.
8. Loss, F. J., Monke, B. H., Hiser, A. L. & Watson, H. E., *J–R* curve characterization of irradiated low-shelf nuclear vessel steels. In *Elastic–Plastic Fracture: Second Symposium, Vol. II—Fracture Resistance Curves and Engineering Applications. ASTM STP 803*, ed. C. F. Shih & J. P. Gudas. American Society for Testing and Materials, Philadelphia, PA, 1983, pp. II-771–95.
9. Kawamura, K., Otsubo, T. & Mori, T., Determination of Nitride and dissolved nitrogen in steel by hydrogen hot extraction. *Trans. ISIJ*, **14** (1974) 347–56.
10. Miglin, M. T., Van Der Sluys, W. A., Futato, R. J. & Domian, H. A., Effects of strain aging in the unloading compliance *J*-test. In *Elastic–Plastic Fracture Toughness Test Methods: The User's Experience, ASTM STP 856*, ed. E. T. Wessel & F. J. Loss, American Society for Testing and Materials, Philadelphia, PA, 1985, pp. 150–65.
11. Feldstein, J. G., Bloom, J. M. & Van Der Sluys, W. A., Repair welding of heavy section steel component in LWRs, EPRI Report No. NP-3614, Vol. 2, July 1984.
12. Cottrell, A. H. & Bilby, B. A., Dislocation theory of yielding and strain ageing of iron. *Proc. Phys. Soc.*, **62A** (1949) 49–62.

13. Houssin, B., Slama, G. & Moulin, P., Strain aging sensitivity of pressure vessel steels and welds of nuclear reactor components. In *1st Int. Seminar on Assuring Structural Integrity of Steel Reactor Pressure Vessels*, ed. L. E. Steele, & K. E. Stahlkopf, Berlin, Germany, August 1979, pp. 57–67.
14. Leslie, W. C. & Rickett, T. L., Influence of aluminium and silicon deoxidation on the strain aging of low-carbon steels, *J. Metals, Trans. AIME*, **197** (1953) 1021–31.
15. Marschall, C. W., Landow, M. P. & Wilkowski, G. M., Effect of dynamic strain aging on fracture resistance of carbon steels operating at light-water-reactor temperatures. Presented at 21st National Symposium on Fracture Mechanics, Annapolis, Maryland, June 28–30, 1988, to be published by ASTM, American Society for Testing and Materials.
16. Brindley, B. J., The effect of dynamic strain-aging on the ductile fracture process in mild steels. *Acta Met.*, **18** (1970) 325–9.
17. Pereira, J. K., Uncertainties that can arise in the application of *J*-integral fracture analysis methods. Presented at OECD Meeting on NDE and Fracture Mechanics in Component Assessment, Wusenlingen, Switzerland, October, 1988, to be published by OECD, Organization for Economic Cooperation and Development.

Int. J. Pres. Ves. & Piping **43** (1990) 301–316

Ductile Crack Growth of Semi-elliptical Surface Flaws in Pressure Vessels

W. Brocks, H. Krafka, G. Künecke & K. Wobst

Bundesanstalt für Materialforschung und -prüfung (BAM),
Unter den Eichen 87, D-1000 Berlin 45, Germany

ABSTRACT

Experiments on ductile crack growth of some axial surface flaws in a pressure vessel have revealed the well-known canoe shape, i.e. a larger crack extension has occurred in the axial direction than in the wall thickness direction. Two tests have been analyzed by finite element calculations to obtain the variation of the J-integral along the crack front, and the stress and strain state in the vicinity of the crack. The local crack resistance depended on the local stress state. To predict ductile crack extension correctly, J_R-curves have to account for the varying triaxiality of the stress state along the crack front.

INTRODUCTION

Crack initiation and stable crack growth in ductile materials are usually described by J-resistance curves which are obtained from standard specimens. Whereas the initiation value, J_i, is found to be independent of the specimen geometry, $J(\Delta a)$-curves may vary with the shape and size of the specimens.[1,2] No criteria exist for applying these curves to surface flaws in components. A leak-before-break analysis requires reliable crack resistance data up to 100% of the remaining ligament, which is beyond any accepted condition of J-control. The analysis also needs a description of how the crack shape will develop during stable growth. It is commonly assumed that the flaw remains geometrically similar, i.e. the aspect ratio a/c is constant while the crack grows. But evidence exists from experiments by Pugh[3] and Milne[4] that the flaw shape may develop quite differently and expand in the longitudinal direction under the surface more than in the wall thickness direction. The consequences for a leak-before-break analysis are evident.

Int. J. Pres. Ves. & Piping 0308-0161/90/$03·50

The present contribution reports on experimental investigations by Wobst and Krafka[5] and elastic–plastic finite element (FE) analyses by Brocks and Künecke[6] of test vessels containing an axial surface crack which have been performed for two different materials and crack configurations to analyze the transferability of J_R-curves from specimens to structures. The variation of the J-integral along the crack front as well as the stresses in the vicinity of the crack tip have been evaluated in order to study the dependency of the local crack resistance on the local stress state. Continuing earlier work by Brocks and Noack,[7] the ratio of the hydrostatic to the von Mises effective stress is introduced to characterize the triaxiality of the stress state. By introducing triaxiality-dependent resistance curves, a more realistic assessment of the local ductile crack growth can be given.

TEST MATERIALS

The test materials were two German standard steels of the types 20 MnMoNi 5 5 and StE 460. The chemical composition and the mechanical properties of these steels are given in Table 1. Crack resistance curves of both steels, which were determined on different types of specimen, are shown in Figs 1 and 2.

The J_R-curve of the steel 20 MnMoNi 5 5 in Fig. 1 was obtained from 20% side-grooved compact tension specimens of thickness $B = 25$ mm (CT25sg) at a temperature of 22 ± 2°C according to the standard test procedures of

TABLE 1

Chemical Compositions (Per Cent Mass) and Mechanical Properties (T-Orientation, 20 ± 2°C, $\dot{\varepsilon} = 2 \times 10^{-4}\,s^{-1}$) of the Investigated Steels

Steel	*C*	*Mn*	*Mo*	*Ni*	*Si*	*Cr*	*V*	*Cu*	*Al*	*P*	*S*	N_2
20MnMoNi55	0·19	1·43	0·49	0·56	0·22	0·14	—	0·07	0·023	0·009	0·009	—
StE 460	0·17	1·52	0·01	0·62	0·28	0·04	0·18	0·03	0·010	0·009	0·009	0·012

			20MnMoNi55		*StE 460*	
			Sheet	*Round*	*Sheet*	*Round*
Lower yield point	R_{eL}	(MPa)[a]	459	466	480	490
Ultimate tensile strength	R_m	(MPa)	604	609	647	647
Elongation	A_{10}	(%)	17	16	19	20
Reduction of area	Z	(%)	63	66	67	64

[a] 1 MPa = 1 N/mm^2.

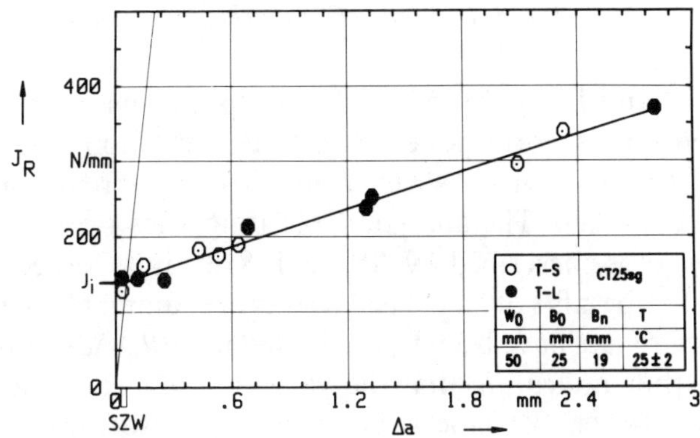

Fig. 1. J_R-curve of the steel 20 MnMoNi 5 5 obtained at a temperature of 22 ± 2°C from 20% side grooved CT25 specimens in T-L and T-S orientation (SZW = stretch zone width).

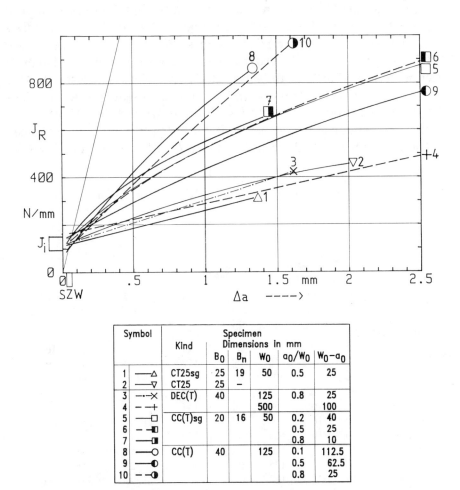

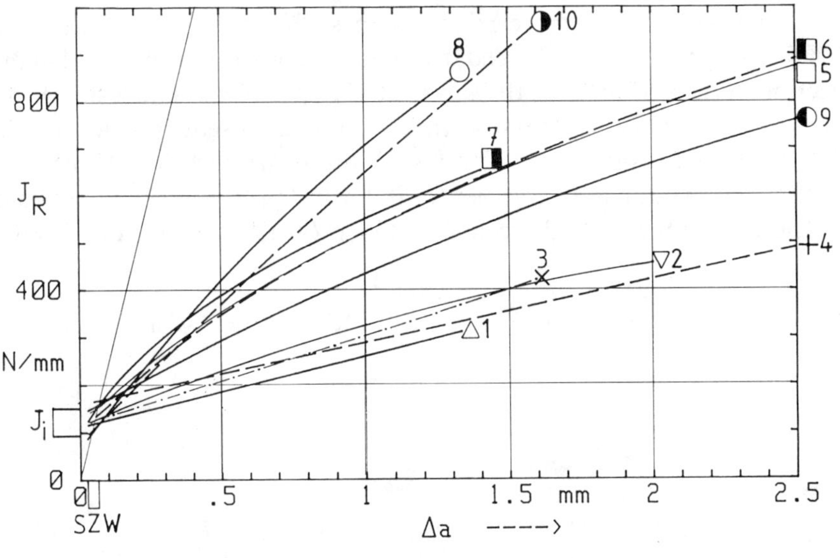

Symbol		Kind	B_0	B_n	W_0	a_0/W_0	W_0-a_0
			Specimen Dimensions in mm				
1	—△	CT25sg	25	19	50	0.5	25
2	—▽	CT25	25	–			
3	-·-×	DEC(T)	40		125	0.8	25
4	– –+				500		100
5	—□	CC(T)sg	20	16	50	0.2	40
6	– –◧					0.5	25
7	—◨					0.8	10
8	—○	CC(T)	40		125	0.1	112.5
9	—◐					0.5	62.5
10	– –◑					0.8	25

Fig. 2. J_R-curves of the steel StE 460 obtained at a temperature of 22 ± 2°C from different specimens, CT25sg, CT25, DECT and CCT, in T-L orientation.

ASTM E 813[8] and DVM-Merkblatt.[9] The specimens were cut in the T-L and T-S orientations, corresponding with the definition given in ASTM E 399,[10] out of that part of the sheet from where the circular inserts for the vessel tests were taken. The analytical blunting lines were calculated from two formulas; those described by ASTM E 813[9] and Cornec, Heerens and Schwalbe.[11] The best fits of all the test results are summarized in Table 2. No significant differences have been found between the $J_R(\Delta a)$-values of the two orientations, thus indicating that the varying stress state along the crack front will cause the crack shape, rather than any anisotropy of the material.

The J_R-curves of the steel StE 460, which were determined on centre-cracked tensile panels (CCT), on double-edge cracked tensile panels (DECT) and on compact tension specimens CT25, are compared in Fig. 2. The blunting line in this figure was also established from an analytical approach by Cornec, Heerens and Schwalbe.[11] More details of these tests are given in Ref. 12. The more gradual slopes of the J_R-curves of the CT and DECT specimens are produced by greater constraints in the ligament direction in these specimens than in the CCT specimens. The result that the slopes of the J_R-curves increase in the demonstrated sequence accords with other published results.[13–17] The results are summarized in Table 2.

Table 2 shows different J_i-values for initiating a stable crack in both steels for differently dimensioned specimens. These J_i-values result from J_R-curves

TABLE 2

Fracture Resistance of Various Specimens; Results from J_R-Curve Testing

Specimen	*B (B_n) (mm)*		*W (mm)*	*a/W*	*J_i (N/mm)*	*dJ/da (N/mm²)*		
						Δα = 0·1	*1·0*	*2·0*
Material 20MnMoNi55								
CTsg	25	(19)	50	0·5	139	85	83	81
CT	25		50	0·5	115	143	149	157
Material StE 460								
CTsg	25	(19)	50	0·5	117	146	153	161
CT	25		50	0·5	122	244	171	91
DECT	40		500	0·8	148	147	140	132
DECT	40		125	0·8	123	183	192	203
CCTsg	20	(16)	50	0·2	141	521	296	217
CCTsg	20	(16)	50	0·5	127	544	308	225
CCTsg	20	(16)	50	0·8	130	742	277	194
CCT	40		125	0·1	94	746	506	239
CCT	40		125	0·5	149	323	265	201
CCT	40		125	0·8	100	592	541	486

B_n = net thickness after side grooving (sg); W = specimen width.

and values for the stretch zone width, SZW, which were determined either as averaged results of several tests or—in a few cases—of a single test according to the single specimen method. The scatter of J_i does not imply that the steels show different initiation values depending on the specimen geometry. All that may be concluded is that different J_i-values have been determined in single tests because the test conditions were not perfectly reproducible with respect to the material state or loading.

EXPERIMENTAL PRESSURE VESSEL TESTS

Figure 3 shows the dimensions of the test vessel and the location of the outer surface notch. The test flaw was located in a round of 650 mm diameter which has been welded into the middle of the cylindrical part of the vessel. The vessel was annealed for stress relief at a temperature of 580 ± 20°C for 90 min after welding.

The flaw was parallel to the axis of the vessel and to the longitudinal rolling direction of the steels 20 MnMoNi 5 5 and StE 460. In the following, the two vessel tests will be called VT1 and VT2 (Table 3). Lengths and depths of the machined notches, which were $2c_0 = 180{\cdot}4$ mm, $a_0 = 15{\cdot}0$ mm for VT1 and $2c_0 = 190{\cdot}7$ mm, $a_0 = 20{\cdot}2$ mm for VT2, respectively, were designed so that initiation and stable crack growth could occur from a 5 mm (VT1) and 7·5 mm (VT2) fatigue crack at a pressure lower than the plastic collapse load.

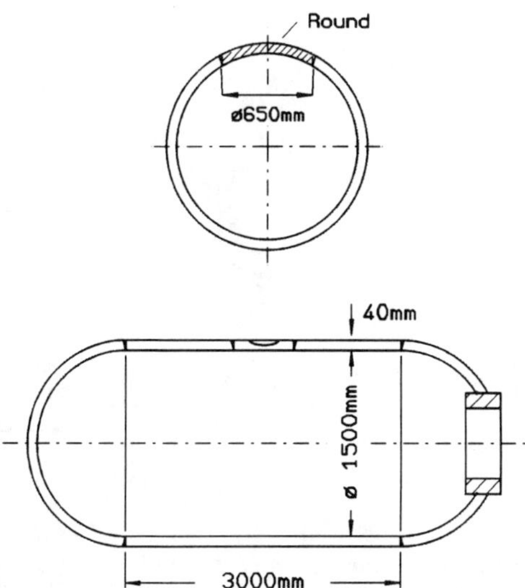

Fig. 3. Test vessel with axial surface flaw.

TABLE 3
Vessel Tests: Geometry, Material Data and Loading

Vessel test		*VT1*	*VT2*
Pressure vessel			
Inner radius, r_i	(mm)	750	750
Length of cylindrical part, L	(mm)	3 000	3 000
Wall thickness t	(mm)	40	39·8
t/r_i		0·053	0·053
Surface flaw			
Length $2c$	(mm)	180·4	192·1
Depth a	(mm)	21·6	28·0
a/c		0·239	0·292
a/t		0·540	0·705
Folias-factor m_F		1·156	1·352
Material (round)		20MnMoNi55	StE 460
Yield strength, σ_y	(MPa)	460	490
Ultimate stress, σ_u	(MPa)	609	647
Hardening exponent of Ramberg-Osgood power law, n		7·1	7·4
Fracture initiation, J_i	(N/mm)	120	110
Internal pressure at initiation, p_i	(MPa)	22·4	20·0
Maximum, p_{max}	(MPa)	24·2	22·36
Calculated yield load, p_y	(MPa)	17·5	21·0
Collapse load p_F	(MPa)	21·2	19·2
p_{max}/p_F		1·14	1·16
p_y/p_F		0·83	1·09
p_{max}/p_y		1·38	1·06

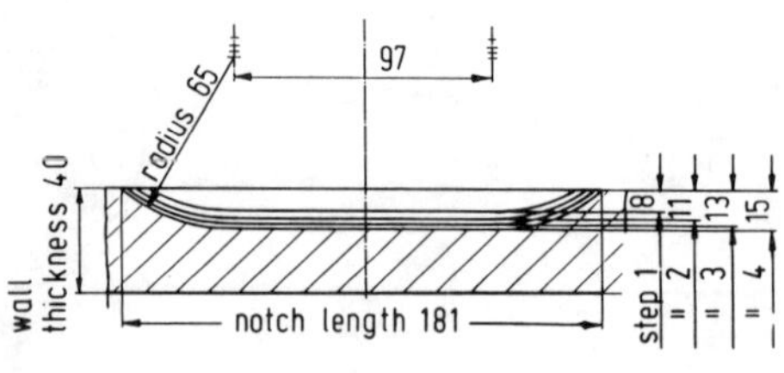

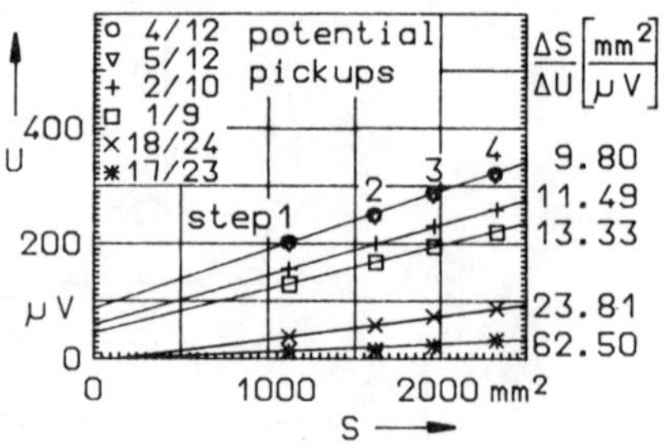

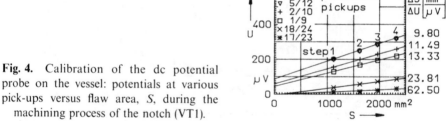

Fig. 4. Calibration of the dc potential probe on the vessel: potentials at various pick-ups versus flaw area, S, during the machining process of the notch (VT1).

This was done by J-assessments based on the Raju and Newman[18] formulas and tables including a small-scale yielding correction.[19] The tests were performed at a temperature of about 21°C following the single specimen procedure.

The global crack growth, fatigue cracking as well as stable crack growth, was measured by the dc potential drop method. The sensitivity of the potential probe was calibrated before by measurements in the course of the notch machining (Steps 1–4 in Fig. 4) showing that there is a linear relationship between the potential, U, and the area, S, of the flaw; but no local crack extension can be measured by this method. Figure 5 shows the potential pick-up positions on the vessel surface and the positions of the clip gages, which measured the notch opening displacements, V, close to the vessel surface at different points due to the internal pressure, for both test configurations.

Figure 6 is a record of the test conditions during fatigue precracking of VT1 by plotting the potential, U, versus the number of cycles, N. The amplitude of the pulsating pressure, $\Delta p = p_U - p_L$, was large, i.e. 15·9 MPa, in the beginning to start the fatigue crack from the machined notch. With increasing fatigue crack growth, the lower value, p_L, was raised to 7·5 MPa, thus reducing the amplitude to 8·9 MPa.

Potential versus notch opening displacement at the clip gage position 1 (see Fig. 5) during the static test VT1 is recorded in Fig. 7. A change of the slope in the U versus V curve indicates the initiation of stable crack growth and thus renders the corresponding value of the total COD at initiation, V_{ti}. The end of the static test phase is denoted by the subscript 'max'. The pressure versus COD curves for both tests are shown in Fig. 8. The pressure at initiation, p_i, follows from V_{ti} in Fig. 7.

At the end of the stable crack growth in VT1 and VT2, the flawed rounds were cut out and broken by tensile loading at a temperature of −120°C. The crack surfaces are displayed in Fig. 9, showing that the intended extensions of the crack have been realized, both with respect to the mean crack extensions and to the crack shapes. The evaluation of the crack shape of VT2 is restricted to one half, only, as the other one has been reserved to investigate the crack tip opening displacement at different crack front positions by cuts perpendicular to the flaw surface. The maximum crack growth does not occur perpendular to the vessel surface but in the axial direction. The local crack growth, Δa, perpendicular to the fatigue crack front, which has been obtained in both tests at p_{max}, can be measured from Fig. 9; it is plotted against the crack front angle, Φ in Fig. 10. Φ is 0° at the deepest point of the crack and $\pm 90°$ at the penetration points of the crack front at the outer surface of the vessel. The maximum Δa along the crack front is reached at $\Phi \simeq 70°$ for VT1 and $\Phi \simeq 80°$ for VT2. The crack

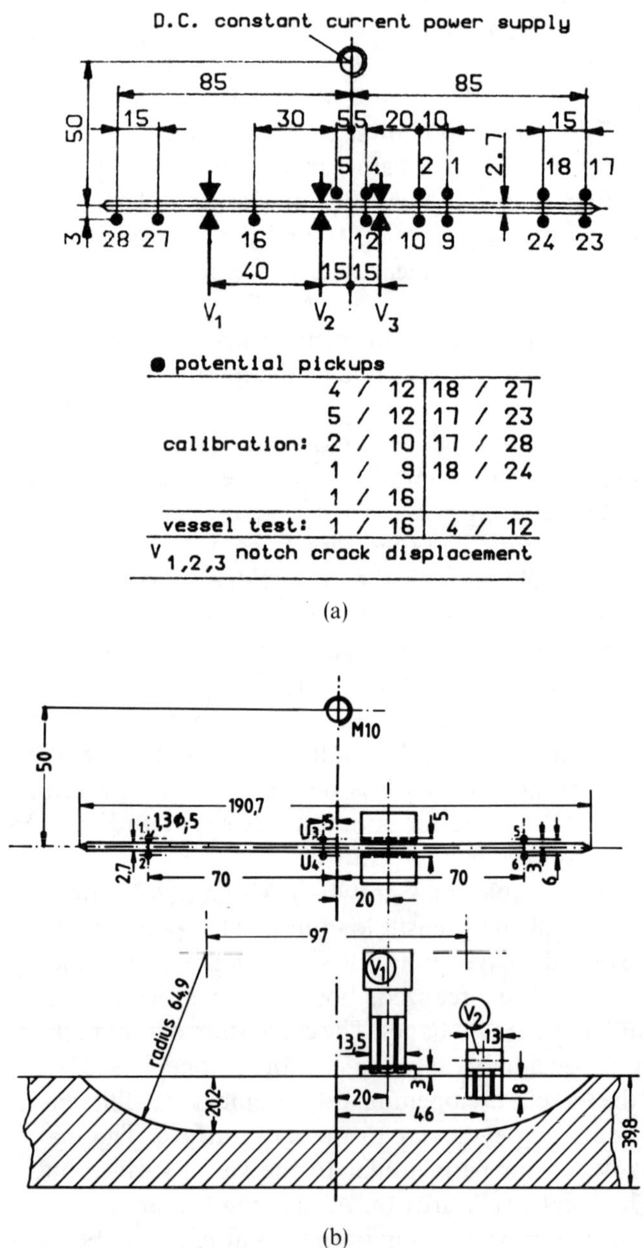

Fig. 5. Clip gage positions for the measurement of notch opening displacements, V, and pick-ups for the dc potential measurement at the surface flaw: (a) VT1; (b) VT2.

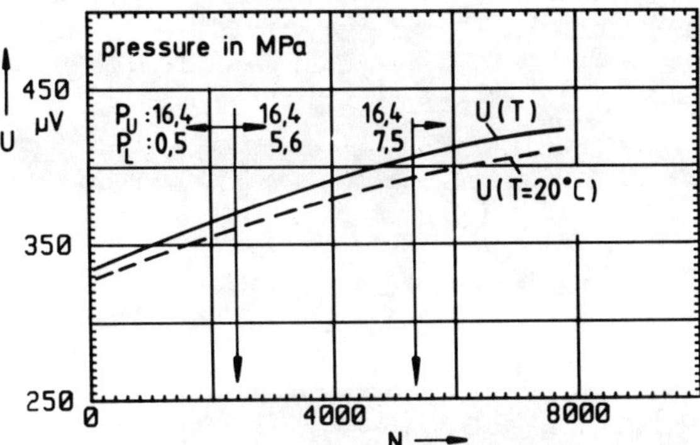

Fig. 6. Control diagram for the first fatigue precracking: potential versus number of cycles, upper and lower pressure, p_U and p_L, in the course of oscillating loading of the vessel (VT1).

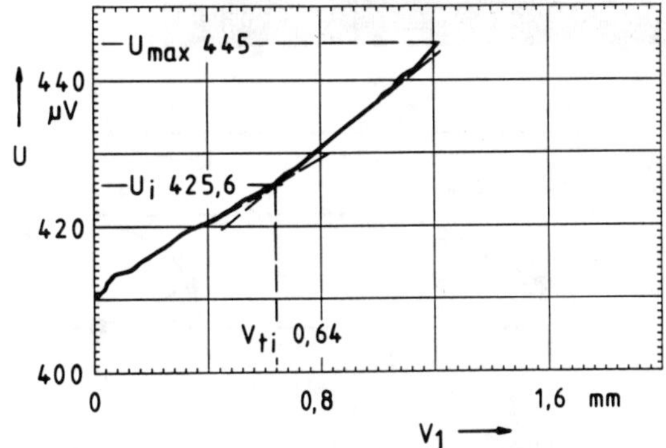

Fig. 7. Static loading of the vessel with the first fatigue crack: measured potential versus notch opening displacement (VT1).

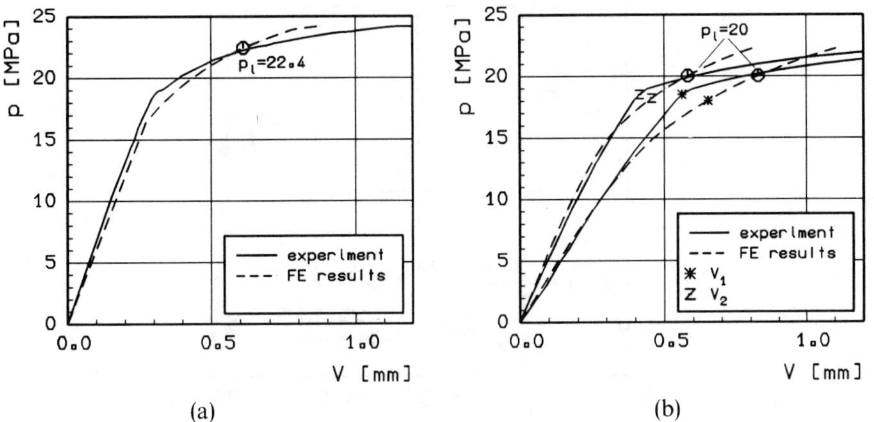

Fig. 8. Static loading of the vessel: internal pressure versus notch opening displacement, comparison of experimental and numerical results: (a) VT1; (b) VT2.

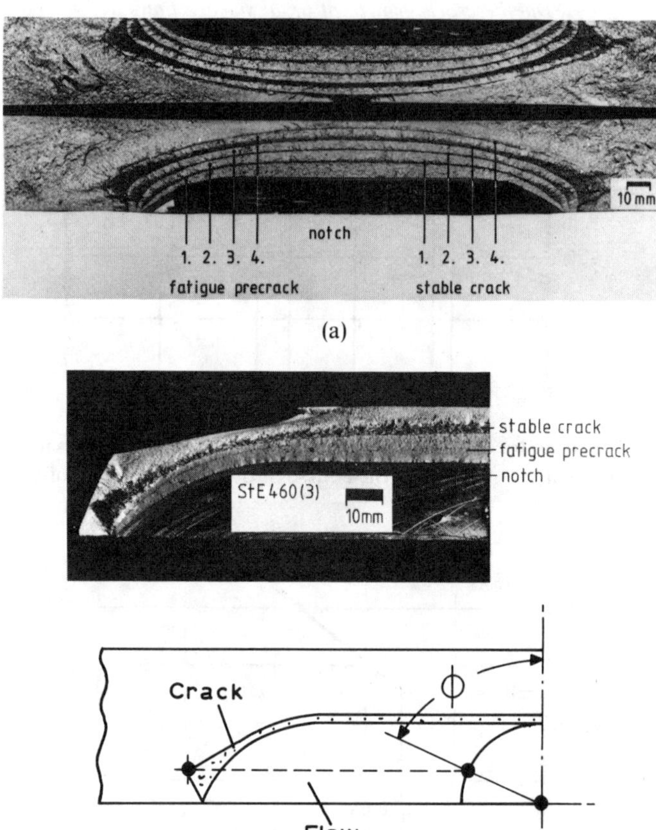

Fig. 9. Crack surfaces of the vessel tests: (a) Alternate phases of fatigue cracking and stable crack growth in VT1 (20 MnMoNi 5 5); (b) fatigue crack and ductile crack in VT2 (StE 460).

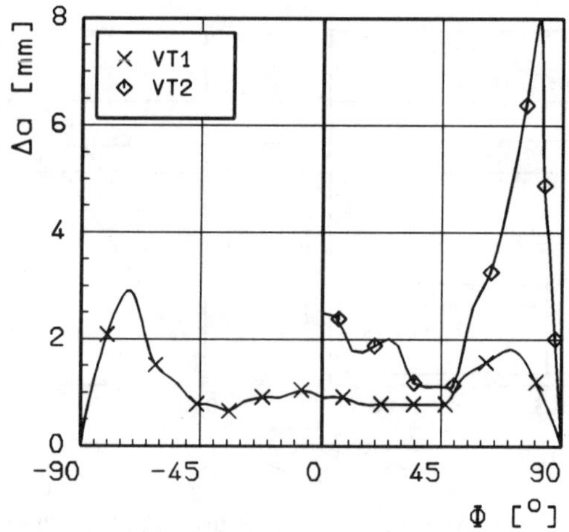

Fig. 10 Variation of stable crack growth along the crack front at p_{max}.

propagation in the axial direction is three to four times larger than the crack growth in wall thickness direction at $\Phi = 0°$. This 'canoe' shape of the ductile crack has been observed before by Pugh,[3] Milne,[4] and other investigators, but with no explanation.

NUMERICAL ANALYSIS OF THE TESTS

Materially and geometrically nonlinear FE analyses of the two structures, VT1 and VT2, were realized with Adina[20] based on the incremental theory of plasticity by von Mises, Prandtl and Reuss and allowing for large strain in the vicinity of the crack. The local J-integral was calculated by a post-processor[21] based on the virtual crack extension method of DeLorenzi.[22] No crack growth was simulated in the FE model. Thus, all the calculated values refer to the configuration of the initial fatigue crack, a_0 and c_0, only. This appeared to be a severe restriction for the analysis of VT2.

The FE calculations, as well as a limit analysis by means of the Folias factor, showed that the initiation of crack growth occurred near or beyond full yielding of the smallest ligament (see Table 3).

Figure 8 shows a comparison of the pressure versus notch opening displacement curves obtained in the vessel tests and in the FE analysis, respectively. The curves indicate a slightly higher plastic limit load in the experiment than in the FE calculation but they coincide at the initiation pressure, p_i. As no crack growth was simulated, the numerical results will necessarily deviate from the experimental data beyond initiation. Consequently, the calculated J-integral will deviate from the 'real' J-values with increasing pressure. This deviation can be eliminated by remembering that J is related to the work of external forces; the latter may be calculated from the area, A, under the load versus load line displacement curve

$$A = \int_0^v \sigma_m(\bar{V})\,d\bar{V} = \frac{r_i}{t}\int_0^V p(\bar{V})\,d\bar{V} \tag{1}$$

Evaluating eqn (1) for both experimental and numerical data, a 'revised' J is obtained by

$$J_{(A)}(\Phi) = J_{(FE)}(\Phi) \times (A_{exp}/A_{FE}) \tag{2}$$

and plotted in Fig. 11 versus the crack front angle Φ for the three pressure values p_i, p_y, p_{max}. This revised J according to eqn (2) will be used for the crack growth analysis. It shows a maximum at $\Phi = 0°$ as in linear elastic fracture mechanics but, with increasing plastification, two additional local maxima develop at $\Phi = \pm 55°$ and $\Phi = \pm 65°$ for VT1 and VT2, respectively. The dashed lines denote the initiation values, J_i, obtained from compact

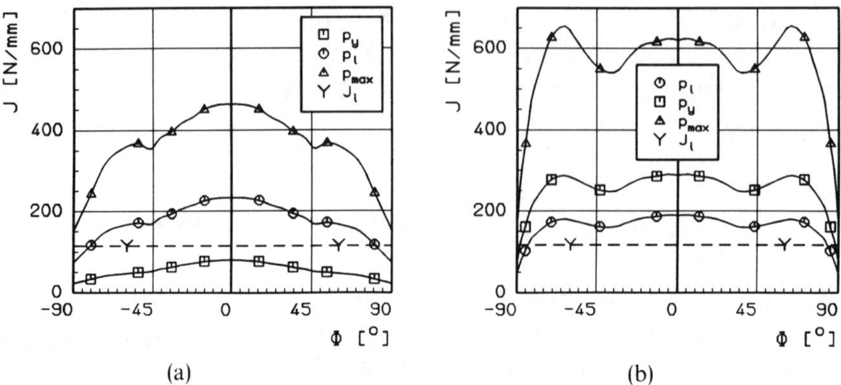

Fig. 11. Variation of J-integral along the crack front from FE calculation, corrected by experimental p versus V data according to eqn (2): (a) VT1; (b) VT2.

specimens for the respective materials. According to these results, some amount of stable crack extension has apparently occurred before the potential curve of Fig. 7 indicated initiation. The variation of CTOD, δ_t, along the crack front is quite similar to that of J (see Fig. 12). This variation of J and CTOD along the crack front does not accord with the variation of stable crack extension, Δa, in Fig. 10.

An attempt to explain the canoe-shape phenomenon has been made by Brocks and Künecke[6] for VT1 by introducing the local ratio of triaxiality of the stress state

$$\chi(\Phi) = \max_r \frac{\sigma_h(r, \theta, \Phi)}{\sigma_e(r, \theta, \Phi)}\bigg|_{\theta=0^\circ} \tag{3}$$

where σ_h and σ_e are the hydrostatic and the von Mises effective stress, r is the distance to the crack tip, and θ is the angle to the ligament in a local polar coordinate system. This ratio has proved to be a parameter characterizing

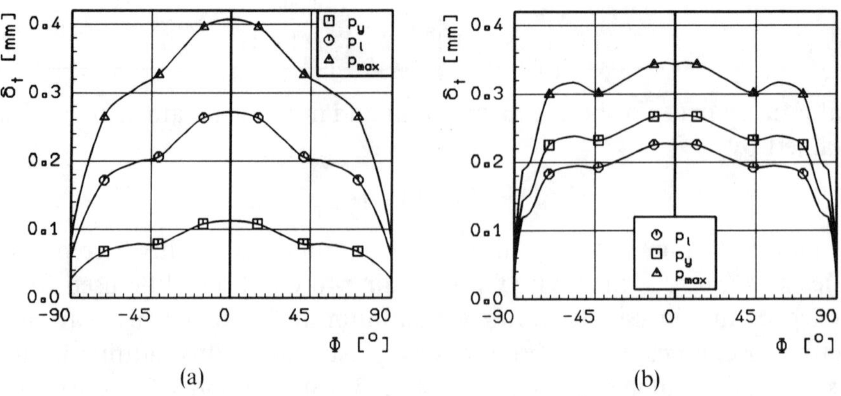

Fig. 12. Variation of CTOD along the crack front: (a) VT1; (b) VT2.

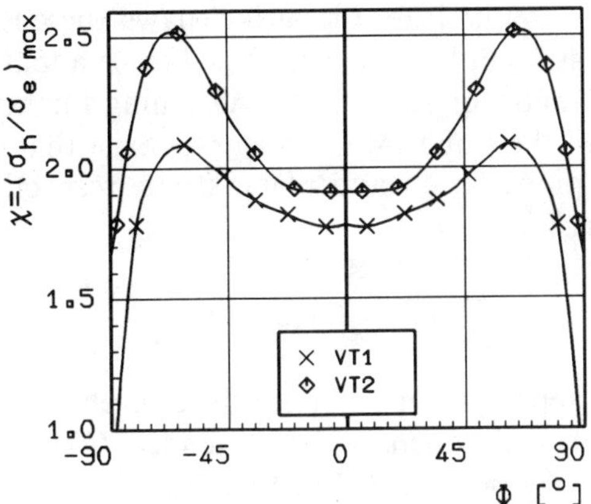

Fig. 13. Variation of the triaxiality of the stress state along the crack front, eqn (3).

the stress state in various specimen geometries[23] and has also been proposed by Kordisch and Sommer.[24] Its physical significance for void growth in elastic–plastic materials is well known from the investigations of McClintock,[25] and Rice and Tracey.[26] The variations of the triaxiality according to eqn (3) along the crack front (Fig. 13) reveal that the maxima of crack growth occur close to the maxima of triaxiality. Another measure of local crack tip constraint, i.e. the ratio $J/(\delta_t\sigma_y)$, has been proposed by Steenkamp.[27] Figure 14 shows that these curves have similar shapes to those of $\chi(\Phi)$. Thus, both quantities seem to be suited to describe the local triaxiality or crack tip constraint that appears to influence the local resistance against ductile crack growth.

In the present paper, the χ-parameter will be used to describe the dependency of the local R-curves on the triaxiality of the stress state. For simplicity, and as no other information about the progress of $\Delta a(\Phi)$ under

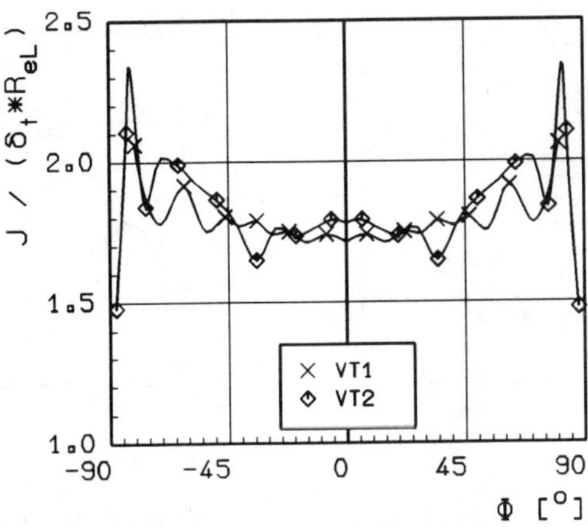

Fig. 14. Variation of crack tip constraint along the crack front.

increasing $J(\Phi)$ was available for the surface flaws, these R-curves have been approximated by straight lines. J_i is assumed to be a material constant not depending on geometry or stress state. Assuming furthermore that upper and lower limits of the slopes of the R-curves exist that correspond to the limiting cases of plane strain and plane stress, respectively, an approximation function is:[6,28]

$$\frac{\Delta J}{\Delta a}(\chi) = A + B\tanh\frac{D-\chi}{C} \tag{4}$$

The resulting $\Delta a(\Phi)$ predictions, by both the conventional method of using J_R-curves of compact specimens and an extended J-concept of χ-dependent local resistance curves, are plotted in Fig. 15.

The conventional concept of a unique characteristic $J(\Delta a)$ overestimates the real crack growth for VT1 at $\Phi = 0°$ in the wall thickness direction up to four times and is not able to predict the maximum crack growth for VT1 and VT2 at $\Phi = 70°$ and $80°$, respectively. Commonly, an overestimation of a critical event is supposed to give a conservative prediction. But this is different with respect to a leak-before-break assessment as the real crack may become unstable due to its axial extension before it penetrates through the vessel wall. Thus, a $J(\Delta a)$ analysis may yield non-conservative estimates. The improved concept of local J_R-curves depending on the triaxiality of stresses gives a much better qualitative and quantitative explanation of the ductile crack growth for VT1. It also gives a qualitatively better prediction of $\Delta a(\Phi)$ for VT2, i.e. it reflects the ratio $\Delta a_{max}:\Delta a_{min}$ obtained in the experiment as well as the approximate position of Δa_{max}. The quantitative results for VT2 are still poor, which might be due to the restriction of the FE analysis which did not account for the geometry changes from the large amount of crack growth.

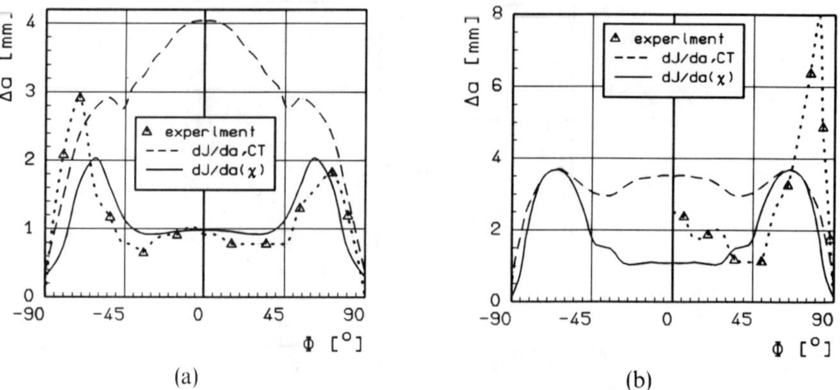

Fig. 15. Variation of stable crack growth along the crack front at p_{max}; test results compared with conventional and improved J-based predictions: (a) VT1; (b) VT2.

ACKNOWLEDGEMENT

The results reported here have been obtained in the course of investigations supported by the Bundesminister für Forschung und Technologie of the Federal Republic of Germany under Contract Number 1500 490.

REFERENCES

1. McCabe, D. E., Landes, J. D. & Ernst, H. A., Prediction of heavy section performance of nuclear vessel steels from surveillance size specimens. *Trans. 7th SMiRT Conf.*, Vol. G, 1983, paper G2/4.
2. Julisch, P. & Stadtmüller, W., Valuation of a ductile vessel rupture by R-curve analysis with CT specimens and wide plate tests. *Trans. 8th SMiRT Conf.*, Vol. G, 1985, paper G 3/1.
3. Pugh, C. E., HSST program quarterly report for Jan. to March 1983. NUREG/CR-3334, Vol. 1, ORNL/TM-8787/V1, NRC Fin. No. B 80119.
4. Milne, I., Notes for EGF task group I exercise in predicting ductile instability: Phase III. Report RL/IM/PL, CEGB, Leatherhead, UK, 1984.
5. Wobst, K. & Krafka, H., Experimental investigation of stable crack growth of an axial surface flaw in a pressure vessel. In *Proc. 14th MPA-Seminar*, Stuttgart, 1988, Vol. 2, paper 41.
6. Brocks, W. & Künecke, G., Elastic–plastic fracture mechanics analysis of a pressure vessel with an axial outer surface flaw. In *Proc. 14th MPA-Seminar*, Stuttgart, 1988, Vol. 2, paper 42.
7. Brocks, W. & Noack, H. D., J-integral and stresses at an inner surface flaw in a pressure vessel. *Int. J. Pres. Ves. & Piping*, **31** (1987) 187–203.
8. ASTM E 813-81, Standard test method for J_{Ic}, a measure of fracture toughness. In *Annual Book of ASTM Standards*. American Society for Testing and Materials, Philadelphia, PA, 1981.
9. DVM-Merkblatt Nr. 2, Ermittlung von Rissinitiierungswerten und Risswiderstandskurven bei Andwendung des J-Integrals. Deutscher Verband für Materialprüfung, Berlin, 1987.
10. ASTM E 399-81. Standard test method for plane-strain fracture toughness of metallic materials. In *Annual Book of ASTM Standards*. American Society for Testing and Materials, Philadelphia, PA, 1981.
11. Cornec, A., Heerens, J. & Schwalbe, K.-H., Bestimmung der Rissaufweitung CTOD und Rissabstumpfung SZW aus dem J-Integral. Berichtsband der 18. Sitzung des DVM-Arbeitskreises Bruchvorgänge, Deutscher Verband für Materialprüfung, Berlin, 1987, pp. 265–79.
12. Aurich, D., Wobst, K. & Krafka, H., J_R-curves of wide plates and CT25 specimens—comparison of the results of a pressure vessel. *Nucl. Engineering & Design*, **112** (1989) 319–28.
13. Garwood, S. J., Effect of specimen geometry on crack growth resistance. *ASTM STP 677*, American Society for Testing and Materials, Philadelphia, PA, 1979, pp. 511–32.

14. Garwood, S. J., Measurement of crack growth resistance of A533B wide plate tests. *Fracture Mechanics, ASTM STP 700*, American Society for Testing and Materials, Philadelphia, PA, 1980, pp. 271–95.
15. Simpson, L. A., Effect of specimen geometry on elastic–plastic R-curves for Zr–2·5% Nb. *Advances in Fracture Research, Proc. 5th Int. Conf. Fracture*, Vol. 2, Cannes, 1981, pp. 833–41.
16. Shih, C. F., German, D. & Kumar, V., An engineering approach for examining crack growth and stability in flawed structures. *Int. J. Pres. Ves. & Piping*, **9** (1981) 159–96.
17. Roos, E., Eisele, U. & Silcher, H., A procedure for the experimental assessment of the *J*-integral by means of specimens of different geometries. *Int. J. Pres. Ves. & Piping*, **9** (1986) 81–93.
18. Raju, I. S. & Newman, C. J., Stress intensity factors for internal and external surface cracks in cylindrical vessels. *Trans. ASME, J. Pressure Vessel Technology*, **104** (1982) 293–6.
19. Brocks, W. & Noack, H.-D., Elastic–plastic *J*-analysis of an inner surface flaw in a pressure vessel. In *Proc. 1986 SEM Fall Conference on Experimental Mechanics*, Keystone, Experimental Mechanics, June 1988, pp. 205–9.
20. Bathe, K. J., ADINA, a finite element program for automatic dynamic incremental nonlinear analysis. Report AE84-1, ADINA Engineering Inc., Watertown, 1984.
21. Matzkows, J., Boddenberg, R. & Kaiser, F., Programm JINFEM, Postprozessor für das Programm ADINA. Report Hb-18-030, IWiS GmbH, Berlin, 1985.
22. de Lorenzi, H. G., On the energy release rate and the *J*-integral for 3D crack configurations. *J. Fracture*, **19** (1982) 183–93.
23. Brocks, W., Künecke, G., Noack, H. D. & Veith, H., On the transferability of fracture mechanics parameters to structures using FEM. In *Proc. 13th MPA-Seminar*, Stuttgart, 1987, Vol. 1, paper 3.
24. Kordisch, H. & Sommer, E., 3D-effects affecting the precision of lifetime predictions. 19th Nat. Symp. on Fracture Mechanics. San Antonio, 1986. ASTM STP 969, 1988, pp. 73–87.
25. McClintock, F. A., A criterion for ductile fracture by growth of holes. *J. Appl. Mechanics*, **35** (1986) 363–71.
26. Rice, J. R. & Tracey, D. M., On the ductile enlargement of voids in triaxial stress fields. *J. Mech. Phys. Solids*, **17** (1969) 201–17.
27. Steenkamp, P. A. J. M., Investigation into the validity of J-based methods for the prediction of ductile tearing and fracture. PhD-thesis, Technical University Delft, 1986.
28. Brocks, W., Veith, H. & Wobst, K., Experimental and numerical investigations of stable crack growth of an axial surface flaw in a pressure vessel. IAEA Specialists' Meeting, Stuttgart, FRG, 1988.

Int. J. Pres. Ves. & Piping **43** (1990) 317–327

A Leak-Before-Break Assessment Method for Pressure Vessels and Some Current Unresolved Issues

J. K. Sharples & A. M. Clayton

UKAEA, Structural Integrity Centre, NRL-Risley, Warrington WA3 6AT, UK

ABSTRACT

The structural integrity diagram, a plot of crack depth against crack length, can be used to investigate a wide range of safety arguments for flawed pressure vessels, including Leak-Before-Break. It enables clear margins to be shown for defects which might exist in the vessels and indicates crack sizes and loadings where the Leak-Before-Break case is valid. The use of this diagram requires a model of crack shape development as a crack grows through the wall of the vessel up to the stage at which the deepest part of the crack breaks through the wall, and this is considered for a number of growth mechanisms. Uncertainties exist, however, in the understanding of crack behaviour relevant to this issue. These uncertainties are reviewed and work programmes underway in the UK aimed at resolving some of them are outlined.

INTRODUCTION

Over recent years, the concept of Leak-Before-Break (LBB) has been considered in establishing safety cases for pressure vessels, particularly in the nuclear industry. The various stages in the development of a LBB argument may be explained with the aid of the diagram shown in Fig. 1. This diagram shows crack depth (a), normalised to the vessel wall thickness (t) as a function of crack length ($2c$). An initial part-through crack would be a point on the diagram. The crack could grow by fatigue, tearing or any other process due to the service load and environment (e.g. stress corrosion cracking), until it reaches some critical depth at which the remaining

Int. J. Pres. Ves. & Piping 0308-0161/90/$03·50

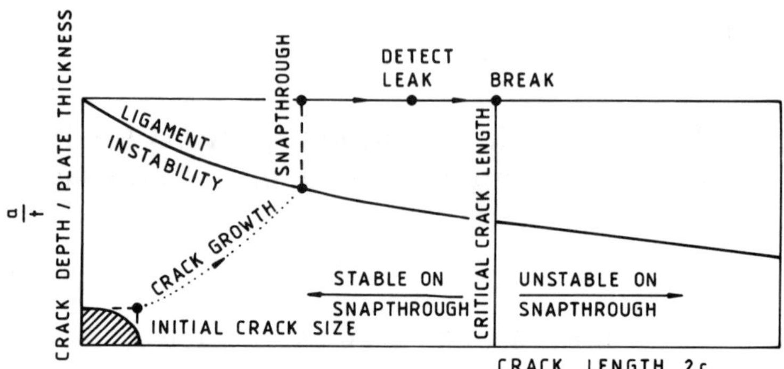

Fig. 1. LBB diagram.

ligament ahead of the crack breaks through the wall. The crack will then continue growing in surface length until there is sufficient opening to cause a detectable leak or until the crack becomes unstable. A LBB argument is therefore aimed at demonstrating that leakage of an appropriate gas or fluid through a crack in the wall of a vessel, could be detected prior to the crack attaining conditions of instability at which rapid crack extension would occur. LBB is one of a number of safety arguments which can all be shown in a diagram of the sort shown in Fig. 1 which is known as a structural integrity diagram.[1]

Consideration of LBB in large pressure vessels (e.g. primary vessels of Magnox and fast reactor nuclear power systems) are in many ways different from that in piping, where much of the research and development work in this field has been concentrated in recent years. In piping, crack or leak detection is often complex (because a vast amount of piping is usually present in a given plant) and there can be local bends with high imposed mechanical stresses. In pressure vessels, however, the loading is usually better defined and crack detection is easier since the vessel often can be contained and leakage sampled. Difficulties arise in developing LBB arguments in pressure vessels because through-wall bending can be present in many situations, due to such features as stiffeners and nozzles, and this bending can result in crack closure on one side of the wall.

LBB arguments can be invaluable in providing satisfactory safety cases for pressure vessels that are difficult to inspect, due to inaccessible or hazardous conditions. However, for cases where it is unacceptable or undesirable for any leak to occur in a vessel, on reliability, economical or health grounds, it is equally important to be able to assess and predict events in the life of a crack up to back surface penetration.

This paper contains a review of the general methods that are used to predict crack growth behaviour in the wall of a pressure vessel, leading to the development of a LBB argument. Present uncertainties surrounding these issues are highlighted and discussed and work programmes within the UK aimed at resolving some of them are outlined.

METHOD OF ASSESSMENT

The CEGB R6 procedure[2] is generally used in the UK for deducing the 'Ligament Instability Curve' and 'Critical Crack Length' in a diagram similar to that shown in Fig. 1. In the R6 method, the ratio of the stress intensity factor (at any point of interest on a crack) to the fracture toughness is denoted by *Kr* and the ratio of the applied load to the plastic limit load is donated by *Lr*. A failure assessment curve is obtained by plotting values of *Kr* against *Lr* which cause crack initiation (or where appropriate, by using a suitably modified toughness, the desired amount of ductile tearing). For a given load, crack dimensions can be adjusted until the (*Kr*, *Lr*) co-ordinates fall on the R6 assessment curve.

A structural integrity diagram, such as Fig. 1, defines bounding defect sizes. These can be very large and have associated with them very high stress intensity factors (or *J*-integral values) even with normal service loads. Where stable tearing has been allowed in defining these bounding defect sizes, care has to be taken in evaluating fatigue crack growth, since it may be enhanced by tearing increments in each cycle. Models of tearing and fatigue interaction have been developed by Chell[3] for such situations.

Important factors to be considered when constructing a structural integrity diagram with a view to establishing a LBB safety argument are:

(a) The choice of material properties (upper bound, lower bound or mean values) can have a marked influence. It is therefore always necessary to carry out a sensitivity study by considering each one of these properties in turn for the different parts of the structural integrity diagram being considered, in order to ensure that the conservatism of the diagram is always maximised for the purpose it is intended.

(b) The possibility of through-wall bending stresses being present in the vessel should be carefully considered and then used in the calculations. These stresses will result in crack opening being reduced on the least tensile side of the vessel wall and depending on the magnitude of the bending component, they can lead to crack closure which would prevent any leakage occurring, thus nullifying a LBB case. Local bending effects in vessels due to bulging should also be taken into account. The extent of crack closure due to bending is currently not sufficiently understood and a programme of work aimed at achieving a better understanding, is currently underway involving a three-dimensional finite element study.

(c) Uncertainties that usually exist in both the magnitude and distribution and the effect on fracture behaviour of residual stresses is an important factor to be taken into consideration. In particular, care should be taken to ensure that any assumptions or approximations relating to these stresses are such that they are conducive to a credible LBB argument (i.e. to ensure pessimism in terms of leakage being predicted prior to the through-thickness critical crack length being attained). In this regard, assuming the residual stresses to be tensile, the limiting LBB argument would be made by excluding them when evaluating the ligament instability curve, but assuming a constant value of material yield stress across the section under consideration when evaluating the critical crack length.

(d) The ability to predict leakage rates through a crack is inevitably an important factor in the LBB argument and although it is currently possible to do this conservatively (i.e. underpredict the leakage), much further knowledge is required to be able to assess flow rates more accurately in realistic cracks. The rate of flow of gas (or fluid) through a crack is known to depend on such aspects as absolute value of crack opening, roughness and waviness of the crack surfaces and area changes of the crack through the wall. These factors which effect the friction loss must be understood in some detail and an extensive experimental programme of work is currently being undertaken in order to achieve this. Ways in which leakage may be detected and monitored during plant operation must also of course be carefully considered.

The above contains a brief review of the general method of assessing LBB in the UK and some of the general problems and uncertainties that currently exist. However, different specific aspects need to be considered for short, intermediate and for long cracks. This is discussed in the following section.

OUTSTANDING UNCERTAINTIES AND WORK IN THE UK AIMED AT RESOLVING THEM

For constant amplitude loading, the part-through-wall crack sizes at which tearing initiation occurs (beyond which there is combined fatigue and tearing) and the crack sizes at which instability occurs (beyond which there is rapid crack growth) can be evaluated by inputting the peak loads into an R6 calculation. The crack sizes are shown in Fig. 2 and split the structural integrity diagram into three regions.

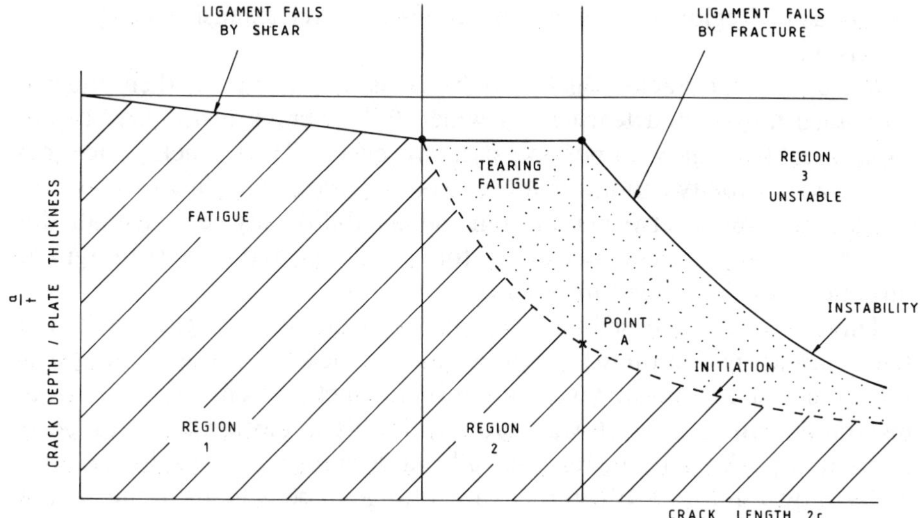

Fig. 2. Structural integrity diagram.

Region 1 is for cracks which remain below the initiation size and only grow by fatigue through the wall. Eventually, the remaining ligament is so small that it can shear under the crack opening. Miller[4] analysed several sets of experiments on the ductile failure of single edge notch plates with thin ligaments under tension to assess the effects of differing surface defect shape and plate thickness and width, defect depth and length. From this work, Miller was able to show that the crack tip opening for the above type of shear to occur is proportional to the width of the remaining ligament, and based on this, Smith[5] has recently deduced that the applied tensile stress, σ_p required for ligament rupture is given by the expression:

$$\sigma_p = \sigma_y \frac{l}{t}\left[\frac{\lambda E t}{4\sigma_y c} + \psi\right] \tag{1}$$

where

σ_y is the yield stress,
l is the ligament width,
t is the wall thickness,
E is Young's Modulus,
c is the crack semi-length,
λ is the constant (of value between 0·3 and 1·3) relating crack tip opening to ligament width, and
ψ is a constraint factor (which is frequently insignificant).

Region 2 is for cracks which undergo combined fatigue and tearing before breakthrough occurs but which finally break through by the same ligament failure as in Region 1. This region can lead to very extensive stable crack extension.

Region 3 is for cracks which may begin as fatigue cracks, then undergo combined fatigue and tearing but which fail by crack instability causing rapid breakthrough with possible dynamic effects. These cracks, once they reach the instability condition, must be considered as unsafe. For very long cracks, the growth between initiation and instability may be very small (due to the rapid rise in stress intensity factor with depth) and a safety margin on initiation may be needed in practice.

This analysis is all based on one peak load value. For a higher peak load, the instability condition will occur at smaller crack sizes. A useful indicator is the crack size associated with initiation in cracks which would just reach instability during through-wall growth. If this is plotted for a range of alternative peak loads, the resulting allowable crack sizes for safe operation are as shown in Fig. 3. When there are a range of peak loads in the normal operation of the plant (i.e. variable amplitude loading) then the largest cycle would be taken to define the initiation and instability sizes, although crack growth could allow for the variable cycling.

Once breakthrough occurs, the through-wall crack may exceed the critical crack length. Again, there will be an initiation length (usually in Region 1) and an instability length (usually in Region 2, but potentially in any region with increased absolute size of the vessel increasing the instability length). Beyond the initiation length, through-wall cracks grow by fatigue and tearing until they reach the size at which instability occurs, beyond which catastrophic failure of the vessel occurs. There will be some intermediate stage after breakthrough and before the full through-wall shape has been developed. This may mean that cracks which might be expected to be

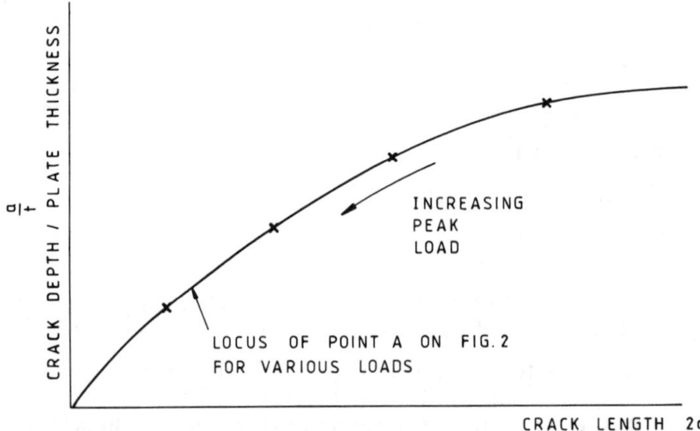

Fig. 3. Demonstration of allowable crack sizes for operation with different peak loads.

unstable on breakthrough are actually stable until some crack length has been developed on the breakthrough face. This usually occurs in a few cycles and thus cannot be relied on for safety arguments.

Overall, then, cracks may take a complex growth plan of the following steps (some of which will be omitted depending on the region in Fig. 2): part-through-wall fatigue, combined fatigue and stable tearing, breakthrough instability, through-wall fatigue, combined fatigue and stable tearing and catastrophic failure.

In order to practically demonstrate the structural integrity diagram shown in Fig. 2 and referred to above, two examples of a plate with a centre-crack loaded in tension have been considered. The first example consists of a 2·5 m wide plate with a 0·06 m thickness and the second one consists of a much wider plate of 50 m with the same thickness. The first example is relevant to wide plate fracture tests being undertaken in the UK and the second example is appropriate to the modelling of a circumferentially oriented crack in a cylinder of large radius to thickness ratio (>100) subjected to an axially applied load. Typical Type 316 stainless steel properties have been used in the calculation. These are:

Young's modulus, $E = 198\,000$ MPa
Poisson's ratio, $\nu = 0{\cdot}3$
Yield stress, $\sigma_y = 234$ MPa
Ultimate stress, $\sigma_u = 593$ MPa
Tearing resistance is given by $J = 0{\cdot}723\,\Delta a^{0{\cdot}607}$ MN/m (Δa in mm).

Applied tensile stress for the plates is about yield stress. The structural integrity diagrams for both examples are contained in Figs 4 and 5 respectively. For completeness, the through-thickness crack initiation and instability lines have also been included in these figures. The two main noticeable differences between the two cases are:

(1) For the 2·5 m plate, through-thickness crack initiation occurs at a crack length somewhat less than that for when initiation occurs at the deepest point of a semi-elliptical surface crack, whereas for the wider plate, initiation occurs at approximately the same crack length (of about 0·2 m).
(2) The 'ligament failure by shear' curve in Region 2 (see Fig. 2) is relatively flat for the 2·5 m plate, whereas for the wider plate the crack depth reduces significantly with crack length.

In addition to specific aspects of each region, there are general uncertainties which apply to all the regions. One problem is that of fracture toughness. For a ductile material, such as stainless steel (appropriate to LBB

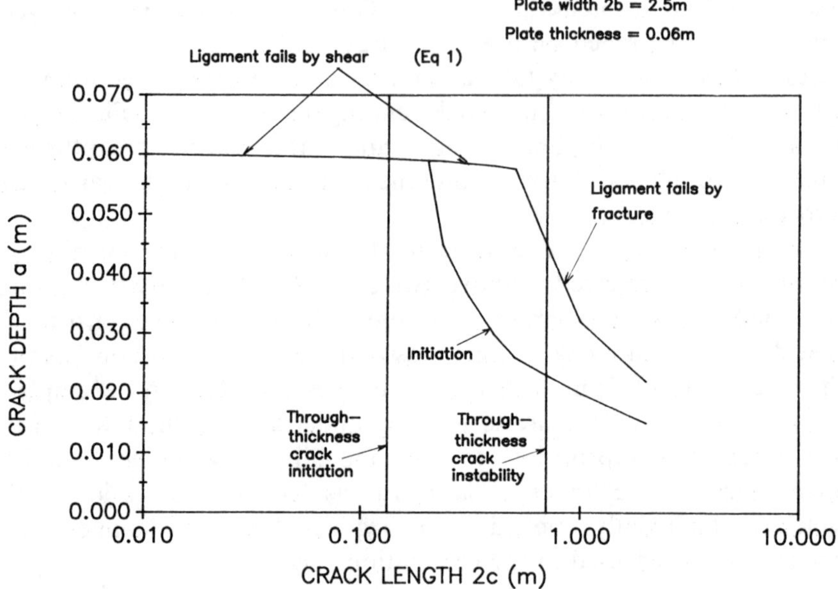

Fig. 4. Structural integrity diagram example for plate width of 2·5 m.

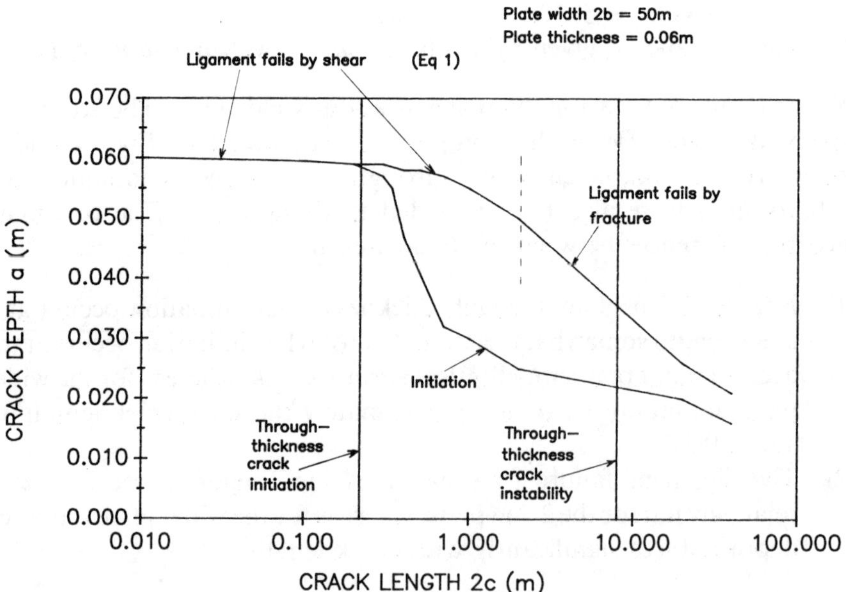

Fig. 5. Structural integrity diagram example for plate width of 50 m.

studies on fast reactor plants), instability may be preceded by a large amount of tearing. In the growth through the thickness, the amount of stable tearing could be between say 5 mm and 25 mm and it may be greater than 1 m for a through-wall crack. These amounts of tearing are far in excess of available materials data from small-scale specimens, and in the case of through-wall cracks, are resulting from a different failure process (plane stress necking ahead of the crack compared with plane strain fracture for the small scale specimens). Nevertheless, the only current option is to use extrapolated data from small-scale tests to evaluate this information.

A further problem in using small-scale fracture data is the fact that only data points should be included which satisfy the conditions of J-controlled growth. The bounds on these limits are given in the R6 document, but they are ill defined and can be extremely restrictive for relatively high work hardening materials such as stainless steel where the amount of J-controlled crack growth would be very small; in some cases less than 1 mm.

Programmes of work are currently being carried out to investigate ductile tearing behaviour in large specimens of plant dimensions. The test programmes are being undertaken on both stainless steel and carbon steel materials,[6] on both plate and cylindrical specimens and by considering both straight fronted through-wall and edge cracks and semi-elliptical surface cracks.

On the selection of the limit load solution for use in an R6 assessment, Miller[7] has compiled a comprehensive compendium for various types of cracks, component geometries and loading conditions. In many instances, several solutions are available for the same geometry and load cases and although guidance is given as to the most appropriate solution to use, there is some doubt in this area. For example, two types of solution have been derived for a semi-elliptical surface crack; that based on 'local' collapse load, at which the ligament goes plastic, and a 'global' collapse load, at which the structure becomes a mechanism. Until recently, there was little evidence to suggest which of these two approaches was the most representative. Miller, considered this further in a later paper.[8] He had previously analysed test results for plates, cylinders and spheres and established a criterion for determining which limit load ('local' or 'global') governed fully plastic instability. He now considered the elastic–plastic regime by using the R6 fracture assessment method to evaluate the J integral. This method implicitly uses the reference stress approximation to evaluate J. In order to calculate the reference stress, the appropriate collapse or limit load must be determined. Miller subsequently compared published results for J integrals with reference stress estimates for plates in tension, cylinders with circumferential defects in tension, and cylinders with axial cracks under pressure. In all the results considered, basing the reference stress on the 'global' collapse load led to much better agreement with calculated J values, than using the 'local collapse load.

A recent fracture test on a Type 316 stainless steel plate, 60 mm thick and 820 mm wide, containing a semi-elliptical surface crack 250 mm long and 42 mm deep, has confirmed Miller's work regarding the suitability of the 'global' limit load solution. It is intended to investigate this further in other large-scale tests scheduled to be carried out.

The final major problem area is that of taking into account the effect of weldments. Assessing defects in welded (or heat affected zone) regions of components, pressure vessels being of no exception, can present two distinct uncertainties; that of which material properties to use (parent plate, weld or heat affected zone) and that associated with residual stresses. Guidelines on dealing with weldments are given in the R6 procedure, aimed at ensuring conservatism in a standard fracture assessment. However, as previously noted, such conservatism is not relevant when constructing a structural integrity diagram to be used in a LBB argument since crack breakthrough to the back surface may be predicted earlier than it would occur in practice. The subject of weldments, therefore, is open to further investigation and a programme of work is being carried out to achieve this for Type 316 stainless steel material relevant to the fast reactor.

CONCLUSIONS

The main conclusions that may be drawn from the work contained in this paper are:

(1) A structural integrity diagram, which can be used to investigate a wide range of safety arguments for flawed pressure vessels, including LBB, can be split up into three regions dependent on crack length.
(2) Different mechanisms are required to be considered for each of the three regions, these mechanisms being: part-through-wall fatigue, combined fatigue and stable tearing, breakthrough instability, through-wall fatigue, combined fatigue and stable tearing and catastrophic failure.
(3) There are general uncertainties which apply to all of the three regions which are principally: the extrapolation of small-scale fracture toughness data to structures, the selection of limit load solution for use in the assessment route and the behaviour of weldments.
(4) The above uncertainties require further investigation and work is ongoing with future work planned within the UK aimed at resolving them.

REFERENCES

1. Clayton, A. M., Multi-legged safety cases for vessels using structural integrity diagrams. In *Innovative Approaches to Irradiation Damage and Fracture Analysis, PVP-Vol. 170*, ed. D. L. Marriot, T. R. Mager & W. H. Bamford. ASME Pressure Vessel & Pipings Conference, Honolulu, July 1989.
2. Milne, I., Ainsworth, R. A., Dowling, A. R. & Stewart, A. T., Assessment of the integrity of structures containing defects. *Int. J. Pres. Ves. & Piping*, **32** (1988) 3–105.
3. Chell, G. G., Fatigue growth laws for brittle and ductile materials including the effects of static modes and elastic–plastic deformation. *Fatigue Engineering Mater. Struct.*, **7**(41) (1984) 237–50.
4. Miller, A. G., The ductile fracture of plates under tension with surface flaws of various shapes. CEGB Report TPRD/B/0341/N83 (1983).
5. Smith, E., The rupture of a thin ligament associated with a part-through crack in a plate subject to tensile deformation. *Int. J. Pres. Ves. & Piping*, **32** (1990) 97–109.
6. Clayton, A. M., Morgan, H. G., Green, D. & Sharples, J. K. Large scale leak before break tests in the UKAEA. IAEA Specialists' Meeting On Large Scale Testing, 25–27 May 1988. To be published in *Nuclear Engineering and Design.*
7. Miller, A. G., Review of limit loads of structures containing defects. CEGB Report TPRD/B/0093/N82 Revision 2 (1987).
8. Miller, A. G., *J* estimation for surface defects. CEGB Report TPRD/B/0811/R86 (1986).

Int. J. Pres. Ves. & Piping **43** (1990) 329–350

Recent Results of Fracture Experiments on Carbon Steel Welded Pipes

G. M. Wilkowski, D. Guerrieri, D. Jones, R. Olson & P. Scott

Battelle, 505 King Avenue, Columbus, Ohio 43201, USA

ABSTRACT

Several pipe fracture experiments were conducted with circumferential cracks in the center of ferritic nuclear pipe welds. These experiments involved either submerged arc or shielded metal arc welds with either through-wall cracks or internal surface cracks. The pipe diameters varied from 940 mm (37 inches) to 152 mm (6 inches), and thickness from 10·9 mm (0·43 inches) to 86·6 mm (3·41 inches). Some of the through-wall and surface-cracked pipe experiments were conducted under constant internal pressure and four-point bending. The test temperature was 288°C (550°F). The results of these experiments are compared with limit-load analyses, the ASME, Section XI, article IWB-3650 criterion, and more elaborate elastic–plastic fracture mechanical analysis.

INTRODUCTION

The objectives of this effort are to evaluate the failure behavior of prototypical carbon steel weldments at light water reactor temperatures and to verify limit-load and elastic–plastic fracture mechanics estimation schemes for cracked pipes. These results are directly related to the development of flaw assessment criteria, such as the proposed ASME, Section XI, article IWB-3650 criteria, as well as for crack stability calculations in leak-before-break criteria.

The carbon steel weld pipe fracture experiments that have been conducted are shown in Table 1. The results of each of these experiments will be briefly

TABLE 1
Carbon Steel Weld Pipe Fracture Experiments

Experiment number	*Material*	*Diameter (mm) (inch)*	*Thickness (mm) (inch)*	*Crack type*[a] *(TWC or SC)*	*Loading*[b] *(B, P + B)*
4141-7	A516 Gr70 SAW	940 (37)	86·6 (3·41)	TWC	B
WJ-1	A106 B SMAW	152 (6)	11·2 (0·44)	TWC	B
4141-8	A106 B SAW	406 (16)	26·2 (1·03)	SC	P + B
4141-9	A106 B SAW	406 (16)	26·2 (1·03)	TWC	P + B

[a] TWC = through-wall crack; SC = surface crack.
[b] B = bending; P + B = pressure plus bending.

reviewed in the following sections. Afterwards, these results will be compared with the net-section-collapse analysis, ASME criteria, and *J*-estimation scheme predictions.

EXPERIMENTAL RESULTS

The results of the four experiments are briefly described below.

Experiment 4141-7—cold-leg SAW pipe

This experiment was conducted on a cold-leg pipe section that was removed from a cancelled pressurized water reactor (PWR). The base metal of the pipe was A516 Grade 70. The weld tested was a shop-fabricated submerged arc weld (SAW).

A through-wall circumferential crack of 37% of the pipe circumference was put in the center of the weld. Figure 1 is a schematic diagram of the cracked pipe cross-section and the instrumentation. Also shown is a stainless steel cladding 3·2-mm (0·12 inch) thick typically used for corrosion resistance. The pipe was tested at 288°C (550°F) in four-point bending without internal pressure. A schematic diagram of the test pipe in the test frame is shown in Fig. 2.

The load versus load-line displacement record is shown in Fig. 3. The significant load drops after the maximum load was reached, corresponding to unstable crack jumps. These crack jumps are believed to be induced by dynamic strain aging[1] rather than from the compliance of the test machine. The moment at crack initiation and the maximum moment are given in Table 2.

The crack growth pattern is shown schematically in Fig. 4, while Fig. 5

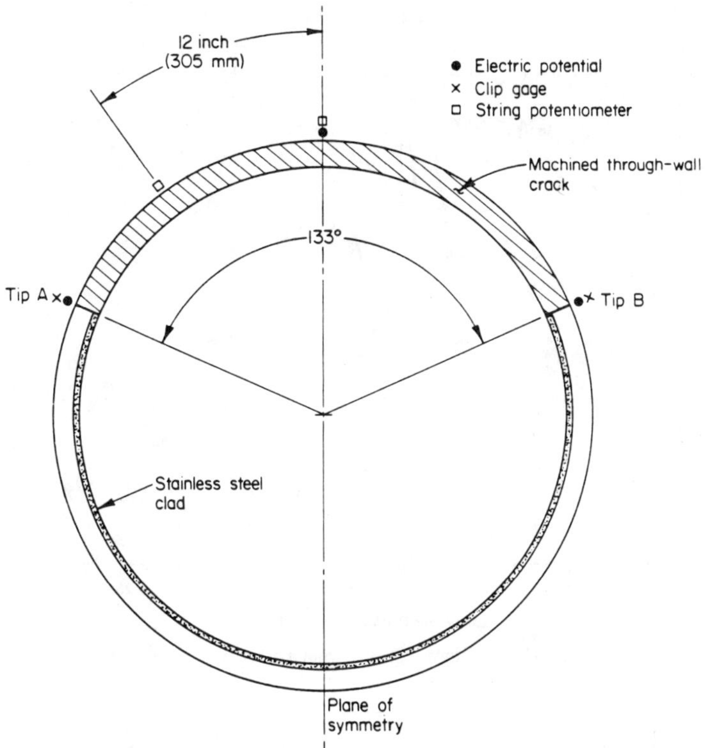

Fig. 1. Schematic diagram of cold-leg pipe cracked cross-section and instrumentation (Experiment 4141-7).

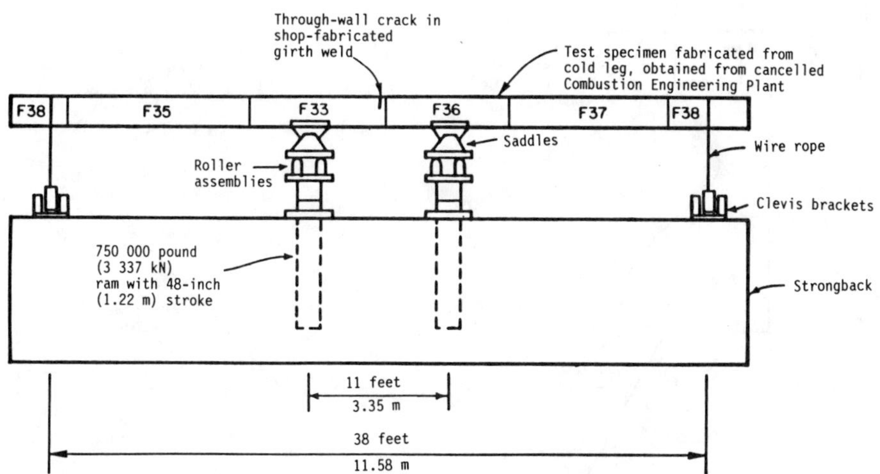

Fig. 2. Schematic diagram of cold-leg pipe in test frame (Experiment 4141-7).

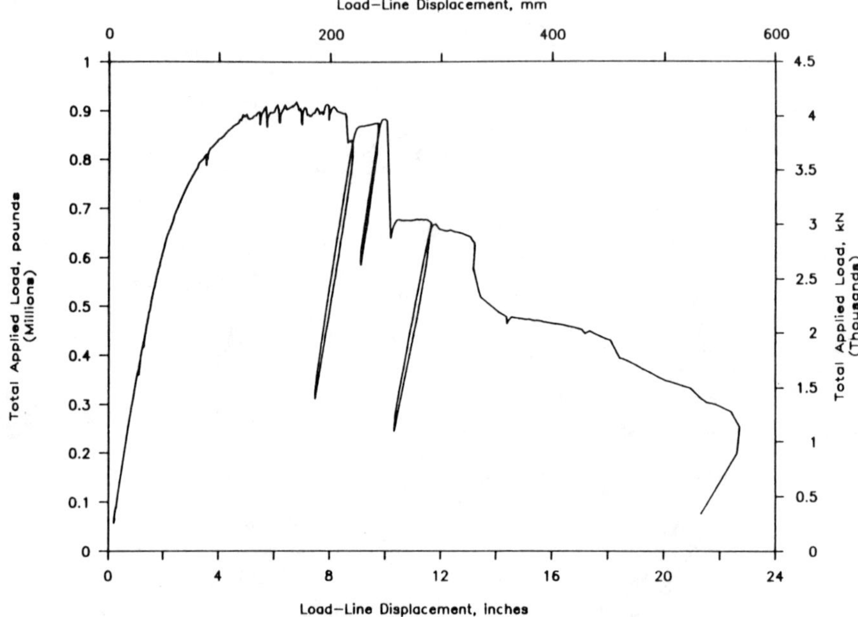

Fig. 3. Load versus load-line displacement record from cold-leg experiment (4141-7).

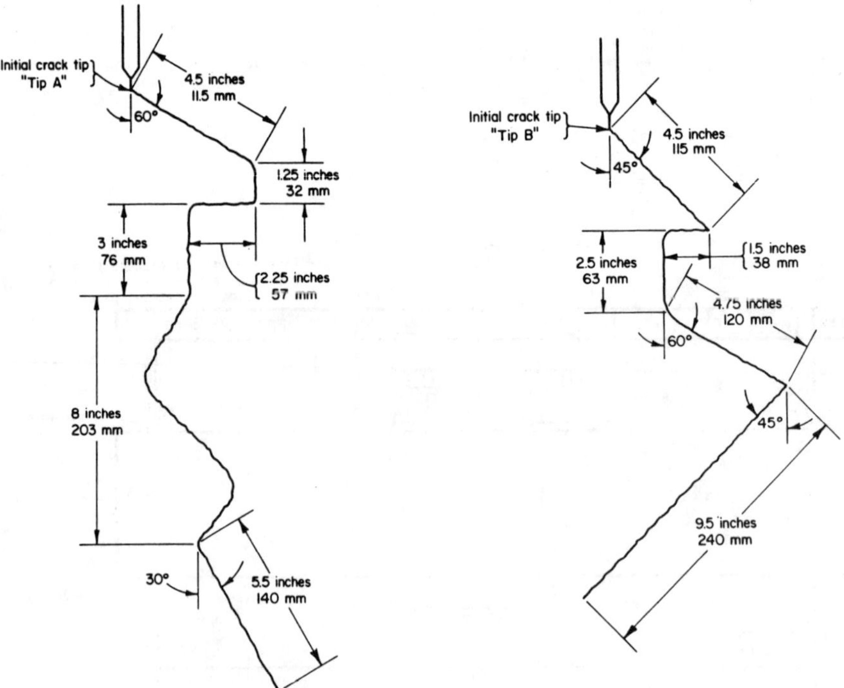

Fig. 4. Schematic diagram of crack growth in cold-leg experiment.

TABLE 2
Summary of Ferritic Weld Pipe Test Results

Experiment number	Outside diameter		Thickness		Flaw dimensions		Pressure		Bending moment			
									Initiation		Maximum	
	mm	*(inches)*	*mm*	*(inches)*	$2c/\pi D$	d/t	*MPa*	*(psi)*	*kN-m*	*(in-kips)*	*kn-m*	*(in-kips)*
4141-7	932·9	(36·73)	86·6	(3·41)	0·37	1·0	0	0	6 424	(56 862)	8 401	(74 358)
WJ-1	168·3	(6·625)	11·2	(0·44)	0·30	1·0	0	0	40·1	(355)	51·3	(454)
4141-9	406	(16)	26·2	(1·03)	0·37	1·0	11·0	1 600	172·9	(1 531)	347·8	(3 078)[a]
4141-9	406	(16)	26·2	(1·03)	0·37	1·0	0	0	N/A[b]		441·3	(3 906)
4141-8	406	(16)	26·2	(1·03)	0·50	0·66	15·5	2 250	422·7	(3 741)	593·8	(5 256)

[a] Estimated.
[b] Data at crack initiation determined when test at pressure.

Fig. 5. Photographs of crack growth in cold-leg experiment.

shows a photograph of the crack growth pattern. Note that the crack grew in a zig-zag fashion.

Experiment WJ-1—A106 Grade B SMAW TWC pipe experiment

This experiment was conducted on a nominal 152-mm diameter Schedule 80 A106 Grade B pipe. The crack was a circumferential through-wall crack in the center of a shielded metal arc weld (SMAW) girth weld. The pipe material was donated to this program from the Duquesne Power Whipjet program.

The main objective of conducting this experiment was to assess the proposed ASME IWB-3650 ferritic pipe flaw evaluation procedure. In the IWB-3650 proposed procedure, the fracture toughness of 70XX shielded metal arc welds is classified along with base metals. The basis for this comes from three *J*-resistance (*J–R*) curve tests. The average J_{Ic} value (given in Table 5-2 of Ref. 2) is 0·126 MJ/m^2 (721 in-lb/in^2). The range of the values was not given. To add to the database, this SMAW pipe weld was tested as part of the Degraded Piping Program.[3]

Pedigree welds fabricated by Duquesne Power Company, Pittsburgh, PA, were available at Battelle. From these welds, one 203-mm (8-inch) nominal diameter Schedule 80 pipe E7018 SMAW weld was used for compact (tension) (C(T)) specimen testing, and one 152-mm (6-inch) nominal diameter Schedule 80 pipe with a similar SMAW was tested in bending with a 30% through-wall circumferential crack. Both C(T) and pipe tests were conducted at 288°C (550°F).

The C(T) specimens were tested with and without side grooves. The *J–R* curves and the ASME reference *J–R* curve for base metal are shown in Fig. 6.

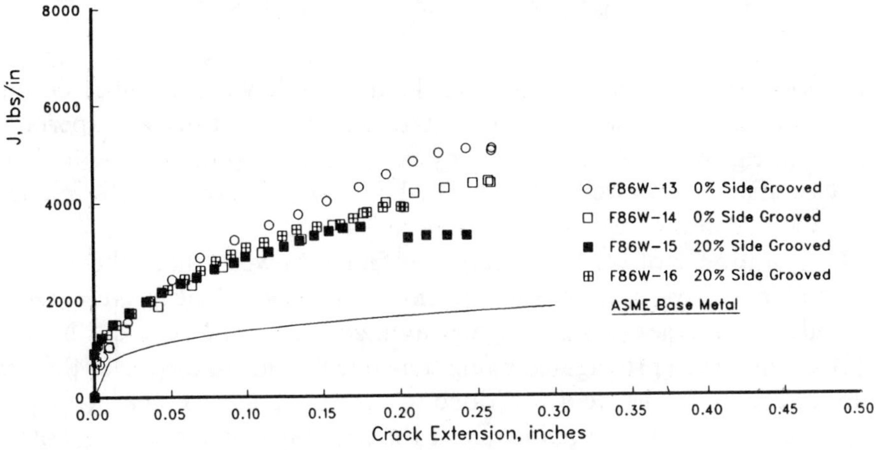

Fig. 6. Whipjet *J–R* curve.

The C(T) specimen *J–R* curves are much higher for the weld than the ASME base metal reference *J–R* curve. Hence, these data substantiate the classification of this E7018 SMAW with the base metal in the IWB-3650 criteria.

The load versus load-line displacement data from the pipe test are shown in Fig. 7. After the maximum load, there were numerous small crack instabilities that can be seen as small drops in load. With each of these crack instabilities, there was also an audible sound. These crack instabilities are believed to be indicative of dynamic strain-aging.[1] If the crack instabilities

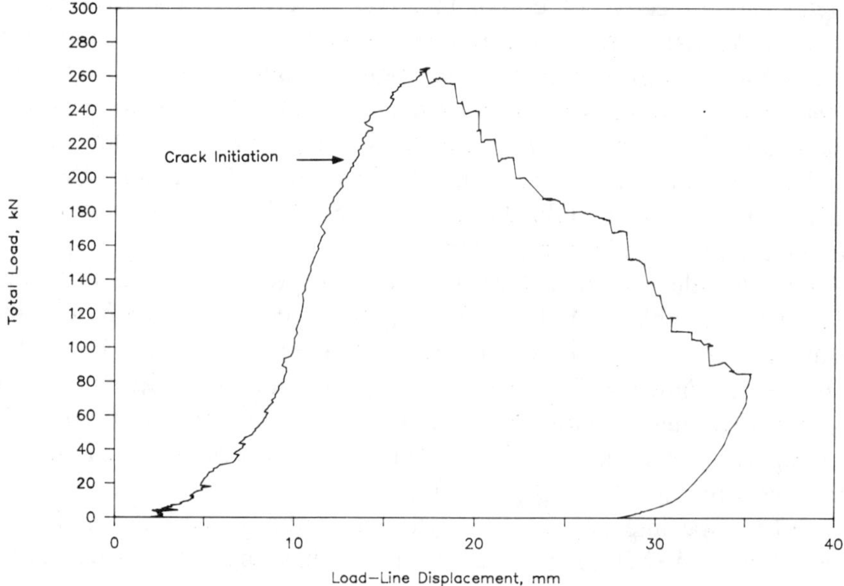

Fig. 7. Whipjet load-displacement curve.

had not occurred, then the maximum load might have been higher since the maximum load corresponded to the start of an instability in this experiment. The crack growth was monitored by the dc electric potential method, and crack initiation was well defined. The bending moment at crack initiation and the maximum moment are given in Table 2.

Observations of the fracture surface showed that the initially circumferentially-oriented crack quickly turned to a slant fracture. On the outside surface, the crack propagated along the toe of the weld. On the inside surface, the crack propagated along the opposite side of the toe of the weld. No sectioning of the weld was done to determine if the crack propagated with the small crack instabilities along the fusion line or if it went completely through the weld metal.

Experiment 4141-8—406-mm diameter surface-cracked SAW pipe

This experiment was conducted on a 406-mm (16-inch) diameter Schedule 100 A106 Grade B pipe procured from a cancelled nuclear plant. The SAW weld procedure was obtained from a US PWR vendor and is given in Appendix C of Ref. 4. This weld was the lowest toughness material tested in this program. The *J–R* curves from 25·4-mm (1-inch) thick 1T, 3T and 9·5T C(T) tests at 288°C (550°F) are shown in Fig. 8. These specimens were made from a similar weld in 25·4-mm (1-inch) thick plate of A516 Grade 70 ferritic steel,[4] but had the plan-form dimensions (i.e. height, width, pinhole diameter, etc.) of 1T, 3T or 9·5T specimens as per ASTM definitions.

To simulate radial growth of a surface crack in a pipe, a single-edge notch (tension) (SE(T)) specimen was tested. Figure 9(a) shows a comparison of the load-displacement records of the SAW and A516 Grade 70 SE(T) specimens. Note that in the weld there was a dynamic crack jump at crack initiation. After this crack arrested, ductile tearing occurred for a small amount of additional crack growth, and then another small crack jump occurred. The calculated *J–R* curves for the weld and A516 Grade 70 SE(T) specimens are shown in Fig. 9(b). The low toughness and possible dynamic crack instability

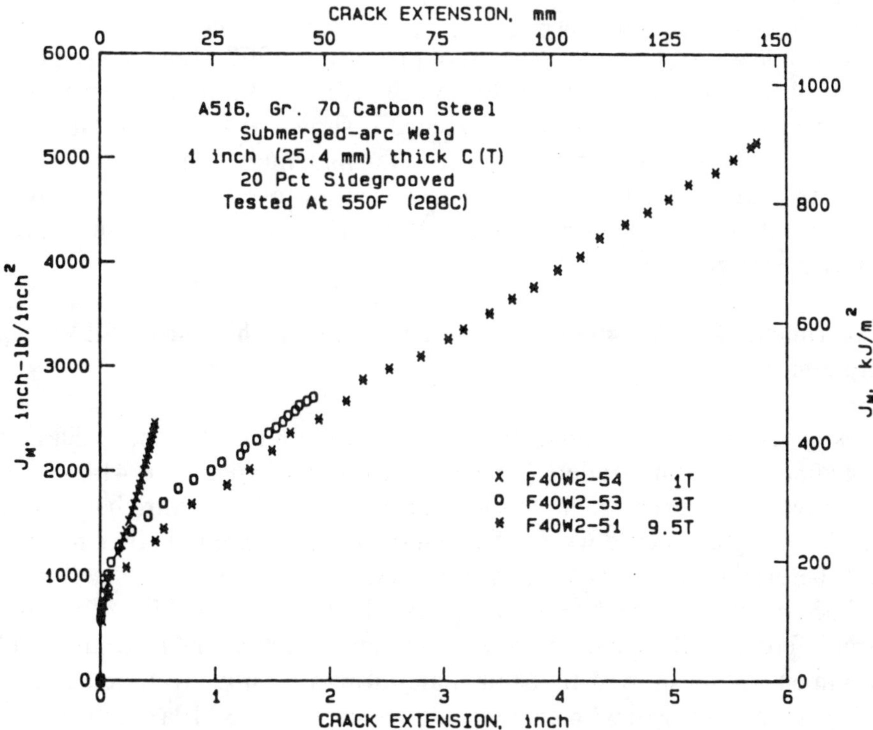

Fig. 8. 1T, 3T and 9·5T *J–R* curves of ferritic SAW.

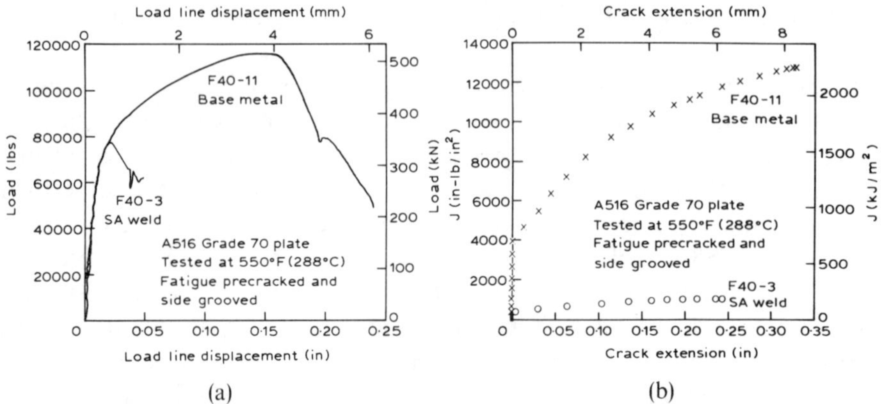

Fig. 9. SE(T) weld and A516 base metal test results. (a) Load–displacement records; (b) *J–R* curves.

at crack initiation made this material of interest for a surface-cracked pipe experiment.

The surface crack geometry was an internal crack with a length of 50% of the circumference and 66% of the pipe thickness. The pipe specimen was fatigue precracked because C(T) specimen results showed that this material was sensitive to notch acuity.

The pipe was tested with an internal pressure of 15·72 MPa (2250 psi), and then loaded in four-point bending until failure. The total bending load versus load-line displacement test record is shown in Fig. 10(a). Note that there was a small load drop just prior to the maximum load. The dc electric potential data showed that this load drop was due to a small crack instability (see Fig. 10(b)). The bending moment at crack initiation and the maximum moment are given in Table 2.

Experiment 4141-9—406-mm (16-inch) TWC carbon steel SAW pipe experiment

This experiment was conducted on the same 406-mm (16-inch) diameter Schedule 100 A106 Grade B pipe as was used in Experiment 4141-8. The objective of this experiment was to gather data on the fracture behavior of this low toughness weld with a through-wall circumferential crack to assess leak-before-break (LBB) fracture analyses.

The circumferential through-wall crack had a length of 37% of the pipe circumference. The pipe specimen was fatigue precracked because C(T) specimen results showed that this material was sensitive to notch acuity.

The pipe was tested with an internal pressure of 11·03 MPa (1600 psi), and then loaded in four-point bending until failure. To conduct this experiment

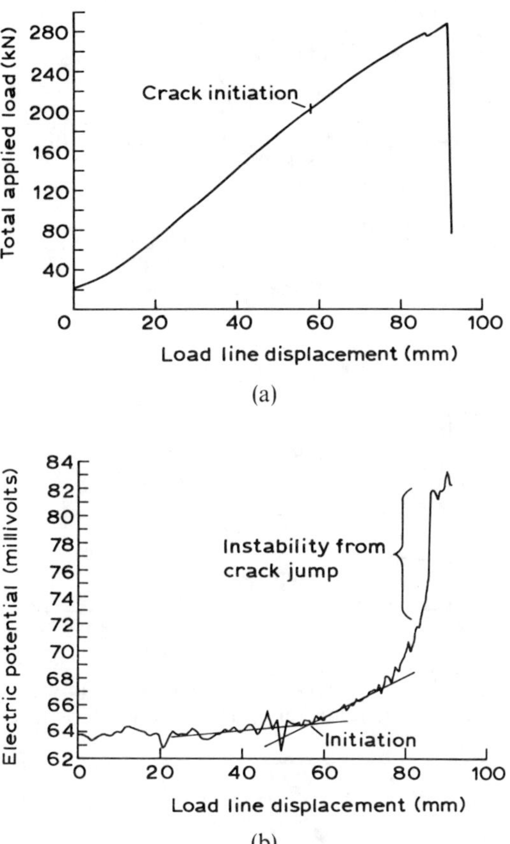

Fig. 10. Experiment 4141-8 pipe test data. (a) Load versus load-line displacement record; (b) dc electric potential data versus load-line displacement.

under pressure and at 288°C (550°F), an experimental high-temperature bladder technique was used. The bladder is essentially an inner tube as in a tire, but a patch is placed between the crack and the bladder to keep the bladder from extruding out the crack. This method was used in 152-mm (6-inch) and 254-mm (10-inch) diameter pipes,[4] but this was the first 406-mm (16-inch) diameter pipe test. The total bending load versus load-line displacement test record is shown in Fig. 11(a). In conducting the experiment, a leak in the bladder occurred after crack initiation. This is indicated in the bending load versus load-line displacement record in Fig. 11(a). Figure 11(b) shows the dc electric potential versus load-line displacement record, which also shows crack initiation and where the leakage occurred. After the leakage was detected and could no longer be overcome by pumping in additional water, the bending load was lowered and the pressure was totally relieved. The temperature of the test specimen

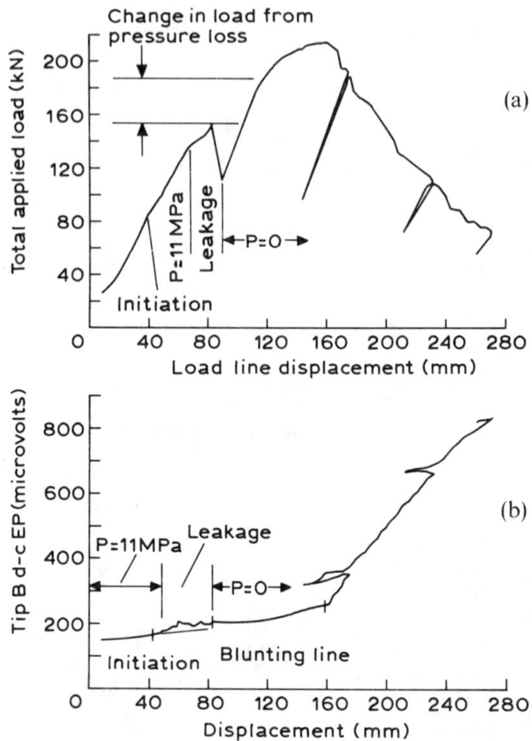

Fig. 11. Experiment 4141-9 pipe test data. (a) Load versus load-line displacement; (b) dc electric potential data versus load-line displacement.

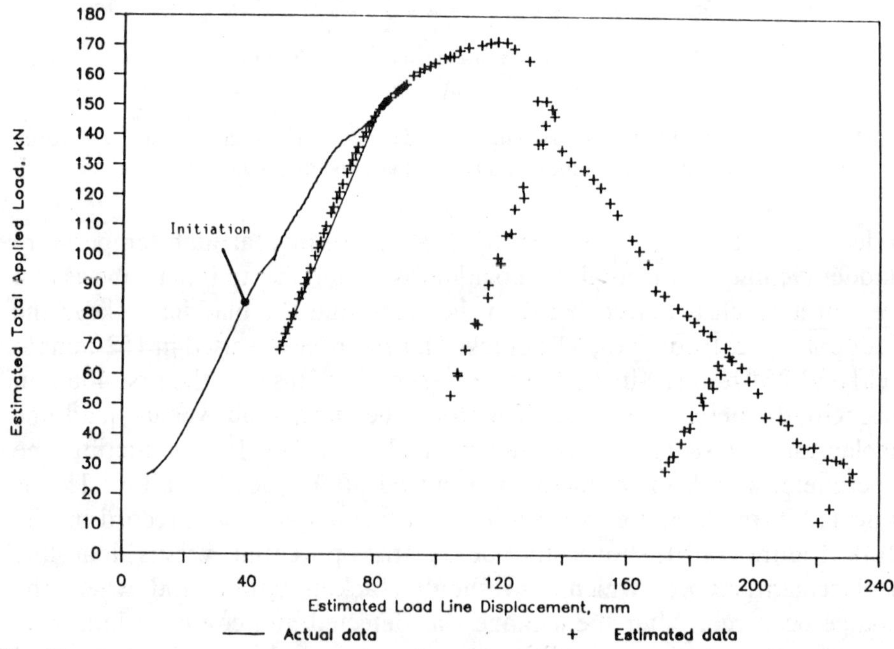

Fig. 12. Estimated load versus load-line displacement record for Experiment 4141-9 if pressure remained constant.

was then re-established and the test continued in four-point bending. Figure 12 shows the maximum load for combined bending and pressure using the change in bending load due to pressure loss in Fig. 11.

NET-SECTION-COLLAPSE, SCREENING CRITERION, IWB-3650, AND EPFM PREDICTIONS

The data from the four ferritic weld tests were analyzed using several methods. These results are summarized below.

Net-section-collapse analysis.

The net-section-collapse analysis[5] is the simplest analysis procedure. This assumes that there is negligible crack growth from crack initiation to the maximum load. The material strength data are used to estimate the flow strength, σ_f, of the material. The different flow stress definitions of the pipe base metals and weld metals are given in Table 3.

The most commonly used definitions of flow stess is the average of the yield and ultimate strength. It was shown that the average of the yield and ultimate strength definition corresponded closely to the 95% confidence level from circumferentially cracked pipe experiments (see Section 4.1 in Ref. 6).

Another definition experimentally derived from pipe fracture experiments is 1·15 times the average of the yield and ultimate strengths. The Degraded Piping Program results showed that this definition, on average, agreed with the circumferentially cracked ferritic and stainless steel pipe experiments, where fully plastic conditions were developed (see Section 4.1 in Ref. 6).

A third definition comes from the ASME IWB-3640 and IWB-3650 flawed pipe analyses. In these approaches, the flow stress is defined in terms of the ASME design stress, S_m. For austenitic steels the ASME flow stress is $3S_m$.[7] For ferritic steels, the ASME flow stress is $2{\cdot}4S_m$.[2] For weld metals, the base metal S_m values are used.

The calculated stresses by the net-section-collapse analysis and the experimental stresses are given in Table 3. In general, this shows that using the weld metal strength will result in overpredicting the failure loads.

For the cold-leg weld test, Experiment 4141-7, the net-section-collapse analysis predicted maximum stress was close to the experimental results when using σ_f equal to the average of the base metal yield and ultimate strengths. This was somewhat surprising due to the large diameter of the pipe where contained plasticity was anticipated. Using the ASME definition of flow stress, this pipe reached 31% higher stress than predicted.

For the whipjet pipe test, WJ-1, the experimental loads were slightly below the loads predicted by the net-section-collapse analysis when using the

TABLE 3
Summary of Ferritic Welded Pipe Results Compared to Net-Section-Collapse Analysis

Experiment number	Material	Flow stress						Experimental data				$(\sigma_b)_{NSC}$ Using						Experimental/predicted NSC^a using		
		$\sigma_{f1}{}^b$		$\sigma_{f2}{}^c$		$\sigma_{f3}{}^d$		σ_a		σ_b		$\sigma_{f1}{}^b$		$\sigma_{f2}{}^c$		$\sigma_{f3}{}^d$				
		MPa	(ksi)	MPa	(ksi)	MPa	(ksi)	MPa	(ksi)	MPa	(ksi)	MPa	(ksi)	MPa	(ksi)	MPa	(ksi)	$\sigma_{f1}{}^b$	$\sigma_{f2}{}^c$	$\sigma_{f3}{}^d$
4141-7	Base (516-70)	359	(52·1)	413	(59·9)	299	(43·4)		0	188	(27·3)	172	(25·0)	198	(28·8)	143	(20·8)	1·09	0·95	1·31
	Weld (SAW)	589	(85·4)	677	(98·2)	N/A			0	188	(27·3)	283	(41·0)	324	(47·1)	N/A		0·66	0·58	N/A
WJ-1	Base (A106B)	440	(63·8)	506	(73·4)	299	(43·4)		0	252	(36·6)	272	(39·5)	313	(45·4)	185	(26·9)	0·93	0·81	1·36
	Weld (SMAW)	492	(71·4)	566	(82·1)	N/A			0	252	(36·6)	304	(44·2)	350	(50·8)	N/A		0·83	0·72	N/A
4141-9[e]	Base (A106B)	423	(61·4)	487	(70·7)	299	(43·4)	34	(5·0)	124	(18·0)	161	(23·4)	192	(27·9)	101	(14·6)	0·81	0·70	1·18
	Weld (SAW)	455	(66·0)	522	(75·8)	N/A		34	(5·0)	124	(18·0)	176	(25·6)	209	(30·4)	N/A		0·75	0·65	N/A
4141-9[f]	Base (A106B)	423	(61·4)	487	(70·7)	299	(43·4)		0	158	(22·9)	203	(29·5)	234	(33·9)	143	(20·8)	0·78	0·68	1·10
	Weld (SAW)	455	(66·0)	522	(75·8)	N/A			0	158	(22·9)	218	(31·7)	251	(36·4)	N/A		0·72	0·63	N/A
4141-8	Base (A106B)	423	(61·4)	487	(70·7)	299	(43·4)	49	(7·1)	212	(30·8)	234	(34·0)	279	(40·5)	191	(27·7)	0·92	0·80	1·09
	Weld (SMAW)	455	(66·0)	522	(75·8)	N/A		49	(7·1)	212	(30·8)	257	(37·2)	304	(44·1)	N/A		0·86	0·74	N/A

[a] Axial stress plus bending stress/(Axial stress plus $(\sigma_b)_{NSC}$).
[b] $\sigma_{f1} = (\sigma_y + \sigma_u)/2$.
[c] $\sigma_{f2} = 1{\cdot}15(\sigma_y + \sigma_u)/2$.
[d] $\sigma_{f3} = 2{\cdot}4S_M$, ASME IWB-3650 definition of σ_f.
[e] Pressure and bending load, maximum load estimated.
[f] Bending only.

average of the base metal yield and ultimate strengths to define the flow stress. Using the ASME definition of flow stress, this pipe reached 36% higher stress than predicted.

For the surface-cracked ferritic SAW pipe test (Experiment 4141-8) the experimental loads were slightly below the predicted loads from the net-section-collapse analysis when using the base metal yield and ultimate strengths to define the flow stress. Using the ASME definition of flow stress, this pipe reached 9% higher stress than predicted. This was somewhat surprising since this was a low toughness weld in a large diameter pipe.

For the through-wall-cracked ferritic SAW pipe test, Experiment 4141-9, the net-section-collapse analysis predicted loads were for both the pure bending and the pressure and bending cases (see Table 3). The experimental maximum load for both of these cases was about 20% below the net-section-collapse analysis predicted loads when using the base metal yield and ultimate strengths to define the flow stress. When using the ASME definition of flow stress, the experimental loads were 10–18% higher than predicted by the net-section-collapse analysis.

In all of these cases, if the weld metal strength is used to define the flow stress, the net-section-collapse analysis predicted loads are higher than the experimental results. In some cases, this difference may be as much as 40%. Hence, the weld metal strength should not be used to define the flow stress in the net-section-collapse analysis.

An important conclusion from these results is that the use of the ASME definition of flow stress and the net-section-collapse analysis appears adequate to conservatively predict the maximum loads for these ferritic welded pipes.

IWB-3650 analysis

The ASME Section XI Committee is in the process of approving a ferritic pipe flaw evaluation criterion. This is known as Article IWB-3650. The technical basis for this work is given in Ref. 3. For circumferential cracked pipe, there are two material classifications, as noted in Table 4.

TABLE 4
Material Classifications of Circumferential Cracked Pipe

Category	*Materials*	*Toughness*
1	Base metals and E70XX SMAWs	$J_{Ic} > 105\ KJ/m^2$ ($600\ in\text{-}lb/in^2$)
2	SAW and non-70XX SMAWs	$J_{Ic} > 61{\cdot}3\ KJ/m^2$ ($350\ in\text{-}lb/in^2$)

A screening criterion is given in this criterion where the user determines if the failure mode will be limit load, elastic–plastic fracture, or linear-elastic fracture. If an elastic–plastic failure mode is indicated, then a simplified stress multiplier procedure is used. The stress multiplier accounts for the lower failure stresses from an elastic–plastic failure and is called a Z-factor. The Z-factors are calculated from the following equations:

$$Z = 1{\cdot}2[1 + 0{\cdot}021A(D - 4)] \tag{1a}$$

for material category 1, and

$$Z = 1{\cdot}35[1 + 0{\cdot}0184A(D - 4)] \tag{1b}$$

for material category 2.

where

$$A = [0{\cdot}125(R/t) - 0{\cdot}25]^{0{\cdot}25} \qquad \text{for } 5 \leq R/t \leq 10 \tag{2a}$$

$$A = [0{\cdot}4(R/t) - 3{\cdot}0]^{0{\cdot}25} \qquad \text{for } 10 < R/t < 20 \tag{2b}$$

D = nominal pipe diameter (in inches)
R = mean pipe diameter
t = thickness.

In the ASME IWB-3650 analysis, the flow stress for ferritic materials is defined as $2{\cdot}4S_m$ in the net-section-collapse analysis. The experimental loads were compared with the predicted loads from the net-section-collapse analysis for this definition of flow stress in Table 3.

For the four ferritic weld experiments conducted in this program, the calculated A, Z, and the ratio of experimental values to ASME predicted failure stresses, using the Z-factors, are given in Table 5. The ratio of experimental to predicted failure stresses ranged from 1·7 to 2·59, without the ASME safety factors. These are conservative predictions in all cases.

TABLE 5
Comparison of Experimental to IWB-3650 Predictions for Ferritic Weld Pipe Tests

Experiment number	*Material category*	R/t	A[a]	D (*mm*)	D (*inches*)	Z[b]	*Experimental/predicted*
4141-7	2 (SAW)	4·88	0·775	933	(36·73)	1·98	2·59
WJ-1	1 (SMAW)	7·03	0·891	152	(6)	1·25	1·70
4141-9	2 (SAW)	7·27	0·901	406	(16)	1·62	1·78
4141-8	2 (SAW)	7·27	0·901	406	(16)	1·62	1·76

[a] Using eqn (2).
[b] Using eqn (1).

Dimensionless Plastic-Zone criterion

The Dimensionless Plastic-Zone criterion was developed in the Degraded Piping Program initially to help determine when limit-load analysis or elastic–plastic fracture analyses are applicable.[8] This methodology was subsequently extended to making predictions in the elastic–plastic range from an emprical relation (eqn (3)).

$$\sigma/\sigma_{nsc} = (2/\pi)\arccos[e^{-C(DPZP)}] \tag{3}$$

where

σ = failure stress
σ_{nsc} = net-section-collapse analysis predicted failure stress
DPZP = Dimensionless Plastic-Zone parameter $= 2EJ_i/(\pi^2\sigma_f^2 D)$ (3a)
E = elastic modulus
$J_i = J$ at crack initiation which may be a valid J_{Ic}
σ_f = flow stress $= (\sigma_y + \sigma_u)/2$ (3b)
D = mean diameter
C = empirical constant: 3·0 for 95% confidence fit of all flaw types; 4·62 for average fit of through-wall crack data; 21·8 for average fit of surface-crack data.

Table 6 gives the comparison of the experimental results to the Dimensionless Plastic-Zone criterion for the four ferritic weld pipe experiments. The 95% confidence coefficient, C, gives predictions that are slightly more conservative than the ASME IWB-3650 analysis. Using the average fit coefficient gives predictions that, on average, are slightly better than the ASME IWB-3650 predictions. However, both the ASME and Dimensionless Plastic-Zone criterion analyses appear to be more conservative than needed.

EPFM analyses

The existing through-wall crack estimation schemes that are available in the NRCPIPE code[10] were compared with the ferritic weld pipe test results.

Table 7 gives the comparisons of the predicted and experimental loads for both crack initiation and at maximum load for Experiment 4141-7, the cold-leg SAW test. Calculations were carried out using both the base metal and weld metal stress–strain curves. In both cases, the weld metal deformation theory J-integral crack growth resistance, J_D–R, curve was used. This J_D–R curve was extrapolated using a power-law method.[9] These results showed that the initiation and maximum load predictions were slightly conservative for all the methods when the base metal stress–strain curve was used. The

TABLE 6
Comparison of Ferritic Weld Experimental Results to DPZP Criterion Predictions of Maximum Loads

Experiment number	σ_f		J_i		*D*		*DPZP*[a]	*Predicted* σ/σ_{nsc}		*Experimental* σ (σ_{nsc})	*Experimental/DPZP predicted*	
	MPa	*(ksi)*	MJ/m^2	*(lb/in)*	*mm*	*(inches)*		*Average*[b]	*95% Confidence*[c]		*Average*	*95% Confidence*
4141-7	359	(52·1)	0·192	(1 095)	933	(36·7)	0·063	0·46	0·38	1·09	2·37	2·87
WJ-1	440	(63·8)	0·105	(580)	168	(6·6)	0·125	0·62	0·52	0·93	1·50	1·78
4141-9	423	(61·4)	0·082	(470)	406	(16)	0·044	0·39	0·32	0·78	2·43	2·00
4141-8	423	(61·4)	0·082	(470)	406	(16)	0·044	0·75	0·32	0·92	1·22	2·87

[a] $DPZP = 2EJ/(\pi^2 \sigma_f^2 D)$.
[b] C = 4·6 for TWC, C = 21·8 for SC in eqn (3).
[c] C = 3·0 for TWC and SC.

TABLE 7
J-Estimation Scheme Predictions Compared to Experiment 4141-7 Results

Analysis method	*Predicted initiation load*				*Experimental/predicted*	
	a		*b*			
	MN	*(klb)*	*MN*	*(klb)*	*a*	*b*
GE/EPRI	2·33	(523)	2·84	(638)	1·34	1·10
NUREG/CR-3464	2·65	(595)	2·73	(614)	1·18	1·14
LBB.NRC	2·91	(654)	3·23	(727)	1·07	0·96
LBB.GE	2·68	(603)	3·40	(765)	1·16	0·92
LBB.ENG	2·61	(587)	3·36	(756)	1·19	0·93
Experimental	3·12	(702)	3·12	(702)	—	—
Analysis method	*Predicted maximum load*				*Experimental/predicted*	
	a		*b*			
	MN	*(klb)*	*MN*	*(klb)*	*a*	*b*
GE/EPRI	3·32	(746)	5·31	(1 194)	1·23	0·76
NUREG/CR-3464	3·91	(879)	5·51	(1 238)	1·04	0·74
LBB.NRC	3·91	(878)	5·71	(1 284)	1·04	0·72
LBB.GE	3·73	(838)	5·69	(1 278)	1·10	0·72
LBB.ENG	3·59	(807)	6·24	(1 403)	1·13	0·65
Experimental	4·09	(918)	4·09	(918)	—	—

[a] Using base metal stress–strain curve and weld J_D–R curve.
[b] Using weld metal stress–strain curve and weld J_D–R curve.

GE/EPRI method is the most conservative of all the *J*-estimation schemes. These predictions are, however, more accurate than the ASME IWB-3650 or Dimensionless Plastic-Zone criterion. If the weld metal stress–strain curve is used, then all the analyses overpredict the experimental loads.

Table 8 gives the comparisons of the predicted and experimental loads for both crack initiation and at maximum load for Experiment WJ-1, the 152-mm diameter SMAW test. In this case, the calculations were carried out using the base metal stress–strain curve. However, two different J_D–R curves were used. Both of these J_D–R curves were from the same weld, but from different C(T) specimens. This was done to assess the sensitivity of the analyses to different *J*–*R* curves. These J_D–R curves were extrapolated using a power-law method.[9] The modified *J*-resistance, J_M–R, curves were also used, and they were extrapolated linearly.[9] These results showed that initiation loads were, in general, conservatively predicted. The maximum

TABLE 8
NRCPIPE Predicted Load/Experimental Load for Pipe Experiment WJ-1 using Power-Law Extrapolation of *J-R* Curves from Different Specimens

Analysis method	*C(T) Specimen No. 16*			*C(T) Specimen No. 13*		
	Initiation load	*Maximum load*		*Initiation load*	*Maximum load*	
		J_D	J_M		J_D	J_M
GE/EPRI	0·79	0·88	0·97	0·67	0·95	0·99
NUREG/CR-3464	1·02	1·09	1·09	0·79	1·09	1·10
LBB,NRC	0·96	1·00	1·04	0·81	1·03	1·04
LBB.GE	0·92	1·02	1·12	0·78	1·11	1·14
LBB.ENG	0·91	1·02	1·12	0·77	1·10	1·14

loads were predicted better using the J_D–R curves. The J_M–R curve predictions of the maximum load were, on average, 4% higher in this case than the maximum load predictions using the J_D–R curves. In general, these predictions are far more accurate than either the ASME IWB-3650 or the Dimensionless Plastic-Zone criterion, but were very slightly nonconservative. If the weld metal stress–strain curve is used, then all the analyses overpredict the experimental loads.

In regard to the sensitivity of the load predictions using the *J–R* curves from the two different specimens for Experiment WJ-1, the initiation loads varied by 15·6% on average, the maximum load using the J_D–R curve varied by 5·4% and the maximum loads using the J_M–R curve varied by only 1·4%. These results show that the initiation load is the most sensitive to variations in the *J–R* curve.

The results from Experiment 4141-9 are compared with the *J*-estimation scheme predictions in Table 9. For this experiment, the experimental results

TABLE 9
Comparison of Experimental Maximum Load to NRCPIPE Predictions for Experiment 4141-9[a]

Analysis method	*Experimental/predicted maximum load*
GE/EPRI	1·258
NUREG/CR-3464	1·021
LBB.NRC	0·995
LBB,GE	1·123
LBB.ENG	1·142

[a] Maximum load data were used where the internal pressure was zero, so these were four-point bending solutions.

consisted only of the maximum load where the pressure was zero. The predictions used the base metal stress–strain curve and the weld metal J_D–R curve. The J_D–R curve was extrapolated in the power-law fashion as in Ref. 9. The comparisons in Table 8 show reasonable predictions for all the methods, with the GE/EPRI being the most conservative and the LBB.NRC method being the least conservative. As shown in Ref. 10, this is the general trend of the analyses for all the through-wall cracked pipe data.

CONCLUSIONS

The results of these experiments and analyses show: (1) that the SMAW welds examined had higher toughness values than the welds used in the technical basis of the ASME Section XI ferritic pipe flaw evaluation procedure, IWB-3650; (2) the failure loads were well above those predicted by the ASME IWB-3650 flaw evaluation procedure; (3) the failure behavior, either in terms of small unstable crack jumps or crack growth in irregular paths, is not considered in the fracture analyses, and (4) variations in J–R curves tend to have a larger effect on the predicted load at crack initiation in pipes rather than on the maximum load predictions.

ACKNOWLEDGEMENTS

The work described in this paper was sponsored by the US Nuclear Regulatory Commission's Office of Nuclear Regulatory Research under contract number NRC-04-84-103 for the Degraded Piping Program—Phase II. The NRC program officer is Mr Michael Mayfield. We thank the NRC for the opportunity to conduct this research and publish these results.

Others at Battelle who also contributed to the results in this paper are: Dr J. Ahmad, Mrs B. Blanton, Dr F. Brust, Mr N. Frey, Mr J. Garrabrant, Mr R. Gertler, Mr N. Ghadiali, Mr G. Kramer, Mr J. Kramer, Mr M. Landow, Dr C. Marschall, Mr M. Oliver, Mr J. Ryan, Mr D. Shoemaker and Mr M. Wilson.

REFERENCES

1. Marschall, C. W., Landow, M. P. & Wilkowski, G. M., Effect of dynamic strain aging on fracture resistance of carbon steels operating at light-water-reactor temperatures. Presented at ASTM 21st National Fracture Symposium, June 1988. ASTM STP 1074, 1990, pp. 339–60.

2. Evaluation of flaws in ferritic piping. EPRI Report NP-6045, prepared by Novetech Corporation, October 1988.
3. Wilkowski, G. M., *et al.*, Degraded Piping Program—Phase II. Sixth Program Report, October 1986–September 1987, Batelle Columbus Laboratories, NUREG/CR-4082, Vol. 6, April 1988.
4. Wilkowski, G. M., *et al.*, Degraded Piping Program—Phase II. Semiannual Report, October 1985–March 1986, Battelle Columbus Division, NUREG/CR 4082, Vol. 4, September 1986.
5. Kanninen, M. F., Broek, D., Marschall, C. W., Rybicki, E. F., Sampath, S. G., Simonen, F. A. & Wilkowski, G. M., Mechanical fracture predictions for sensitized stainless steel piping with circumferential cracks. Final Report, EPRI NP-192, September 1976.
6. Wilkowski, G. M., *et al.*, Degraded Piping Program—Phase II. Semiannual Report, April 1986–September 1986, Battelle Columbus Laboratories, NUREG/CR-4082, Vol. 5, April 1987.
7. Evaluation of flaws in austenitic steel piping. (Technical basis document for ASME IWB-3640 analysis procedure), prepared by Section XI Task Group for Piping Flaw Evaluation, EPRI Report NP-4690-SR, April 1986.
8. Wilkowski, G. M., *et al.*, Degraded Piping Program—Phase II. Semiannual Report, October 1984–March 1985, Battelle Columbus Laboratories, NUREG/CR-4082, Vol. 2, July 1985.
9. Wilkowski, G. M., Marschall, C. W. & Landow, M. Extrapolation of C(T) specimen *J–R* curves for use in pipe flaw evaluations. Paper presented at ASTM 21st National Fracture Mechanics Symposium, June 1988. ASTM STP 1074, 1990, pp. 56–84.
10. Wilkowski, G. M., *et al.*, Degraded Piping Program—Phase II. Final Program Report, March 1984–January 1989, Batelle Columbus Laboratories, NUREG/CR-4082, Vol. 8, March 1989.

Int. J. Pres. Ves. & Piping **43** (1990) 351–366

Strength Behaviour of Flawed Pipes under Internal Pressure and External Bending Moment: Comparison between Experiment and Calculation

D. Sturm & W. Stoppler

Staatliche Materialprüfungsanstalt (MPA), University of Stuttgart, Pfaffenwaldring 32, D-7000 Stuttgart 80, Germany

ABSTRACT

Experimentally determined failure curves for pipes weakened by surface longitudinal or circumferential defects, were compared with results calculated with the aid of engineering approximation methods. Considering the scatter bands of the mechanical properties and the geometrical dimensions, then by use of the engineering approximation methods, one can make only rough estimates of the load bearing behaviour.

INTRODUCTION AND OBJECTIVES

The design of the primary cooling systems of pressurized water reactors in the Federal Republic of Germany accords with the KTA-Rule 3201.2[1] under which both the design and accident conditions are taken into account. Details concerning the permissible defect size are contained in the KTA-Rule 3201.3.[2] However, fracture mechanics analyses and defect size determination by non-destructive testing form the basis of an individual evaluation of every important piping component.

Extensive investigations have shown that under near-operational conditions on pipes similar to constructional units ($Di \times t = 760 \times 52$ mm) made of materials of high upper-shelf impact energy, the critical defect dimensions determined by fracture mechanics, even under accident conditions, are always several times larger than those of the smallest defect which is detectable by non-destructive testing (Fig. 1).[3]

In the present paper, it is shown how the currently used engineering

Int. J. Pres. Ves. & Piping 0308-0161/90/$03·50

Exclusion of rupture of pipes with longitudinal flaw

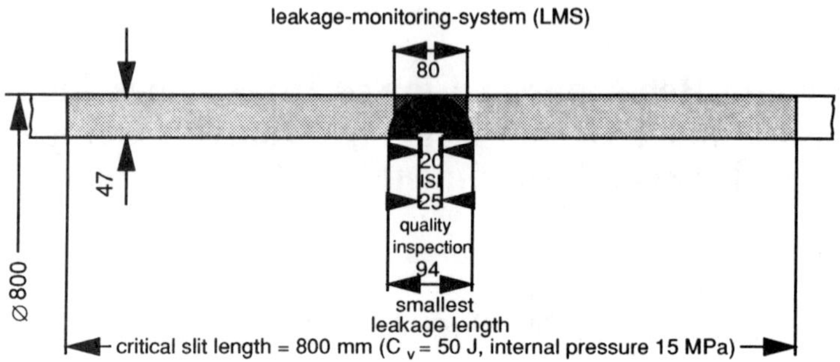

Exclusion of rupture of pipes with circumferential flaw

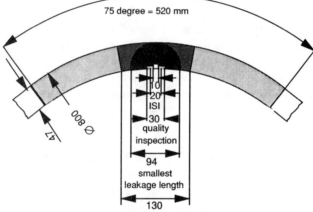

critical slit length for a bending moment M_b= 5 MNm (C_v= 50 J)

75 degree = 520 mm

10

20

ISI

30

quality inspection

94

smallest leakage length

130

leakage-monitoring-system (LMS)

Ø 800

47

Fig. 1. Permissible, detectable and critical through-wall lengths for longitudinal and circumferential cracks in pipes.

approximation methods are able to accomplish the calculation of the critical size of longitudinal and circumferential defects. The effect that the scatter of the wall thickness and the diameter, as well as the mechanical properties, can have on the calculated load-bearing capacity is illustrated. Since the numerical methods of calculation based on the finite element method are normally only applied if the critical loading (e.g. the internal pressure or external bending moment at crack initiation or instability), is to be determined for a specified assumed defect geometry under complex loading a corresponding comparison is dispensed with here, because of the high calculational cost. These results also depend on the degree of scatter of the input data.

PIPES, MATERIALS, TEST RESULTS

For the tests, pipes principally of 800 mm diameter and 47 mm wall thickness were available. These dimensions are close to the main cooling piping of a 1300 MWe pressurized water reactor (PWR). They were made of ferritic constructional steel and whilst having equivalent yield and tensile strength values they had different upper-shelf notch impact energy values ($C_v = 200$ J or $C_v = 50$ J).[4,5] The pipes were tested under the conditions prevailing in a PWR, particularly with respect to internal pressure and temperature ($p_i = 15$ MPa and temperature = 300°C). Some of the pipes were also subjected to additional monotonic loading by an external bending moment in a four point bending test. Before testing, the pipes were provided

TABLE 1
Material Properties and Pipe Dimensions

	Unit	*Material*			
		20 MnMoNi 55		*NiMoCr-melt*	
		Value	*Standard deviation*	*Value*	*Standard deviation*
Guarantee values					
Yield strength $R_{p0.2}$ 300°C	MPa	≥392	—	—	—
Ultimate strength R_m 300°C	MPa	≥530	—	—	—
Measured values					
Yield strength $R_{p0.2}$ 300°C	MPa	428	10	417	43
Ultimate strength R_m 300°C	MPa	605	17	622	25
Outer diameter	mm	797·9	3·8	793·9	4·5
Wall thickness	mm	47·2	0·4	47·2	0·4
Notch depth	mm	—	0·1	—	0·1

with mechanically-introduced longitudinal or circumferential defects. In this manner the load at crack initiation or at instability associated with a specific defect geometry could be determined. The statistically evaluated material properties and pipe dimensions together with the guarantee values and the calculated standard deviations are given in Table 1. The material properties were taken perpendicular to the flaw direction. The data from the pipe tests are tabulated in Tables 2 and 3. In both the pipes with longitudinal defects and those with circumferential defects, a clear influence of the toughness, expressed as upper-shelf notch impact energy, on the load-bearing behaviour can be recognized.

TABLE 2
Test Results, Pipes with Longitudinal Defects

Experiment number	*Dimensions (mm)*	*Flaw*			*Temperature (°C)*	*Failure pressure (MPa)*	*Nominal stress (MPa)*
		Type	*Length (mm)*	*Depth (mm)*			
BVZ 010		S	650	47·2	20	23·8	187
BVZ 011	Outer	S	1 102	47·2	20	14·8	117
BVZ 012	diameter	S	1 105	47·2	20	14·4	113
BVZ 022	800	A	782	38·3	305	21·9	173
BVZ 030		A	1 500	36·2	300	19·5	155
BVZ 080	Wall	A	1 500	36·2	17	20·4	164
BVZ 070	thickness	I	700	38·2	265	22·4	177
BVZ 060	47·2	A	1 500	36·0	305	18·0	143
BVS 010	Length	S	800	47·2	155	17·5	137
BVS 020	2 500/	A	709	37·3	320	14·8	117
BVS 030	5 000	A	1 100	35·0	305	13·1	103
BVS 042		A	709	38·3	245	16·8	132

BVZ: Material 20 MnMoNi 55 (CVN upper shelf impact energy $C_v > 150$ J).
BVS: NiMoCr–melt (CVN upper shelf impact energy $C_v \sim 50$ J).
Flaw type: A, outer surface crack; I, inner surface crack; S, through-wall crack, slit.

TABLE 3
Test Results, Pipes with Circumferential Defects

Experiment number	*Dimensions (mm)*	*Flaw*			*Temperature (°C)*	*Internal pressure (MPa)*	*Maximum bending moment (kNm)*
		Type	*Angle (degrees)*	*Depth (mm)*			
BVZ 090		O	—	—	20	15·2	12·110
BVZ 161	Outer	S	20	47·2	130	0·0	13·910
BVZ 091	diameter	S	60	47·2	20	0·0	9·650
BVZ 100	800	S	60	47·2	5	15·0	8·970
BVZ 160		A	20	26·0	233	14·6	13·110
BVZ 110	Wall	A	60	39·4	220	13·4	7·050
BVZ 120	thickness	A	60	20·0	195	13·3	11·030
BVZ 130	47·2	I	60	20·0	235	15·2	11·300
BVZ 150		A	90	20·0	267	15·7	9·300
BVZ 140		A	120	20·0	248	15·2	8·300
BVS 090	Length	O	—	—	250	14·8	11·500
BVS 060	5 000	S	60	47·2	140	15·6	5·510
BVS 110		A	20	36·0	247	15·3	8·500
BVS 102		A	20	20·0	235	15·1	11·240
BVS 070		A	60	20·0	232	16·0	6·600
BVS 080		A	120	20·0	253	14·8	5·550

BVZ: Material 20 MnMoNi 55 (CVN upper shelf impact energy $C_v > 150$ J).
BVS: NiMoCr-melt (CVN upper shelf impact energy $C_v \sim 50$ J).
Flaw type: O, without crack; A, outer surface crack; I, inner surface crack; S, through-wall crack, slit.

CALCULATIONAL METHODS

The bases of the engineering approximation methods for the calculation of the failure pressure or the maximum tolerable bending moment are failure theories, which take into account the linear elastic or partially plastic behaviour of the material, and the results from tests which serve to calibrate the geometrical factors. In addition to the tensile strength, to some extent the toughness values, such as notch impact energy, fracture toughness or crack resistance, are also required.

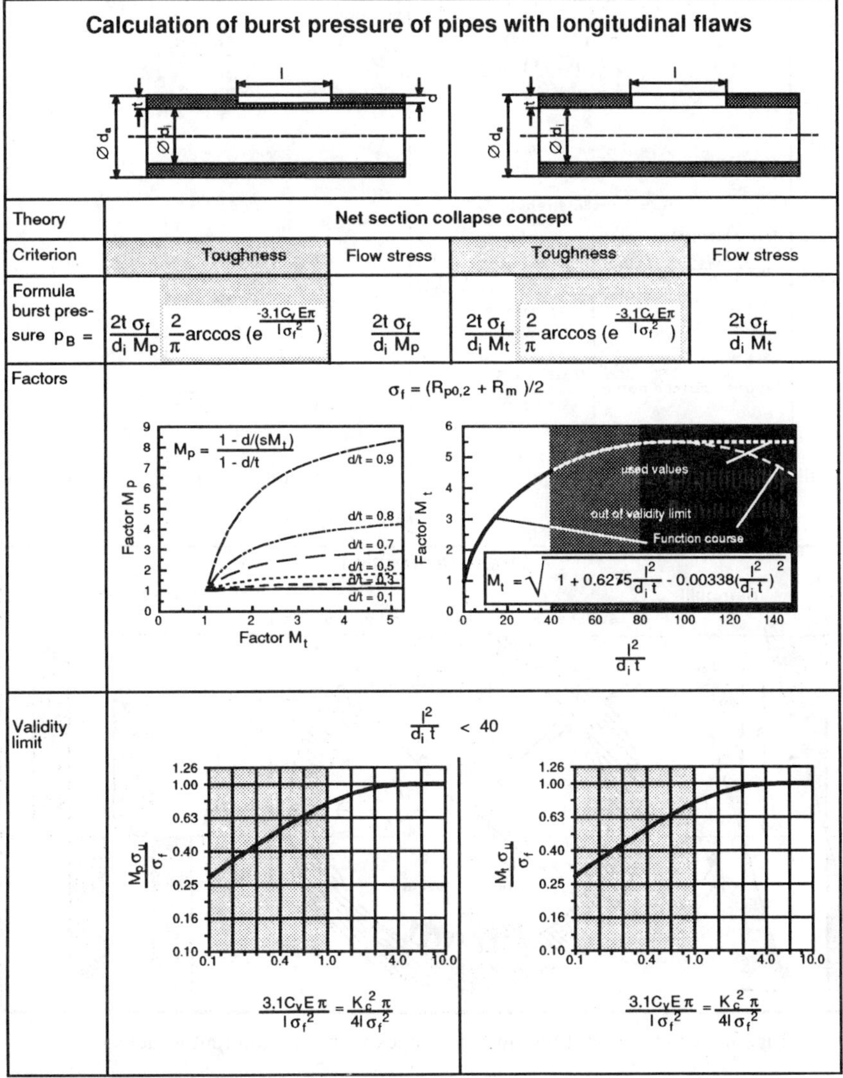

Fig. 2. Calculational methods for pipes with longitudinal defects.

Calculation of the failure moment of pipes exposed to internal pressure and external bending

Theory	Bernoulli "Moment"-method		Net section collaps concept	
Criterion	Tensile strength		Flow stress	
	Surface flaw	Through-wall crack	Surface flaw	Through-wall crack
Formula $M_b =$	$\frac{I_{\hat{x}}}{a}\left(R_m - \frac{A_w}{A_n} p_i\right) - (\hat{y} + b)\, A_w\, p_i$		$2\, \sigma_f\, r_m^2\, t\, (2\, k \sin\beta - f \sin\alpha)$	
Factors	$I_{\hat{x}} = I_x - \hat{y}^2 A_n - \left\lvert A_f + \frac{\sin 2\alpha}{2}(R^2 - r^2)\right\rvert \frac{R^2 + r^2}{4}$ $I_x = \frac{\pi}{4}(r_a^4 - r_i^4)$ $\hat{y} = \frac{2}{3}\frac{(R^3 - r^3)\sin\alpha}{A_n}$ $A_a = \pi r_a^2$ $A_i = \pi r_i^2$ $A_f = (R^2 - r^2)\,\mathrm{arc}\,\alpha$ $A_n = A_a - (A_i + A_f)$ R_m = Tensile strength		$\beta = \frac{\pi - f\alpha}{2} - \frac{\pi\, r_m\, p_i}{4\, t\, \sigma_f}$ $\sigma_f = \frac{R_m + R_{p\,0.2}}{2}$ Variation $R_{eH} \leq \sigma_f \leq R_m$ $r_m = r_i + t/2$ $f = d/t$	
	Internal surface notch $b = \frac{2}{3}\frac{(R^3 - r^3)\sin\alpha}{A_w}$ $R = (r_i + d)$ $r = r_i$ $A_w = A_i + A_f$ $a = \hat{y} + r_a$ External surface notch $b = 0$ $A_w = A_i$ $R = r_a$ $r = (r_a - d)$ for $r_a \cos\alpha < r$ $a = \hat{y} + r$ for $r_a \cos\alpha \geq r$ $a = \hat{y} + R$	$b = 0$ $A_w = A_i$ $R = r_a$ $r = r_i$ for $\alpha < 90°$ $a = \hat{y} + r$ for $\alpha \geq 90°$ $a = \hat{y} + R$	for $\beta + \alpha \leq \pi$ $k = 1$ for $\beta + \alpha > \pi$ $k = 1 - f$ $\beta = \pi + \frac{1}{k}\left(\frac{f\alpha - \pi}{2} - \frac{\pi\, r_m\, p_i}{4\, t\, \sigma_f}\right)$	$f = 1$

Stress Distribution

Fig. 3. Calculational methods for pipes with circumferential defects.

The best known calculational methods for pipes with longitudinal surface and through-wall defects together with their bounding conditions, are shown in Fig. 2. Both expressions are based on the assumption of plastic collapse in the net cross-section where for one expression the flow stress σ_f[4,6–8] and for the other in addition the toughness of the material has to be inserted as the failure criterion. By contrast, for the calculation of the load-bearing behaviour of pipes with surface or through-wall defects in the circumferential direction, essentially two engineering calculational methods find application (Fig. 3), which are based either on the assumption of plastic collapse in the net cross section[5,9–11] or on the classical bending theory.[5,9,12] Toughness or fracture mechanics property values are not taken into consideration in the above. The range of applicability of these calculational methods therefore remains strictly limited to ductile materials of high notch impact energy.

RESULTS

Pipes with longitudinal defects

Using the two calculational equations for the net-section-collapse shown in Fig. 2 (based on toughness (T) or based on flow stress (F)) the load-bearing curves for pipes of 800 mm diameter and 47 mm wall thickness were calculated, specifically for surface notches of various depths (notch depth ratios of $d/t = 0{\cdot}25, 0{\cdot}5$ and $0{\cdot}75$) and also for through-wall slits (leak-before-break curves, which separate the leak regime from the catastrophic failure regime). These curves are shown graphically in Fig. 4.

In the application of the toughness dependent method, the upper shelf values of $C_v = 50$ J and $C_v = 200$ J were put into the equation as the notch toughness energy. The curves for the high notch impact energy value (upper bound of the scatter band) lie distinctly above the curves for the low toughness (lower bound of the scatter band).

The curves calculated by the flow stress dependent method—in this method of calculation a deformable material of high upper-shelf notch impact energy is assumed—up to a defect length of about 400 mm, are equal in shape and magnitude to the upper bound curve, calculated by the toughness dependent equation. For greater defect lengths, the curves from the flow stress method stay at a significantly higher stress.

Comparison of the experimentally determined curves (Figs 5 and 6) with the calculated ones (toughness dependent) shows that the calculated curves lie below (about 30%) the experimental ones. This means that for the range of pipe dimensions investigated, the calculated curves lie on the side of safety (are conservative) for both ranges of toughness.

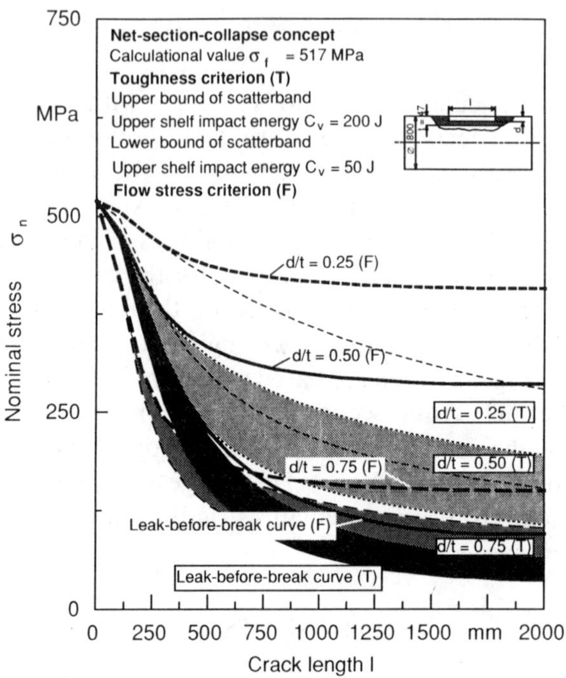

Fig. 4. Calculated load-bearing curves for pipes containing longitudinal flaws and with different values of notch impact toughness, using the plastic collapse method.

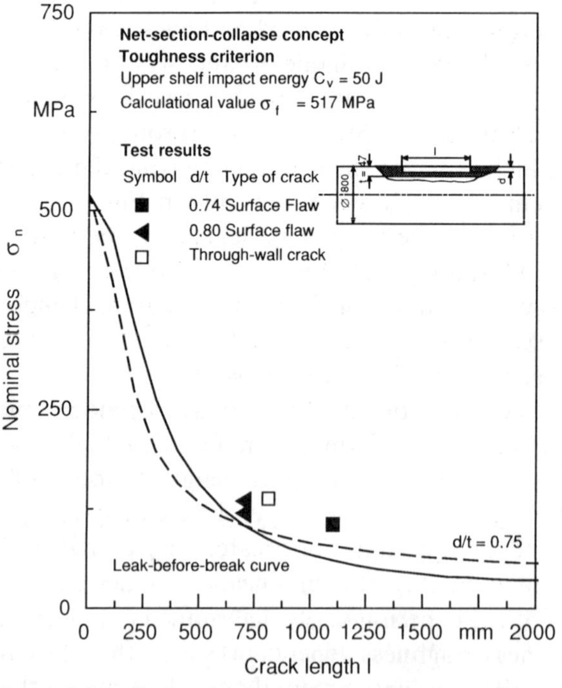

Fig. 5. Comparison of the calculated curves (plastic collapse, toughness dependence) with the test results on pipes of low toughness material (NiMoCr-melt) containing longitudinal flaws.

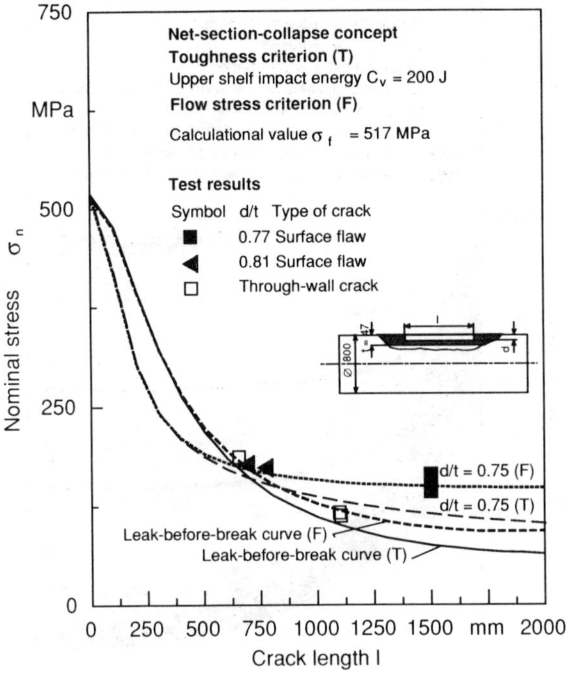

Fig. 6. Comparison of the calculated curves (plastic collapse, toughness or flow stress dependence) with the test results on pipes of high toughness material (20MnMoNi55) containing longitudinal flaws.

As an example, in Fig. 6 the calculated curves using the flow stress dependent method are compared with the curves based upon the experimental results of the pipes of the high toughness material; good agreement was found. In calculating the curves, the choice of failure criterion to be inserted into the equations is significant. In the literature the flow stress, the mean of the yield and ultimate tensile strength, has found application as a suitable value for the failure stress. The test results described here confirm this application, using a flow stress of 517 MPa.

Flow stress values that lie between the yield and ultimate strength are also used as failure criteria. The influence of the choice of the magnitude of the failure stress, the scatter in diameter and wall thickness on the load-bearing capacity, and the leak-before-break curves are shown in Fig. 7. The curves were calculated by the flow stress dependent equation. The diameter was inserted with a scatter band of $\pm 1\%$ of the nominal diameter of 800 mm and the wall thickness was taken with a scatter band of $\pm 15\%$ of the nominal wall thickness of 47 mm according to DIN 1629.[13] In Fig. 9 it is clearly shown that the influence of the scatter of diameter and wall thickness is of minor importance compared with the influence of the magnitude of the chosen failure stress value.

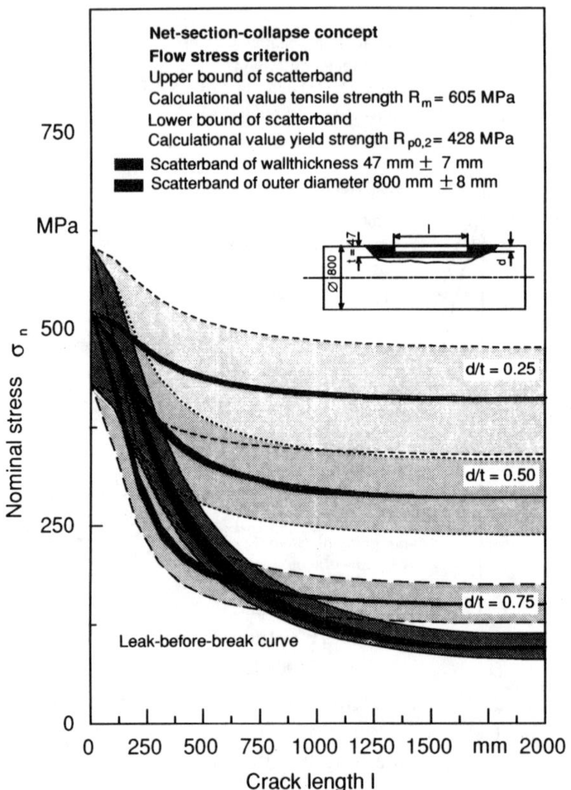

Fig. 7. Influence on the load-bearing curves of the failure criterion and the scatterbands in diameter and wall thickness using the plastic collapse criterion on pipes with longitudinal flaws.

Pipes with circumferential defects

For the calculation of the load-bearing behaviour of pipes with defects in the circumferential direction, which are loaded by internal pressure and by an additional external bending moment, the equations shown in Fig. 3 (moment method (M) and net-section-collapse concept (P)) were used. Surface notches of various depths ($d/t = 0{\cdot}25$, 0·5 and 0·75) and the through-wall slit were considered. The slit curve represents the leak-before-break curve, which separates the leak regime from the catastrophic failure regime. A diameter of 800 mm and a wall thickness of 47 mm were taken as the nominal dimensions. The graphical representation of these curves is given in Fig. 8. In the equations it is assumed that the material is ductile. In both equations the flow stress $\sigma_f = 517$ MPa was taken as the failure stress. The curves calculated on the basis of net-section-collapse lie considerably higher than those determined by the moment method.

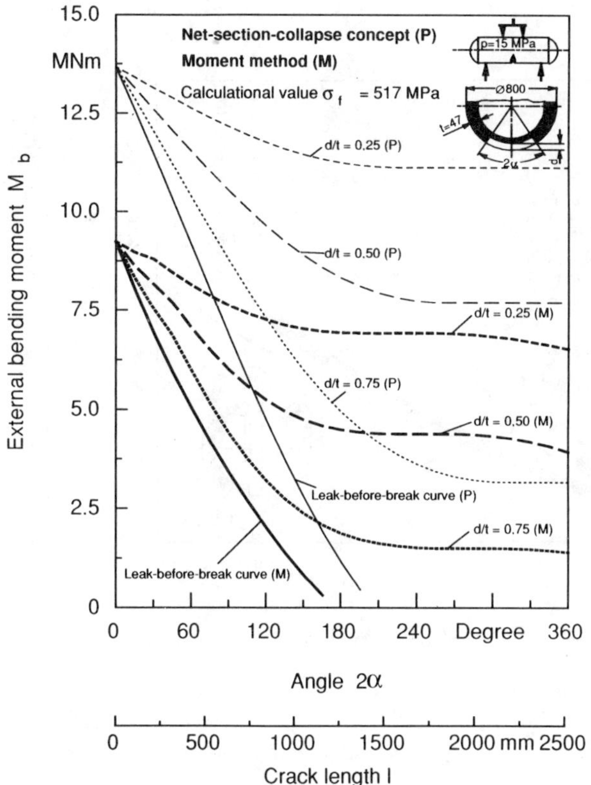

Fig.8. Calculated load-bearing curves for pipes with circumferential defects using net-section collapse and moment methods.

Comparison of the experimentally determined curves (Figs 9–12) with the calculated ones shows that the use of the net-section-collapse concept for the tubes of the lower toughness material (Fig. 9), leads to calculated values for the tolerable bending moment or critical defect lengths that are too high. For pipes of the high toughness material (Fig. 10) this statement is also valid for circumferential surface notches. The experimentally determined leak-before-break curve shows good agreement with the calculated one. In contrast to this, the moment method gives values which lie below the experimentally determined ones (Figs 11 and 12), especially if the yield strength $R_{p0 \cdot 2}$ instead of the flow stress σ_f is taken as the failure criterion (Fig. 13).

The influence of the failure criterion on the path of the load-bearing and leak-before-break curves may be seen in Fig. 13. By taking the test results into consideration, in high toughness pipes a failure criterion up to the ultimate tensile strength R_m can be adopted; for tubes of the lower toughness material, the failure criterion adopted must not exceed the flow stress σ_f.

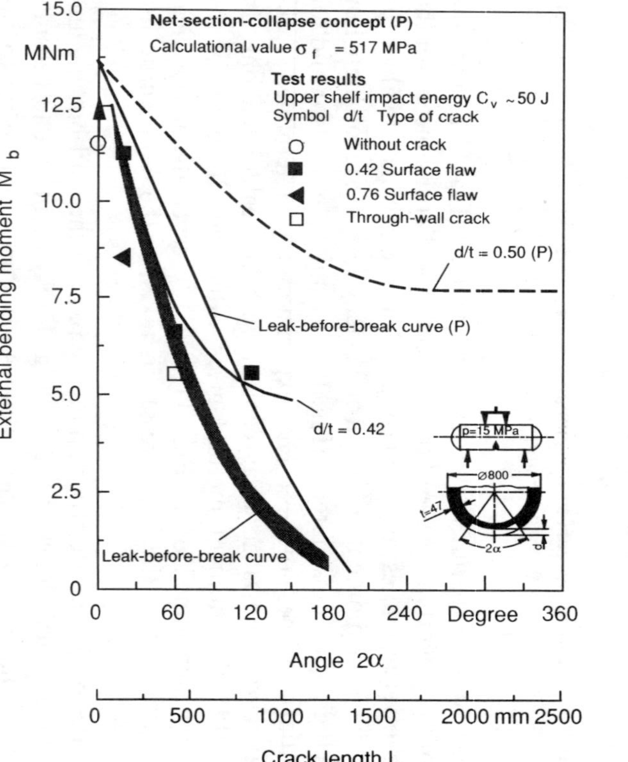

Fig. 9. Comparison of the calculated curves (net-section collapse) with the test results for pipes of low toughness material containing circumferential cracks.

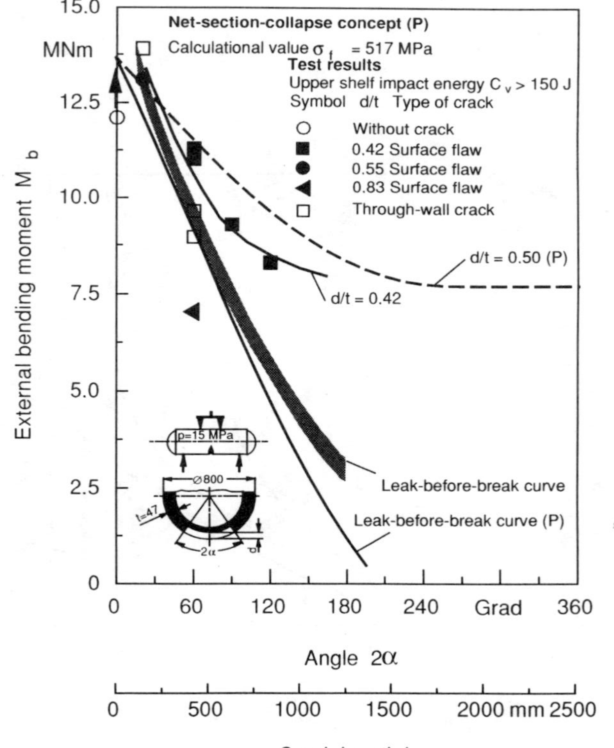

Fig. 10. Comparison of the calculated curves (net-section collapse) with the test results for pipes of high toughness material containing circumferential cracks.

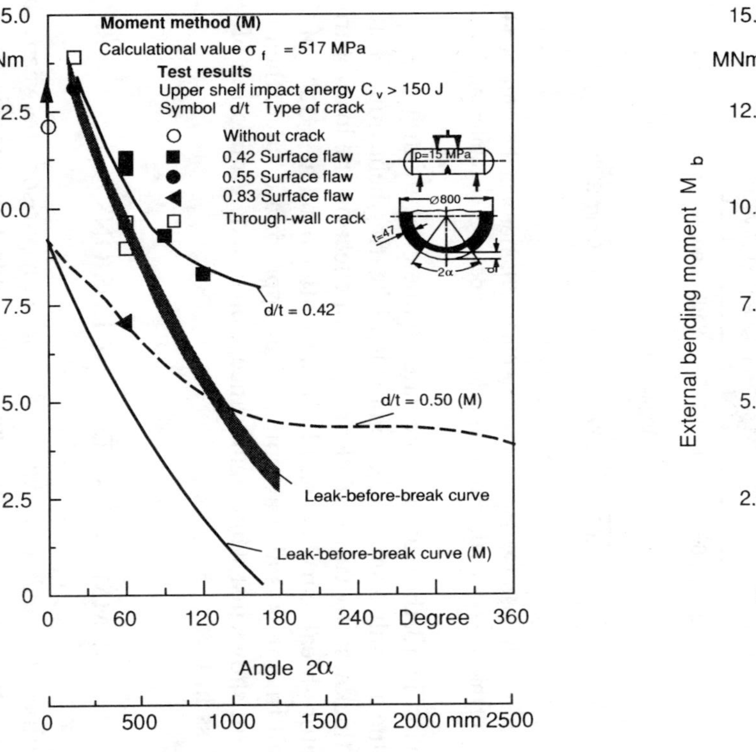

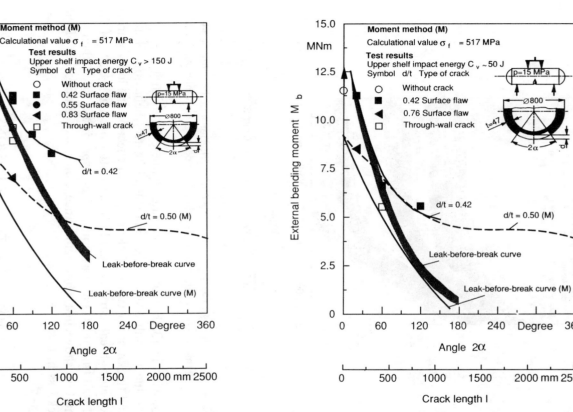

Fig. 11. Comparison of the calculated curves (moment method) with the test results for pipes of high toughness material containing circumferential cracks.

Fig. 12. Comparison of the calculated curves (moment method) with the test results on pipes of low toughness material containing circumferential cracks.

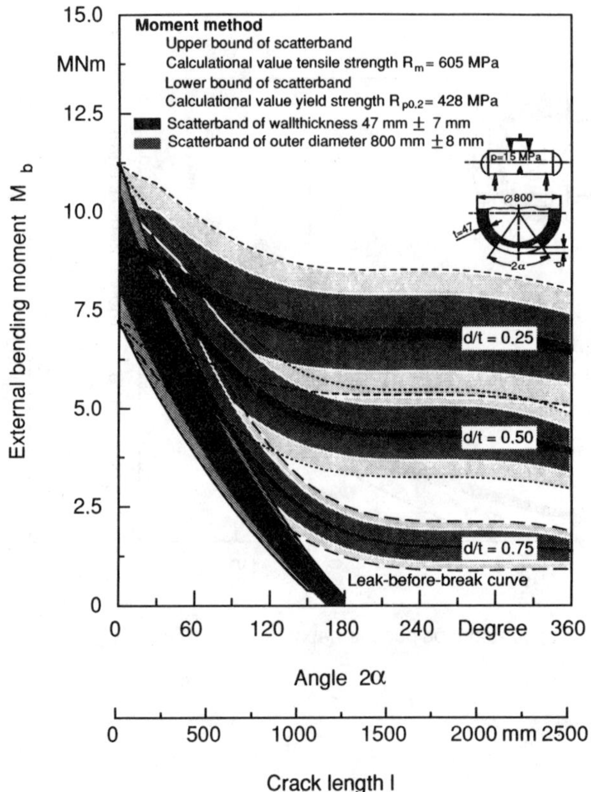

Fig. 13. Influence on the load-bearing curves of the failure criterion and the scatterbands in diameter and wall thickness using the moment method for pipes with circumferential cracks.

Also shown in Fig. 13, the scatter of the outer diameter of about $\pm 1\%$ of the nominal value has only a small effect on the load-bearing behaviour. Scatter in the wall thickness, by contrast, clearly has an effect on the load-bearing and leak-before-break curves (Fig. 13). If the scatter band at $\pm 15\%$ of the nominal wall thickness is also accepted as an appropriate value, then this representation shows just what significance a precise determination of the wall thickness has.

SUMMARY AND CONCLUSIONS

Tests on pipes formed the basis for the comparison between the experimental and calculated failure loads. The outer diameter was 800 mm and the wall thickness, 47 mm. Artificial defects of various dimensions lying in the longitudinal or circumferential direction were introduced into the pipes.

The failure loads were calculated with the aid of engineering approximation methods which were developed for ductile materials.

The effect of the scatterband of the material properties, the dimensions and the failure criterion ($R_{p0 \cdot 2} < \sigma_f < R_m$) on the calculated failure loads was determined. To accomplish a realistic prediction for leak or break, in the evaluation of calculational methods the scatterbands of input values (material properties, pipe and defect dimensions) must be taken into account.

Calculated results (failure loads) which lie conservatively with respect to safety are obtained by the use of guaranteed minimum values

—for pipes with a longitudinal defect based on calculation of plastic collapse (toughness criterion);

—for pipes with a circumferential defect based on calculation using the moment method.

This statement is also valid for smaller pipe dimensions than the main coolant piping described here.[9]

The defect dimensions allowed by the regulatory codes are several times smaller than the critical defect sizes determined experimentally; hence a sufficient margin of safety always exists.

More precise statements concerning critical defect geometry, including crack initiation and instability behaviour, and accounting for the scatter of the input data, e.g. mechanical properties, are only possible with the aid of advanced means of calculation based on the finite element method,[14–16] which leads to correspondingly higher calculational costs.

REFERENCES

1. Sicherheitstechnische Regel des Kerntechnischen Ausschusses KTA Nr.: 3201.2 Komponenten des Primärkreises von Leichtwasserreaktoren. *Teil 2: Auslegung, Konstruktion und Berechnung. Fassung 3/84.* Carl Heynmanns Verlag KG, Köln, 1984.
2. Sicherheitstechnische Regel des Kerntechnischen Ausschusses KTA Nr.: 3201.3 Komponenten des Primärkreises von Leichtwasserreaktoren. *Teil 3: Herstellung. Fassung 12/87.* Carl Heynmanns Verlag KG, Köln.
3. Bartholome G. & Bieselt, R. W., The application of Leak Before Break concepts on nuclear piping of KWU-plants. Paper presented at the IAEA Meeting on Recent Trends in the Development of Primary Circuit Technology, Madrid, 25–28 November 1985.
4. Sturm, D. & Stoppler, W., *Forschungsvorhaben Phänomenologische Behälterberstversuche—Versuche zum Traglast—und Bruchverhalten von Rohren mit Längsfehlern—150 279, Phase 1.* MPA, Stuttgart, FRG, July 1985. Distributed by Fachinformationszentrum (FIZ), Karlsruhe.

5. Sturm, D. & Stoppler, W., *Forschungsvorhaben Phänomenologische Behälterberstversuche—Versuche zum Traglast—und Bruchverhalten von Rohren mit Umfangsfehlern—150 279, Phase 2.* MPA, Stuttgart, December 1987.
6. Hahn, G. T., Sarrate, M. & Rosenfield, A. R., Criteria for crack extension in cylindrical pressure vessels. *Int. J. Fracture Mech.*, **6** (1969) 187–210.
7. Eiber, R. J., Maxey, W. A., Duffy, A. R. & Atterbury, T. J., Investigation of the initiation and extent of ductile pipe rupture. Final Report Task 17, BMI-1866 (1966) and BMI-1908 (1971).
8. Maxey, W. A., Fracture initiation, propagation and arrest. 5th Symp. on Line Pipe Research, Houston, TX, 1974, American Gas Association Catalogue No. L30174.
9. Absicherungsprogramm zum Integritätsnachweis von Bauteilen. Zusammenfassender Bericht mit Bewertung, MPA, Stuttgart, FRG, February 1989.
10. Kanninen, M. F., Broek, D., Marschall, C. W., Rybicki, E. F., Sampath, S. G., Simonen, F. A. & Wilkowski, G. M., Mechanical fracture predictions for sensitized stainless steel piping with circumferential cracks. EPRI NP-192, Final Report, September 1976.
11. Rodabough, E. C., Comments on the Leak-Before-Break concept for nuclear power plant piping systems. US Nuclear Regulatory Commission Report Nureg/CR-4305, July 1985.
12. Szabo, J., *Einführung in die Technische Mechanik.* Springer-Verlag, Berlin, 1961, pp. 91–135.
13. DIN 1629. Nahtlose kreisförmige Rohre aus unlegierten Stählen für besondere Anforderungen. Technische Lieferbedingungen. Beuth Verlag GmbH, Berlin, October 1984.
14. Forschungsvorhaben, Versuchsbegleitende Rechnungen und Analysen zu Behälterversuchen mit dem Programm DAPSY, RS 478, GRS-A-640, Gesellschaft für Reaktorsicherheit GRS, Köln, August 1981.
15. Forshungsvorhaben, Analytische Tätigkeiten—Bruchvorgänge in Behältern und Rohrleitungen, Abschlußbericht RS 477, GRS-A-1343, Gesellschaft für Reaktorsicherheit GRS, Köln, May 1987.
16. Analytische Begleituntersuchungen, Nachrechnung der Versuchsbehälter BVZ 060 und BVZ 070, SDK-Bericht Nr.3509-01. *SDK Ingenieurunternehmen für spezielle Statik, Dynamik und Konstruktion*, Lörrach, April 1987. Distributed by GRS, Köln.

Int. J. Pres. Ves. & Piping **43** (1990) 367–377

Fracture Behaviour of Stainless Steel Pipes Containing Circumferential Cracks at Room Temperature and 280°C

C. Maricchiolo, P. P. Milella & A. Pini

ENEA/DISP, Via V. Brancati 48, 00144 Rome, Italy

ABSTRACT

The paper presents the experimental results of a research programme on fracture behaviour of austenitic stainless steel and TIG welds in pipes containing circumferential through-wall cracks at room temperature and 280°C. Pipes were loaded in pure bending using a four-point bend test method. The diameter of the pipes under investigation was 168 mm and 324 mm, with a thickness varying from 10 to 17 mm.

As opposed to the behaviour of carbon steel pipes, it is found that the Net Section Collapse (NSC) criterion predicts the moment of instability. Crack mouth opening displacements (COD) and collapse moments calculated using the GE-EPRI engineering approach show a rather high scatter with respect to experimental results.

1 INTRODUCTION

The use of the double ended guillotine break (DEGB) as a design criterion may not necessarily lead to a real improvement of the safety of a nuclear power plant. It requires, in fact, the installation of a large number of pipe whip restraints to prevent the possible occurrence of large movements of the pipes assumed to have severed, and the crushing of other pipes and components. It also requires the use of heavy barriers to cut off high pressure water flows and avoid jet impingement.

Besides the economics of such a design choice, the use of restraints and barriers results in a considerable loss of accessibility to the piping systems

Int. J. Pres. Ves. & Piping 0308-0161/90/$03·50

during the life of the plant and additional irradiation doses to the workers for their removal, where possible, during scheduled outages.

This has led the Directorate for Safety and Protection (DISP) of the Comitato Nazionale per la Ricerca e lo Sviluppo dell'Energia Nucleare e delle Energie Alternative (ENEA) to consider new and more realistic design criteria such as the Leak-Before-Break (LBB) concept.

To substantiate the applicability of LBB to PWR nuclear power plants, in 1981 ENEA/DISP undertook a research programme to improve knowledge regarding the fracture behaviour of circumferentially through-wall and part-through cracked pipes.[1] Within the framework of the ENEA research programme two different steels have been studied: A 106 B carbon steel and type 316 stainless steel. Some results on carbon steel pipes were already published.[2]

The present paper summarizes the results obtained on stainless steel pipes containing circumferential through-wall cracks, tested at room temperature and 280°C.

2 EXPERIMENTAL PROCEDURE

The experiments considered in this study were conducted using the four-point bend method schematically shown in Fig. 1. During the test the load was quasi-statically increased under displacement control, until the maximum load was reached. Because of the low compliance of the test rig, unstable crack propagation never occurred.

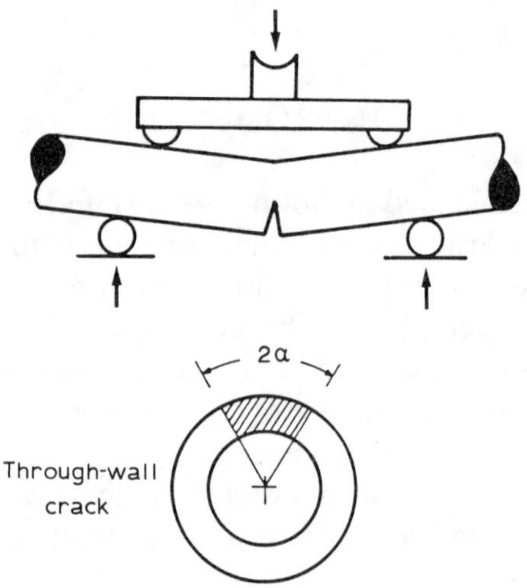

Fig. 1. Schematic diagram of four-point bend test method.

The pipes contained through-wall circumferential cracks of various size, introduced by EDM (Electric Discharge Machining) at the notch tip of milled slots, about 3 mm wide; the flaw tip radius was about 0·1 mm. Fatigue pre-cracking was not applied, because the high toughness of the steel causes large blunting of the flaw tip before the propagation of the crack.

The crack mouth opening displacement (COD) was measured by means of a clip gage mounted at the centre of the flaw. In some of the experiments four crack opening gages were placed, evenly spaced, along half of the crack on the outer surface of the pipes to determine the crack shape and measure the leak area.

The Direct Current Electric Potential (DC-EP) method was utilized to detect the onset of the crack propagation.

Tubes with two different diameters were tested: 168 and 324 mm, with a thickness varying from 10 to 17 mm. Table 1 presents the test matrix of all the experiments conducted so far.

TABLE 1

Summary of ENEA Fracture Experiments on Stainless Steel Pipes

Test code	*Outer diameter (m)*	*Thickness (m)*	*Crack angle (°)*	*Temperature (°C)*
Material: SS 316				
123I60	0·323 9	0·015 8	60·00	RT
121I40	0·323 9	0·016 1	40·00	RT
122I50	0·323 9	0·016 2	50·00	RT
124I90	0·323 9	0·016 4	90·00	RT
125I180	0·323 9	0·016 4	180·00	RT
Material: SS 316L				
L6B1I135	0·168 3	0·010 2	133·00	RT
L6B2I135	0·168 3	0·010 3	135·00	RT
L6C1I180	0·168 3	0·010 5	180·00	RT
L6D1I180	0·168 3	0·010 6	182·00	RT
L6A2I90	0·168 3	0·010 7	90·00	RT
L6A1I90	0·168 3	0·010 9	90·00	RT
L63II140	0·168 3	0·014 2	140·00	280
L62II60	0·168 3	0·014 3	64·00	280
Material: SS 316L-W				
L6C1IS135	0·168 3	0·010 6	133·00	RT
L6D1IS90	0·168 3	0·010 7	90·00	RT
L6A1IS90	0·168 3	0·010 7	90·00	RT
L6A2IS135	0·168 3	0·010 7	135·00	RT
L6F1IS180	0·168 3	0·010 8	180·00	RT
L6F2IS180	0·168 3	0·011 0	180·00	RT

TABLE 2
Summary of Material Properties

Material specification	*Test temperature (°C)*	*Young's modulus (MPa)*	*Yield strength (MPa)*	*Ultimate strength (MPa)*	J_{IC} *(MN/m)*
SS 316L	RT	198 000	240	560	0·750
SS 316	RT	198 000	358	636	0·524
SS 316L-W	RT	198 000	432	619	0·249
SS 316L	280	176 000	171	440	0·372

3 MATERIAL PROPERTIES

Three types of stainless steel were analysed; namely, AISI 316, AISI 316 Low Carbon (316 L) and Tungsten Inert Gas (TIG) weld on 316 L. Welds were performed according to ANSI B 31.7 and ASME Sections III and IX.

The material characterization programme involved several tests on compact tension (CT) and tensile specimens cut out of the pipes; the tests were conducted in accordance with the ASTM standards. Table 2 summarizes the average values of the mechanical properties measured, that were used in the fracture mechanics evaluation of the experiments.

4 ANALYSIS OF THE EXPERIMENTAL RESULTS

Two different methods of analysis were used to assess the fracture behaviour of defected pipes and are compared in this study: the Net Section Collapse (NSC) criterion and the GE-EPRI engineering approach.

4.1 Net Section Collapse criterion

The Net Section Collapse (NSC) criterion is based on the limit load analysis which assumes that failure occurs when the stress on the pipe cross-section reaches the flow stress of the material. In the case of a pipe with a circumferential through-wall crack under pure bending, the limit moment is given by[3]

$$\mathrm{NSCL} = 2\sigma_f R^2 t\{2 \sin[(\pi - \alpha)/2] - \sin \alpha\} \tag{1}$$

where σ_f is the flow stress (herein assumed as the average of the yield and ultimate strengths), R and t are the mean radius and the thickness of the pipe, and α is defined in Fig. 1.

To compare the experimental results from different materials and geometries, the applied remote stress, σ_{rem}, was calculated for all the tests as the experimental failure moment divided by the pipe section modulus

$$\sigma_{rem} = M_{exp}/\pi R^2 t \tag{2}$$

The NSC criterion estimate of the remote stress at failure may be calculated as

$$\sigma_{rem,NSC} = 2\sigma_f\{2 \sin [(\pi - \alpha)/2] - \sin \alpha\}/\pi \tag{3}$$

The comparison between the experimental results and the NSC predictions are shown in Fig. 2, where the remote stresses are normalized to the flow stress values; the solid line represents the NSC estimation.

The NSC criterion seems particularly effective in predicting the fracture behaviour of pipes made of 316 L stainless steel, while it overestimates, by 10–20%, the load carrying capability of pipes of 316 stainless steel, as well as that of the welding material. However, these discrepancies decrease as the crack size increases.

This behaviour is completely different from that of carbon steel pipes, whose experimental resistance was always greater than that predicted by the NSC criterion; at least at room temperature where dynamic strain ageing did not occur.[2]

To establish the applicability of the NSC criterion, a simple rule based on the evaluation of the dimension of the plastic zone ahead of the crack tip was proposed by Battelle Laboratories:[4] if the dimension of the plastic radius is greater than the distance between the crack tip and the neutral axis, then the fully-plastic condition is reached and the NSC criterion is applicable. Using

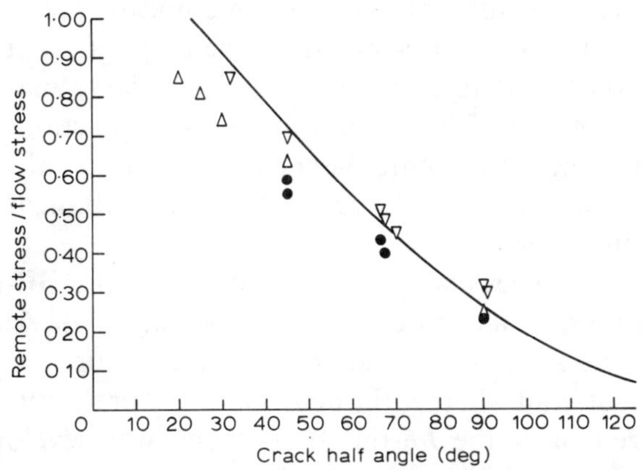

Fig. 2. Experimental results (open and closed symbols) and NSC moment prediction (solid line) versus crack half angle. ——, NSCL; ●, 316-L-W; ▽, 316-L; △, 316.

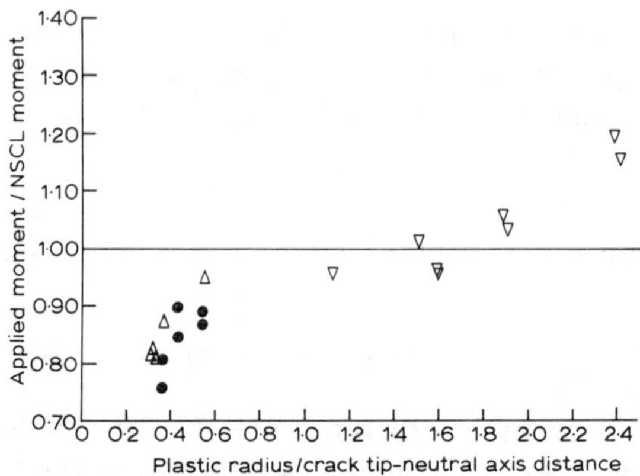

Fig. 3. Ratio of applied moment to NSC moment versus extension of plastic zone. ●, 316-L-W: ▽, 316-L; △, 316.

Irwin's equation of the plastic radius, in pure bending the previous condition becomes

$$EJ_{IC}/[\pi\sigma_f^2(\pi - \alpha)R] > 1 \tag{4}$$

where E and J_{IC} are the Young's modulus and the toughness of the material, respectively.

Figure 3 shows the comparison of the experimental results and the NSC prediction versus the ratio defined in eqn (4). The estimation of the failure load is poor in all the experiments where the above ratio is less than 0·6, i.e. for the tests on the lowest toughness material and small cracks.

4.2 GE-EPRI engineering approach

The GE-EPRI engineering approach is a method to predict the fracture behaviour of cracked structures based on the J-integral approach. So far different pipe geometries subjected to different loads were analysed by means of finite element analysis. The solution in terms of functions to be used to calculate fracture mechanics parameters such as J-integral, crack opening displacement (COD) and crack tip opening displacement (CTOD) are available in tabular form.[5,6]

The material properties play an important role on the GE-EPRI method predictions. At present, a debate exists whether the engineering or the true stress–strain curve should be used for the calculation of the Ramberg-Osgood coefficients and whether the deformation-J or the modified-J values should be utilized. Since the J-estimation scheme was developed under the assumption of small-scale deformation of the overall structure, the use of the engineering curve along with the deformation-J values seems more

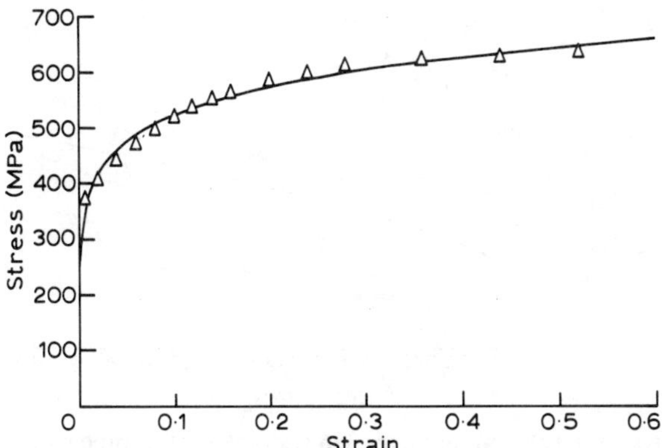

Fig. 4. Engineering stress–strain curve for 316 L stainless steel, at RT. $\alpha = 3{\cdot}042$; $n = 7{\cdot}6507$; $\sigma_0 = 358$ MPa; $\varepsilon_0 = \sigma_0/E$; $E = 198000$ MPa. ——, Best fit; △, test.

consistent with the theory. Furthermore, from the safety viewpoint, this choice is to be preferred, because it results in a prediction of lower limit loads.

As an example, the application of the method is presented for a 168 mm diameter pipe of 316 L material, with a through-wall crack of 90° (test L6A1I90). Figure 4 shows the engineering stress–strain curve of the material with the coefficients of the Ramberg-Osgood equation derived from a best fit to the experimental data. The material *R*-curve is reported in Fig. 5 (open triangles); since the crack growth measured on the CT specimens was always small, the data points were best-fitted by the following equation:

$$J_R = A da^B \tag{5}$$

The R-curve extrapolated data were always used in the GE-EPRI analysis. Finally, Fig. 6 shows the results of the *J–R* analysis using the GE-EPRI engineering approach.

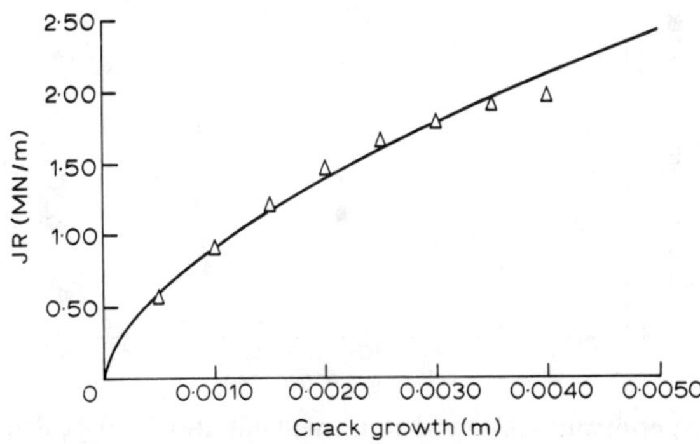

Fig. 5. *J*-Resistance curve for 316 L stainless steel, at RT. $JR = A \times \text{Growth}^B$. $A = 62{\cdot}90$, $B = 0{\cdot}6133$. ——, Best fit; △, test.

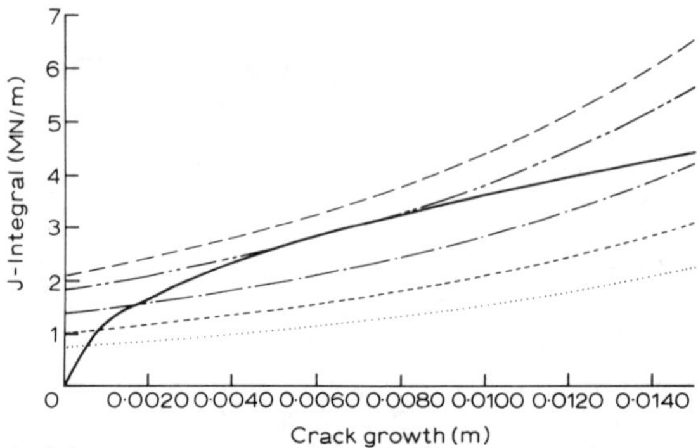

Fig. 6. *J–R* analysis for 168 mm diameter 316 L stainless steel pipe with a 90° through-wall crack, at RT. ———, $M = 0{\cdot}5468 \times 10^{-1}$ MNm; —··—, $M = 0{\cdot}5340 \times 10^{-1}$ MNm; —·—, $M = 0{\cdot}5084 \times 10^{-1}$ MNm; ----, $M = 0{\cdot}4828 \times 10^{-1}$ MNm; ·······, $M = 0{\cdot}4572 \times 10^{-1}$ MNm (initiation); ——, *R*-curve.

Results of all the experiments are plotted in Figs 7, 8 and 9. Figure 7 shows the bending moment (BEND) and crack mouth opening displacement (COD) at initiation, as the ratio of the experimental value to the calculated one, versus the crack angle. It can be seen that there is a large scatter in the results. TIG welds, in particular, seem to behave rather differently from the theoretical expectations, as far as the COD is concerned, while for the bending moment at initiation the experimental results appear to be close to the calculated values. It is not clear whether the discrepancies are to be ascribed to the uncertainty in the potential drop technique, to the material characterization of TIG welds or to the GE-EPRI method or to a combination of all these factors.

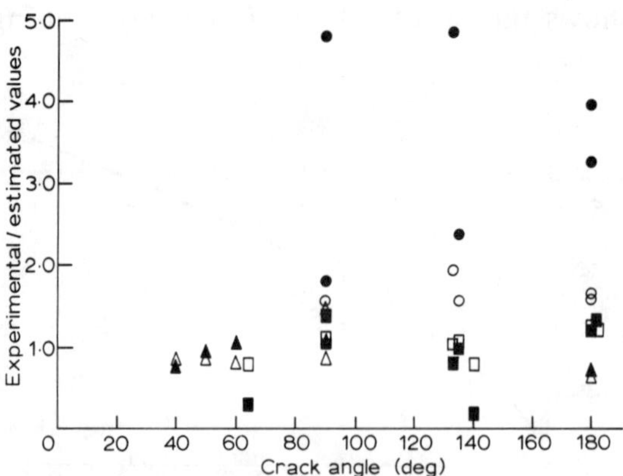

Fig. 7. Ratio of experimental COD and moment at initiation to the calculated values versus crack angle. ●, COD 316 L-W; ■, COD 316L; ▲, COD 316; ○, bend. 316 L-W; □, bend. 316 L; △, bend. 316.

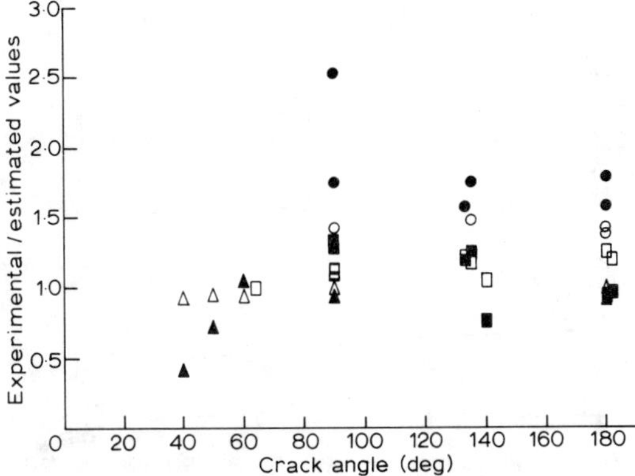

Fig. 8. Ratio of experimental COD and moment at maximum load to the calculated values versus crack angle. Key as Fig. 7.

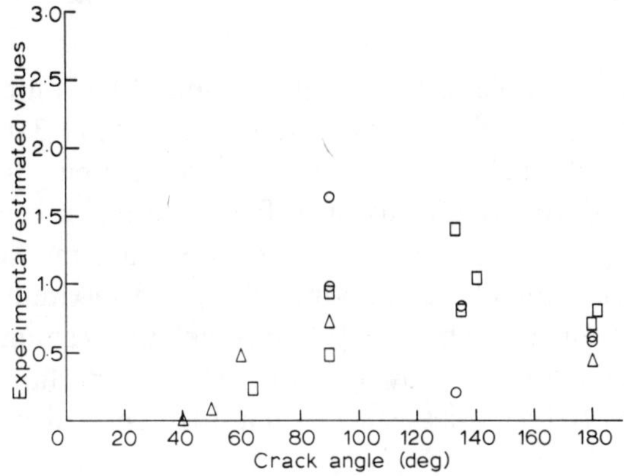

Fig. 9. Ratio of measured stable crack growth to calculated values versus crack angle. ○, Growth 316 L-W; □, growth 316 L; △, growth 316.

Figure 8 is the same plot as Fig. 7 referring to COD and bending moment (BEND) at maximum load. Again, the results are spread out although the scatter is much smaller than in Fig. 7. Finally, Fig. 9 shows the subcritical crack growth, in terms of the ratio of experimental measurements to calculated values versus the crack angle.

5 CRACK SHAPE

Four crack opening gages were used in all the 316 stainless steel pipes and in the 316 L pipes tested at 280°C, to monitor the flaw opening shape during the loading phase.

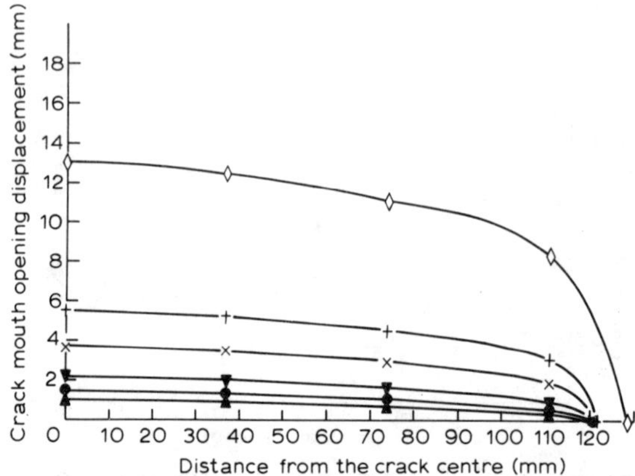

Fig. 10. Crack opening on the surface of a 324 mm diameter 316 stainless steel pipe for various loading steps at RT. ◇, $M = 0{\cdot}3855$ MNm; +, $M = 0{\cdot}3354$ MNm; ×, $M = 0{\cdot}2957$ MNm; ▼, $M = 0{\cdot}2398$ MNm; ●, $M = 0{\cdot}1987$ MNm; ▲, $M = 0{\cdot}1620$ MNm.

Figure 10 is an example of the results obtained. The plot pertains to the Experiment 124I90, namely a 324 mm diameter pipe of 316 stainless steel, with a through-wall crack of 90°. The crack mouth opening displacements (COD) are plotted versus the distance from the crack centre for several values of the applied moment, up to the maximum moment. Subcritical crack propagation commenced at about 80% of the maximum load reaching a value of about 8 mm at each crack tip; the crack tip opening was very large.

A different behaviour is shown in Fig. 11, that pertains to a 168 mm diameter pipe of 316 L stainless steel with a through-wall crack of 64° (Test

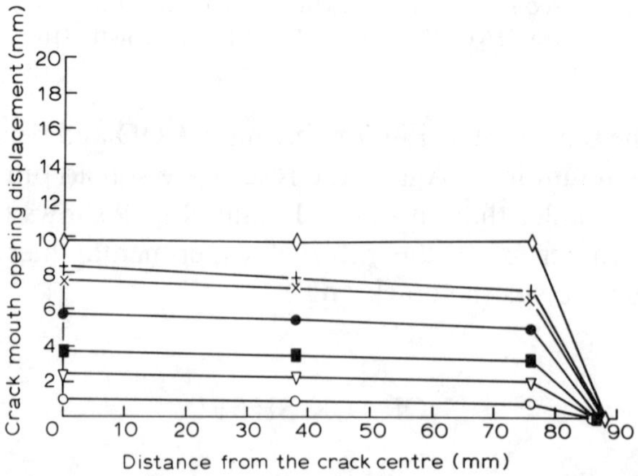

Fig. 11. Crack opening on the surface of a 168 mm diameter 316 L stainless steel pipe for various loading steps at 280°C. ◇, $M = 0{\cdot}0647$ MNm; +, $M = 0{\cdot}0645$ MNm; ×, $M = 0{\cdot}0628$ MNm; ●, $M = 0{\cdot}0584$ MNm; ■, $M = 0{\cdot}0534$; ▽, $M = 0{\cdot}0493$; ○, $M = 0{\cdot}0416$.

L62II60), tested at 280°C. At variance with all the other tests, for this smaller pipe the subcritical crack growth was practically negligible, while the crack tip blunting appeared to be larger.

6 DISCUSSION AND CONCLUSION

Pipe fracture assessment using the Net Section Collapse (NSC) criterion appears to be effective for ductile materials such as 316 L stainless steel, or for a combination of material toughness and crack size which results in the development of a plastic enclave ahead of the crack tip that extends throughout the ligament. The use of the Battelle screening criterion is recommended to judge whether the NSC criterion can be fully applied.

Predictions of crack mouth opening displacement (COD) and bending moment, both at initiation and collapse, using the GE-EPRI method may be different from experimental results; however, the method generally gives conservative predictions of the failure loads. A proper characterization of the mechanical properties of the material is needed.

The crack opening area associated with through-wall cracked pipes of stainless steel material loaded under pure bending is fairly approximated by the ellipse-shaped crack model. Large size flaws develop subcritical crack growth before the maximum moment is reached. This stable growth may be completely absent in pipes with small cracks.

REFERENCES

1. Milella, P. P., Outline of nuclear piping research conducted in Italy. *Nucl. Engrg. Des.*, **98** (1987) 219–29.
2. Maricchiolo, C. & Milella, P. P., Fracture behaviour of carbon steel pipes containing circumferential cracks at room temperature and 300°C. *Nucl. Engrg. Des.*, **111** (1989) 35–46.
3. Kanninen, M. F., Broek, D., Marschall, C. W., Rybicki, E. F., Sampath, S. G., Simonen, F. B. & Wilkowski, G. M., Mechanical fracture predictions for sensitized stainless steel piping with circumferential cracks. EPRI NP-192, September 1976.
4. Wilkowski, G. M., *et al.*, Degraded Piping Program Phase II. NUREG/CR-4082, Vol. 2, July 1985.
5. Kumar, V., German, M. D. & Shih, C. F., An engineering approach for elastic–plastic fracture analysis. EPRI NP-1931, July 1981.
6. Kumar, V., German, M. D., Wilkening, W. W., Andrews, W. R., deLorenzi, H. G. & Mowbray, D. F., Advances in elastic–plastic fracture analyses. EPRI NP-3607, August 1984.

Int. J. Pres. Ves. & Piping **43** (1990) 379–397

Leak-Before-Break Verification Test and Evaluations of Crack Growth and Fracture Criterion for Carbon Steel Piping

Y. Asada

Tokyo University, 3-1 Hongo 7-Chome, Bunkyo-ku, Tokyo 113, Japan

K. Takumi

Nuclear Power Engineering Test Center, Shuwa Kamiyacho Building, 3–13 Toranomon 4-Chome, Minato-ku, Tokyo 115, Japan

N. Gotoh

Nuclear Plant Maintenance Service Department, Hitachi Ltd, Saiwai-cho 3-1-1, Hitachi-shi, Ibaraki-ken 317, Japan

T. Umemoto

Ishikawajima-Harima Heavy Industries Co. Ltd, Isogoku, Yokohama-shi 235, Japan

K. Kashima

Central Research Institute of Electric Power Industries, Komae-shi, Tokyo 201, Japan

ABSTRACT

A proving test on the integrity of carbon steel piping in light water reactors (LWRs) was planned by the Nuclear Power Engineering Test Center (NUPEC) as a four-year verification test program; it was completed at the end of March 1989. The objective of this proving test was to demonstrate the validity of the Leak-Before-Break (LBB) concept for high quality carbon steel piping under actual plant conditions.

This paper briefly describes the results of material property tests, fracture behavior tests, LBB verification tests, numerical analyses of pipe fracture behavior and evaluation of flaw growth and fracture criterion. From these results, LBB has been verified and a fracture criterion has been developed for carbon steel piping in LWRs.

1 INTRODUCTION

Structures, systems, and components in nuclear power plants which are important to safety should be designed to accommodate dynamic effects, including the effects of pipe whipping, discharging fluid, and decompression waves associated with postulated instantaneous pipe ruptures.

According to recent progress in fracture mechanics, it is now possible to establish that Leak-Before-Break (LBB) conditions exist, and a nuclear power plant can be safely shut down upon detection of the fluid leakage.

Under this situation, in Japan, a four-year program named 'proving test on the integrity of carbon steel piping in LWRs' was conducted by NUPEC to demonstrate the validity of the Leak-Before-Break concept for high quality carbon steel piping under actual plant conditions. The program was successfully completed at the end of March 1989.

This report presents the results of each R&D activity in this program, and consists of the following;

(1) Material property tests
(2) Fracture behavior tests
(3) LBB verification tests
(4) Evaluation of flaw growth and fracture criterion.

2 MATERIAL PROPERTY TESTS

These tests were performed to obtain material property data for carbon steel base metal and weldment required for the LBB assessment of carbon steel piping.

Materials used for the test were JIS (Japanese Industrial Standard) STS42 and STS49, SFVC2B, and SGV42 (equivalent to ASTM A333, Gr. 6, ASTM A226 Cl.4 and ASTM A516 Gr.60, respectively) and their welds of submerged arc welding (SAW) and shielded metal arc welding (SMAW). Their chemical compositions are shown in Table 1. The test matrix is presented in Table 2.

TABLE 1
Chemical Composition and Heat Treatment of the Test Materials (wt%)

	C	*Si*	*Mn*	*P*	*S*	*Heat treatment*	*Equivalent*
STS 42 (6-inch)	0·16	0·28	1·30	0·020	0·014	—	ASTM A333 Gr. 6
STS 42 (16-inch)	0·16	0·13	1·14	0·032	0·015	620°C × 1·5 h FC	
STS 49 (26-inch)	0·21	0·28	1·14	0·026	0·012	625°C × 1·5 h AC	ASTM A333 Gr. 7
SFVC 2B	0·20	0·25	1·16	0·004	0·004	610 ~ 625°C × 7·75 h FC	ASTM A266 Cl. 4
SGV 42	0·15	0·22	1·11	0·017	0·003	625°C × 3 h FC	ASTM A516 Gr. 60
Weld metal							
SMAW	0·07	0·63	1·08	0·015	0·004	—	—
SAW	0·13	0·02	1·93	0·010	0·009	625°C × 1·4 h FC	—

FC = Furnace cooling; AC = atmospheric cooling.

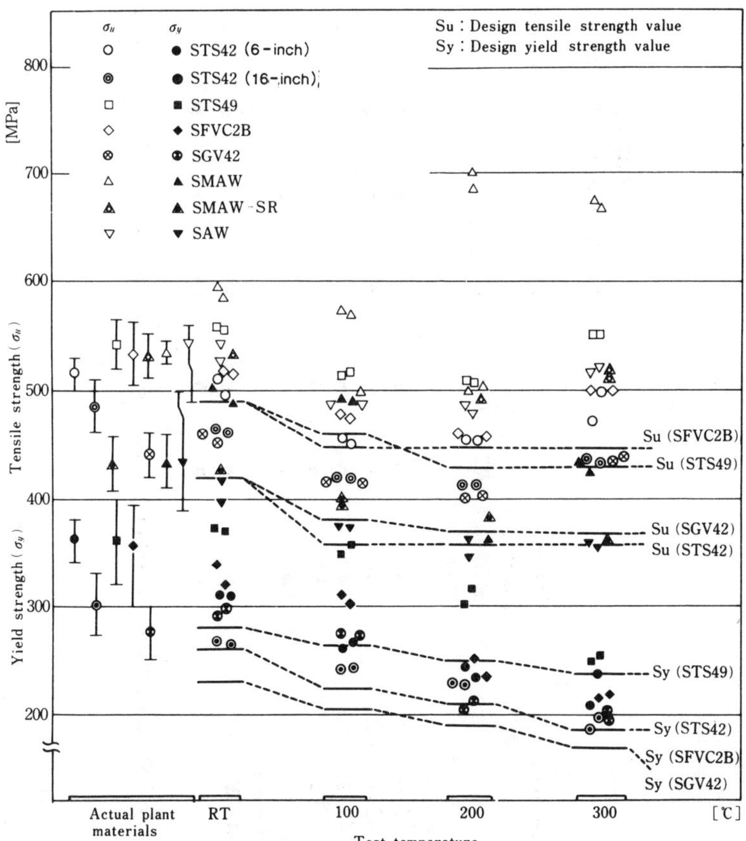

Fig. 1. Tensile properties of test materials.

TABLE 2
Material Property Test Matrix

Material	*Charpy impact transition curve*	*Tensile test*				*Fracture toughness test*				*Center-cracked panel test*			
		RT	*100°C*	*200°C*	*300°C*	*RT*	*100°C*	*200°C*	*300°C*	*RT*	*100°C*	*200°C*	*300°C*
STS 42 (6-inch)	●	●	●	●	●	●	—	—	●	●	—	—	●
STS 42 (16-inch)	●	●	●	●	●	●	●	●	●	●	●	●	●
STS 49 (26-inch)	●	●	●	●	●	●	●	●	●	—	—	—	—
SFVC 2B	●	●	●	●	●	●	—	●	●	●	—	—	●
SGV 42	●	●	●	●	●	●	—	●	●	—	—	—	—
Weld metal													
SMAW	●	●	●	●	●	●	●	●	●	●	—	—	●
SAW	●	●	●	●	●	●	●	●	●	—	—	—	—

SMAW = Shielded metal arc welding; SAW = submerged arc welding; RT = room temperature.

The results of the material property test are summarized as follows.

(1) The tensile tests, Charpy V-notch impact tests, fracture toughness tests, and center-cracked panel tests provided data on the mechanical properties, toughness, and fracture behavior of the carbon steel material.

(2) The results of the tensile and Charpy tests indicated that these data can be taken as being representative of the properties of carbon steel piping used in actual plants. The values of ultimate tensile strength (σ_u) and yield strength (σ_y) obtained with the test materials are shown in Fig. 1. The figure also gives the values of the design tensile strength (S_u) and the design yield strength (S_y) of the steel specified in Notification No. 501 of the Ministry of International Trade and Industry (hereinafter referred as Notification). In addition, the values of the σ_u and σ_y of the steel used in the actual plants are plotted. As shown in Fig. 1, it is recognized that mechanical properties of the test material are almost average or lower-than-average values compared with the value of steel of the same specification used in the actual plants.

(3) Compared with the base materials, the SMAW and SAW weld metals were found to have a higher tensile strength and a similar fracture toughness. (*J*-integral resistance curves are shown in Fig. 2.)

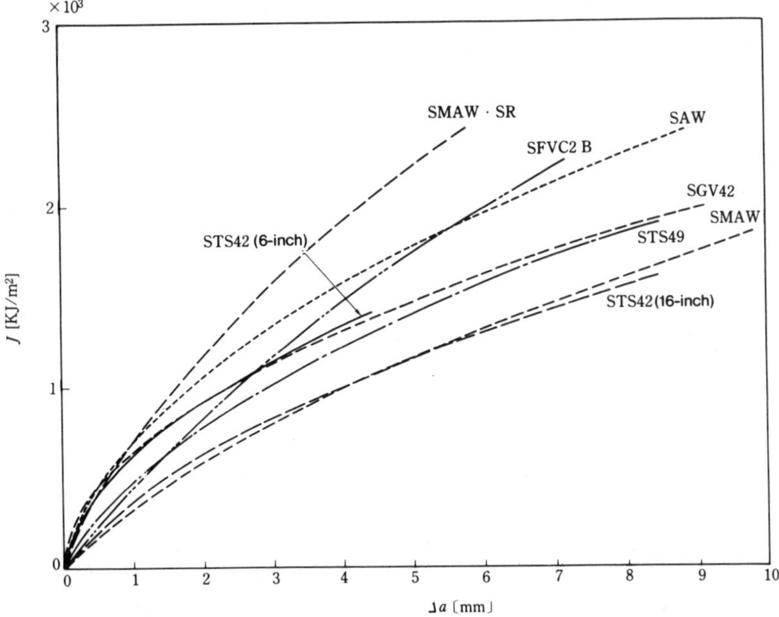

Fig. 2. J–Δa curves for the pipe materials at 300°C.

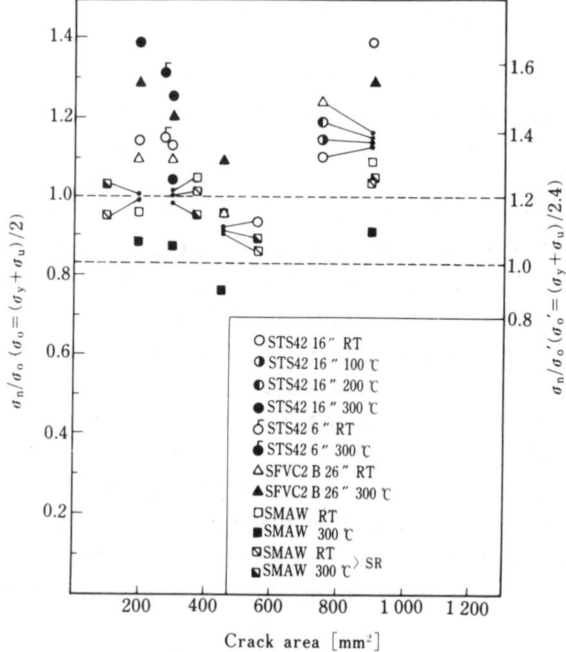

Fig. 3. Net section stress at the maximum nominal load.

(4) The result of a center-cracked panel test is summarized in Fig. 3 in a relation between a normalized net section stress and the area of the crack, where the net section stress (σ_n) was normalized using a value of the flow stress of the material of the following two definitions.

$$\sigma_0 = (\sigma_y + \sigma_u)/2{\cdot}0 \quad (1a)$$

or

$$\sigma_0 = (\sigma_y + \sigma_u)/2{\cdot}4 \quad (1b)$$

As is shown in Fig. 3, values of σ_n/σ_0 are almost 1·0 or above when a flow stress of the definition by eqn (1a) is used, but when using eqn (1b) a reasonable conservatism is observed for all materials irrespective of specimen geometries and test temperatures.

3 FRACTURE BEHAVIOR TESTS

The fracture behavior tests were performed on 6- and 16-inch (1 inch = 2·54 cm) STS42 straight pipes. The pipes tested are base material and weldment by SMAW. Both the 6- and 16-inch pipe specimens were notched in the circumferential direction and subjected to four-point bending loads

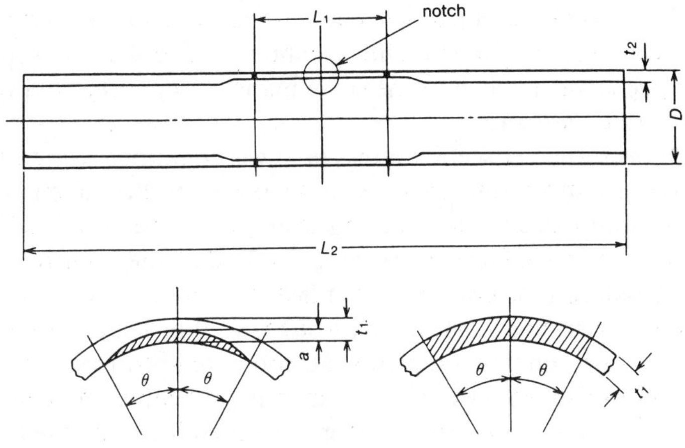

	for low compliance test		for high compliance test	
	6-inch	16-inch	6-inch	16-inch
L_1 [mm]	500	450	300	1,200
L_2 [mm]	2 800	3 950	2 160	3 200
D [mm]	165.2	406.4	165.2	406.4
t_1 [mm]	8.8	18.9	9.8	19.0
t_2 [mm]	11.0	21.4	11.0	21.4
a [mm]	4.4	9.5	4.9	9.5

2θ : notch angle

(a)

(b)

Fig. 4. Configuration of high and low compliance test specimens for (a) bending and (b) tension.

with both low and high compliance. The configuration of the specimens and the dimensions of their initial notches are shown in Fig. 4. In addition, the 6-inch pipe specimens notched in the same manner were subjected to tensile tests with high compliance.

In the 26 tests, the net section stress (σ_n) was computed using equations based on the net stress collapse criteria proposed by Kanninen *et al.*[1]

The maximum loads in the bending tests were evaluated based on net stress collapse criteria. The evaluation showed that the occurrence of the maximum bending moment was identified by net section stress criteria, regardless of notch type (surface or through-wall) or temperature.

The maximum bending moments at room temperature for surface-notched pipes are shown in Fig. 5. The maximum bending moment on the ordinate of this Fig. 5 indicates dimensionless values obtained through division of the maximum bending moment M by moment Mo ($=4\sigma_n R^2 t$), where Mo is the collapse moment of a pipe without cracks. These values were obtained through the following equation, shown in Fig. 5 by the solid line:

$$\frac{M}{Mo} = \cos\frac{\theta}{4} - \frac{1}{4}\sin\theta \tag{2}$$

In the tensile tests for surface notch pipes, dimensionless values for the plastic collapse load were obtained through division of the maximum tensile load P by the tensile load Po ($=2\pi\sigma_n Rt$), where Po is the collapse load of a

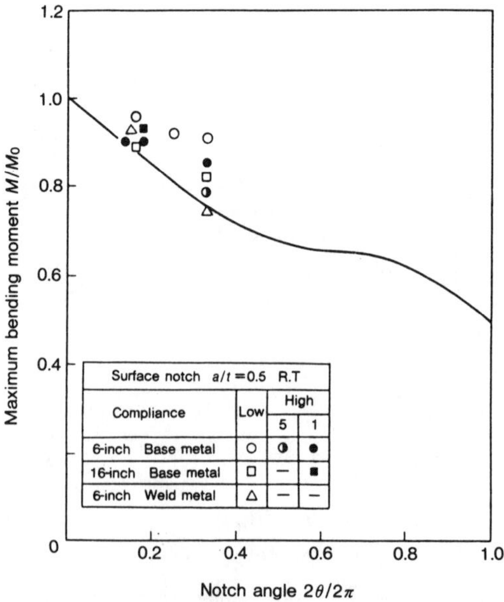

Fig. 5. Maximum bending moments for surface notched pipes.

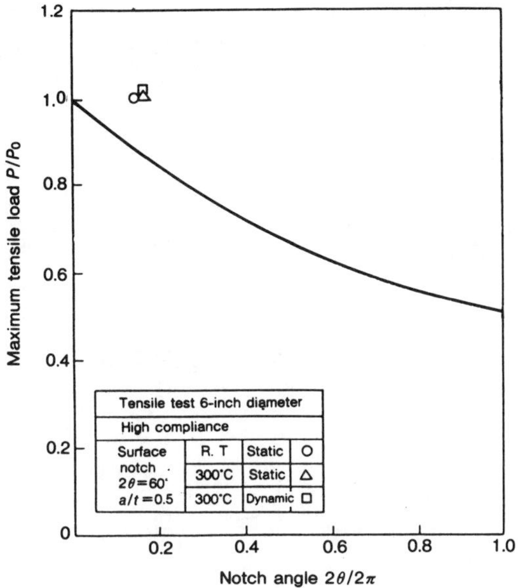

Fig. 6. Maximum tensile loads for surface notched pipes.

pipe without cracks. These values were obtained through the following equation, and are shown in Fig. 6 by the solid line at $a/t = 0{\cdot}5$.

$$\frac{P}{Po} = \pi - (a/t)\theta - 2\sin^{-1}\left(\tfrac{1}{2}(a/t)\sin\theta\right) \tag{3}$$

The test values were consequently larger than the calculated values (see Fig. 6).

The tests elucidated the fracture behavior of 6- and 16-inch carbon steel piping with cracks. The fracture mode conforms with the net stress collapse criterion within the scope of the test.

4 LBB VERIFICATION TESTS

The LBB verification tests were conducted with specimens of three straight pipes, an elbow and a tee. The specimens were subjected to a cyclic bending load and were pressurised with high high temperature water to the pressure–temperature conditions of the actual plants.

After a crack grew through the pipe wall and steam leakage was detected, several additional cyclic bending loadings were applied and crack growth behavior was observed. A schematic diagram of the LBB verification test is shown in Fig. 7, and the specimen dimensions and test conditions are presented in Table 3.

TABLE 3
Test Condition of LBB Verification Test

Specimen	*Specimen no.*	*Material*	*Dimension of the specimen*				*Load condition*		
			Outer diameter (mm)	*Pipe wall thickness (mm)*	*Notch length (mm)*	*Notch depth (mm)*	*Max load* ($\times 10^3$ *N*)	*Min load* ($\times 10^3$ *N*)	*A number of cycle*[a]
Straight pipe	TU-01 (double notches)	STS42	165·2	9·4	105·0	7·9	100–193	25	425
		SMAW	165·2	9·8	105·0	6·9			
	TU-02	SMAW	165·2	9·7	104·6	8·14	178	30	1 425
	TU-03	STS42	165·2	11·1	102·4	5·5	195	34	2 302
Elbow	EL-01	SFVC2B	165·2	15·9	100·3	12·9	67–72	22–25	15 870
Tee	TE-01	SFVC2B	165·2	12·2	50·3	10·1	51–71	8	502

[a] Up to leakage.

Water environmental conditions:

Temperature: 288–289°C; Internal pressure: 794–833 MPa; Flow rate: 0·071 kg/s; Conductivity: 0·006 5–0·024 μS/mm; Dissolved oxygen concentration: 5·8–8·5 ppm.

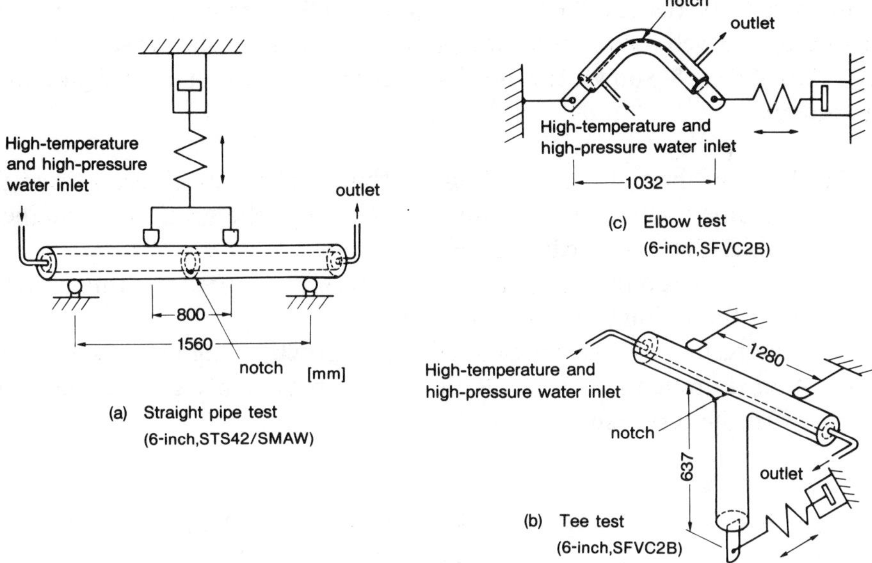

Fig. 7. Schematic diagram of LBB verification test.

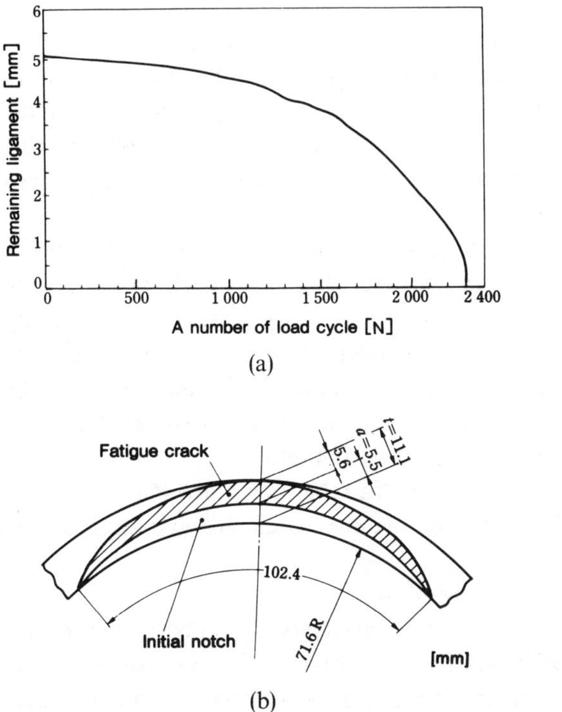

Fig. 8. (a) Crack growth as a function of load cycle using DCP for the straight pipe test TU-03 and (b) a sketch of the fracture surface for the straight pipe test TU-03.

In these tests, the resultant through-wall cracks remained stable and the shapes agreed well with engineering predictions. Test results are summarized as follows (Fig. 8 shows the resultant crack growth of a straight pipe specimen).

(1) In all of the tests, it was observed that a fatigue crack grew through the pipe wall allowing steam leakage while the specimen remained stable. The test verified LBB.
(2) It was also confirmed that the crack grew at a higher rate through the wall than along the inner pipe surface.
(3) The results of measurement of crack growth using ultrasonics (UT) and dc potential drop (DCP) were in reasonably good agreement with the test results.

5 EVALUATION OF FLAW GROWTH AND FRACTURE CRITERION

5.1 Crack growth analysis

5.1.1 Stress intensity factors for surface cracks

Stress intensity factors, K, for a circumferential surface crack in a pipe were computed using the three-dimensional finite element method (FEM) (MARC Program was used), and then compared with those obtained from the Newman–Raju solution[2] to evaluate the applicability of the simple solution to surface cracked pipe.

The results of the K-value analysis at the deepest point in the crack are shown in Table 4. In the case where the crack was small, the Newman–Raju solution agreed well with the finite element method giving a value that approximates the pipe's stress intensity factor and was not greatly influenced by the pipe's curvature. In the case where the crack was deep, the Newman–Raju solution yielded a higher K-value, and by extension, more conservative results, than the finite element method.

The analysis concluded that the Newman–Raju solution could be used for estimating K-values for fatigue crack growth analysis in a pipe.

5.1.2 Crack growth analysis

The fatigue crack growth analysis of carbon steel pipes under operating conditions of boiling water reactor (BWR) and pressurized water reactor (PWR) plants in Japan was performed for various surface crack shapes and amounts of crack growth.

Fatigue crack growth was assumed to be governed by Paris' power law[3]

TABLE 4
Results of *K*-Value Analysis (at the Deepest Point)

Case no.	Diameter	Thickness (t) (mm)	Crack angle (degrees)	Crack depth (t)	Tensile load ($\sigma_t = 98\ MPa$)		Bending load ($\sigma_b = 98\ MPa$)	
					FEM	Newman–Raju	FEM	Newman–Raju
1			8·8	0·2	6·98	8·03	6·08	7·78
2			60	0·25	11·3	11·7	9·73	11·4
3	165·2			0·8	23·1	(28·3)	20·7	(26·4)
4		11·0	120	0·25	11·9	12·4	10·3	12·1
5				0·8	30·6	(42·3)	26·7	(39·7)
6			60	0·25	14·5	16·7	13·2	16·5
7	406·4			0·8	31·8	(44·8)	29·3	(42·9)
8		21·4	120	0·25	19·4	17·5	17·0	17·4
9				0·8	50·7	(70·5)	45·2	67·9
10	660·4	30·7	60	0·25	20·6	20·2	18·7	20·0

Note: solutions in parentheses are obtained from non-applicable crack size for the formula.

with respect to the initial semi-elliptical surface crack depth (a) and surface length ($2c$).

The Newman–Raju solution[2] was used to obtain the stress intensity factor. The period for the calculation was taken as 80 years (the plant life, 40 years, multiplied by a safety factor of two).

In all cases, the cracks did not grow completely through the wall, even in the late stages of life. The deepest crack growth was estimated to be lsss than 40 per cent of the pipe wall thickness.

5.2 Unstable fracture analysis

For the LBB evaluation, the conditions for unstable fracture need to be established for the loading conditions experienced by the carbon steel pipes. The applicable fracture criteria were the elastic–plastic fracture mechanics criterion (J-integral/Tearing Modulus (J/T) criterion) and plastic collapse criterion (net stress collapse criterion). The following items present a summary of the unstable fracture analyses.

5.2.1 Development of evaluation method and procedure

Development of the evaluation method is needed to determine failure criteria for carbon steel pipes, including pipes with a wide variation in outer diameter, especially larger than 16-inch, and the variety of materials, based on the proving test data (6- and 16-inch STS42 base, and SMAW pipe test data).

The fracture evaluation method should be developed considering the following as the basic principles:

(i) A simplified engineering method.
(ii) The result must always be conservative (but not overly conservative).
(iii) The evaluation must be applicable to large outer diameter pipes (especially over 16-inch diameter pipes).

For the J/T criterion such simple methods as the R6 procedure[4] and the full-plastic solution (German–Kumar solution) proposed by GE/EPRI[5] are chosen through comparison of the resultant fracture loadings using these methods with the results of finite element analysis.

The fracture criteria for carbon steel pipe under various conditions were screened by the R6 procedure, including both fracture criteria.

Finally, the fracture load under various conditions was estimated using the proposed correction factor (G-factor, to be described below), to the plastic collapse load. A flow diagram of the evaluation procedure is shown in Fig. 9.

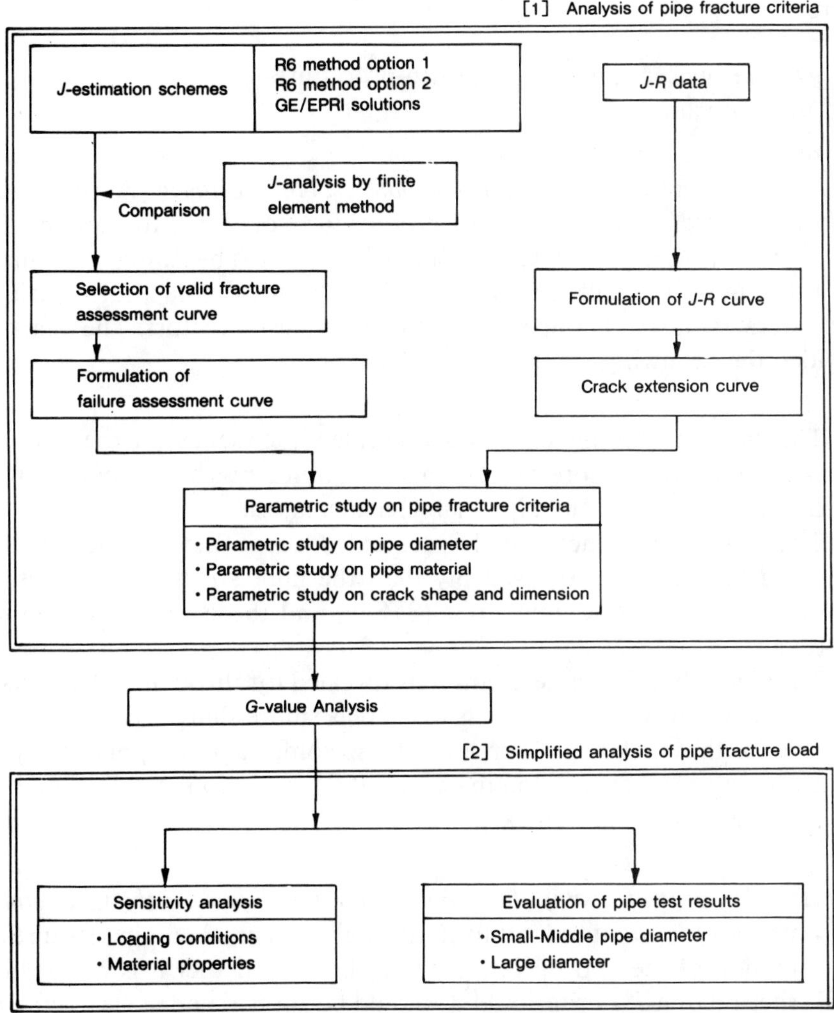

Fig. 9. Flow diagram of fracture assessment.

5.2.2 *Unstable fracture criteria*

5.2.2.1 Determination of failure assessment curve (FAC). FEM and some simplified analyses were performed varying the major factors likely to influence the fracture criteria. The comparisons of FEM and some simplified methods concluded that the R6 procedure (option 2) gave the most appropriate FAC out of the simplified methods.

The following formula was developed as the best approximation of the R6 procedure (option 2) through application of the Ramberg–Osgood constants.

$$K_r = (\alpha L_r^{n-1})^{-0{\cdot}5} \qquad (\alpha = 4{\cdot}412, \quad n = 3{\cdot}309) \tag{4}$$

where

K_r, L_r are two parameters of the R6 procedure,
α, n are constants.

5.2.2.2 Sensitivity analysis for unstable fracture criteria. A sensitivity analysis was performed using the R6 procedure (option 2) to examine the influence of various parameters on fracture criteria. The Zahoor formula[6] (for a through-wall crack) and the Newman–Raju formula[2] (for a surface crack) were employed for determining stress intensity factors. The analysis provided the following results.

- The diameter parameter analysis revealed a tendency for the fracture stress to decrease more sharply than the limit stress based on net stress collapse criteria as the pipe diameter increases.
- The material parameter analysis revealed a tendency for the fracture stress to decrease more sharply than the limit stress in the cases of materials with large flow stress (STS49) and those with low fracture toughness, J_{ic} (SMAW).
- The crack shape parameter analysis revealed for through-wall cracks a tendency for the fracture stress to decrease more sharply than the limit stress as the crack angle increases. It also confirmed the applicability of a fracture mechanism dominated by the net section stress criteria in the case of a surface crack.

5.2.3 Development of a simplified analysis method based on the G*-Factor*
Assessment of the fracture load for a through-wall cracked pipe requires a determination of the appropriate criteria, that is, whether it is net section stress criterion or a J/T criterion that should be applied under the operating conditions of the pipe. This study adopted an approach using the G-factor to facilitate this process of criteria screening and pipe fracture load assessment. The G-factor is expressed by the following equation:

$$G\text{-factor} = \text{plastic collapse load/fracture load (maximum load)} \quad (5)$$

If the G-factor is equal to one, the fracture load will be equal to the plastic collapse load, and the net section stress criterion would therefore be applicable. If the G-factor is greater than one, the fracture load will be less than the plastic collapse load, and the J/T criterion would be applicable. In this equation, the maximum fracture load can be obtained from the R6 procedure (option 2) and the plastic collapse load can be calculated by

TABLE 5
G-Factor for Typical Pipe Crack

Reacter	*Material*	*Diameter (mm)*	*Thickness (mm)*	*Half initial angle θ° (deg)*					
				10	*20*	*30*	*40*	*50*	*60*
BWR		60·5	8·7	1·00	1·00	1·00	1·00	1·00	1·00
		114·3	11·1	1·00	1·00	1·00	1·00	1·00	1·00
		165·2	11·0	1·00	1·00	1·00	1·01	1·03	1·04
	STS42	216·3	18·2	1·00	1·00	1·00	1·02	1·04	1·05
	SFVC2B	267·4	18·2	1·00	1·00	1·02	1·06	1·09	1·10
		318·5	21·4	1·00	1·00	1·04	1·08	1·11	1·13
		355·6	23·8	1·00	1·00	1·06	1·10	1·13	1·14
		406·4	26·2	1·00	1·01	1·08	1·12	1·15	1·17
		457·2	23·8	1·00	1·04	1·11	1·16	1·19	1·22
	STS49	508·0	26·2	1·00	1·03	1·10	1·15	1·18	1·21
	SFVC2B	609·6	31·0	1·00	1·05	1·12	1·17	1·21	1·24
		660·4	33·3	1·00	1·06	1·13	1·19	1·22	1·25
PWR	STS49	406·4	21·4	1·00	1·00	1·05	1·10	1·13	1·16
	SGV42	711·2	31·0	1·00	1·04	1·11	1·17	1·21	1·25
		762·0	33·0	1·00	1·05	1·12	1·18	1·23	1·26

Membrane stress $Pm = 0{\cdot}34Sm$.

simple theoretical equations. The *G*-factors exhibit the following trends (Table 5).

- The *G*-factor is equal to one for small-diameter pipes (less than 6-inch) irrespective of the crack angle.
- The *G*-factor is close to one, even for large-diameter pipes, when the crack angle is small.
- The *G*-factor increases as the wall thickness decreases.

The *G*-factor was approximated by the following expression for more convenient use.

$$\begin{aligned} G &= 1 && (2 \le D < 6) \\ &= (0{\cdot}692 - 0{\cdot}0115D) + (0{\cdot}188 + 0{\cdot}0104D) \times \log_{10}(\theta) && (6 \le D \le 30) \end{aligned} \quad (6)$$

where

D = pipe diameter (inches)

The following are the results of the sensitivity analysis concerning the influence of loading conditions and material properties on the *G*-factor.

Influence of loading conditions:

—The *G*-factor increases slightly with increases in membrane stress (*Pm*). This tendency becomes significant as the crack angle or pipe diameter increases.

Influence of material constants:

—For various pipe diameters, there is little difference between the *G*-factor obtained by using the Notification values and that obtained by using the measured values.

—The fracture bending stress estimated from the Notification values is less than that estimated from the measured values. The analysis based on the Notification values therefore yields more conservative fracture load values. In addition, the degree of conservation can be assessed by use of the flow stress.

CONCLUSION

NUPEC's four-year verification test program on the integrity of carbon steel piping in LWR plant was successfully completed, and the following conclusions were obtained.

(1) Material properties and fracture behavior were fully characterized, and it was confirmed that the net section stress criterion is applicable to carbon steel pipe fracture.

(2) From the results of LBB verification tests using pipes, an elbow and a tee under LWR operating conditions, crack growth behavior of all components was stable at leakage and after leakage. LBB was verified for carbon steel piping.

(3) A fracture criterion that is comprehensively applicable to all carbon steel materials, and even large diameter pipes, was developed by using the R6 procedure (option 2) and *G*-factor relating the plastic collapse load and the maximum fracture load.

REFERENCES

1. Kanninen, M. F., Broek, D., Marschall, C. W., Rybicki, E. F., Sampath, S. G., Simonen, F. A. & Wilkowski, G. M., Mechanical fracture predictions for sensitized stainless steel piping with circumferential cracks. EPRI NP-192, 1976.
2. Newman, J. R. Jr & Raju, I. S., Analysis of surface cracks in finite plate under tension and bending loads. NASA Technical Paper 1578, 1979.
3. Paris, P. & Erdogan, F., A critical analyses of crack propagation laws. *Transactions of the ASME, Series D*, **85** (1963) 528–34.

4. Harrison, R. P., Loosemore, K., Milne, I. & Dowling, A. R., Assessment of the integrity of structures containing defects. CEGB Report No. R/H/R6- Rev. 2, 1980.
5. Kumar, V. & German, M. D., Elastic and fully plastic solutions for through-wall cracks in cylinders. EPRI NP-3607 (Advances in elastic–plastic fracture analysis), 6-1, 1984.
6. Zahoor, A., Gamble, R. M., Mehta, H. S., Yukawa, S. & Ranganath, S., Evaluation of flaws in carbon steel piping. EPRI NP-4824M, Projects 1757-51, 2457-1, -2 Final Report, 1986.

Int. J. Pres. Ves. & Piping **43** (1990) 399–411

Measurement of Leak-Rate Through Fatigue-Cracks in Pipes under Four-Point Bending and BWR Conditions

T. Isozaki, K. Shibata, H. Shinokawa*
& S. Miyazono

Department of Reactor Safety Research, Japan Atomic Energy Research Institute,
Tokai-mura, Ibaraki-ken 319-11, Japan

ABSTRACT

Leak-rate tests were performed using 114 mm and 165 mm (4 and 6 in) diameter, schedule 80 pipes made of austenitic stainless steel SUS304 and carbon steel STS42. Each pipe contained a through-wall fatigue crack and was mounted on a four-point bending machine of 400 kN maximum loading. Tests were done under a pressure of 7 MPa, with a subcooling temperature. The leak rate was measured by a Venturi flow meter and a differential pressure transducer attached to the pressure vessel. Comparisons of the effect of pipe material, diameter and crack angle were made. This paper shows that from a Leak-Before-Break viewpoint, the stainless-steel pipe is superior to the carbon-steel one, and that the pipe with the larger diameter is better than the one with the smaller diameter. No unstable fracture was observed in the tests.

NOTATION

$2a$	Crack length (m)
L	Applied jack load (N)
M	Applied moment (Nm)
M_c	Net section collapse moment (Nm)
p	Pressure (MPa)
P_b	Bending stress on the elastic basis (MPa)
P_m	Axial membrane stress due to internal pressure (MPa)

* Present address: Isogo Engineering Center, Toshiba Corporation, 8 Shinsugitacho, Isogo-ku, Yokohama 235, Japan.

R Mean radius (m)
R_z German Standard for the surface roughness (μm)
S_m Lesser of $2\sigma_y$ or σ_u (MPa)
t Thickness of pipe (m)
W Leak rate (kg/min)
Z Section modulus (m^3)

θ_c Critical half-crack angle (degrees)
θ Half-crack angle of test pipe (degrees)
Θ Nondimensional half-crack angle of test pipe $= \theta/\theta_c$
σ_b Bending stress (MPa)
σ_f Flow stress (MPa) $= (\sigma_y + \sigma_u)/2$
σ_u Ultimate tensile strength (MPa)
σ_y Yield strength (MPa)
σ_Y Equivalent yield strength (MPa) $= (3\sigma_y + \sigma_u)/4$

1 INTRODUCTION

Leak-rate studies play an important role in Leak-Before-Break (LBB) technology. The present leak-monitoring device can detect leaks as small as 4 kg/min within an hour after the leak occurring. LBB technology would progress greatly if it could be shown that unstable fractures occur only at leak rates greater than 4 kg/min or in other words, that no unstable fracture occurs at less than that rate, on the premise that no stress-corrosion crack, erosion, corrosion and wall thinning are observed. In this way, the leak rate can be seen as the prime indicator of through-wall cracks in pressure boundary components. Consequently, the Japan Atomic Energy Research Institute (JAERI) has commenced a leak-rate study; its aim is to determine the through-wall crack length through which 4 kg/min leakage occurs, and to verify that no unstable fracture occurs at this specified leak rate.

To date, the following four types of leak-rate studies have been performed:

(1) through an artificial slit,[1-10]
(2) in a pipe containing an intergranular stress-corrosion-crack,[1]
(3) in a pipe containing a fatigue-crack,[9,11] and
(4) a computer study to obtain the leak rate using a critical flow model.[1,12-14]

This paper describes leak-rate tests on pipes containing a fatigue-crack with or without four-point bending loads under boiling water reactor (BWR) conditions. Test pipes were fabricated from stainless steel SUS304 and carbon steel STS42, into which through-wall fatigue cracks were introduced with a cyclic four-point bending load. Test parameters were crack angle, bending load, pipe materials and diameter, 114 mm and 165 mm (called 4 in. and 6 in. pipes in this paper). By introducing two types of nondimensional parameters to represent the applied moment and initial crack angle, the 4 kg/min leak rate limit was obtained. The region where LBB is accepted is shown by the area between the net section collapse criteria[15] and the leak-rate limit line.

2 TEST APPARATUS AND PROCEDURE

Figure 1 shows the leak-rate test apparatus. It is composed of a 4 m^3 pressure vessel with a 1 m i.d., 400 kW heater, 0·2 m^3 pressure vessel with 0·25 m i.d., nitrogen gas supply, Venturi flow meter and 400 kN bending machine. When a small or no bending load is applied to the pipe, the crack opening displacement (COD) is considered small. In this case, the test pipe was pressurized without any heat source by nitrogen gas supply, by closing the motorized valves MV1, MV2, MV8 and MV9, and by opening MV3. The leak rate was then precisely measured by differential pressure transducers mounted at intervals of 500 mm vertically on the 0·2 m^3 vessel wall and by the Venturi flow meter. When the bending load was applied to the test pipe, however, it became difficult to keep the pressure in the test pipe constant. In this case, the test pipe was connected to a 4 m^3 vessel by closing the valves MV8 and MV9, and by opening MV1, MV2 and MV3. The leak rate was then measured by two differential pressure transducers mounted 985 mm apart on the 4 m^3 vessel and by the Venturi flow meter.

Figure 2 shows a 4 in. (114 mm) test pipe. It was covered with glass wool to a thickness of 50 mm for insulation. The test pipe was mounted on the lower

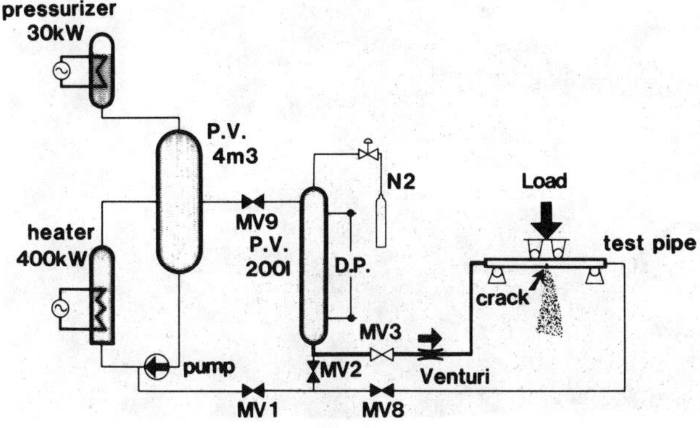

Fig. 1. JAERI's leak-rate test apparatus.

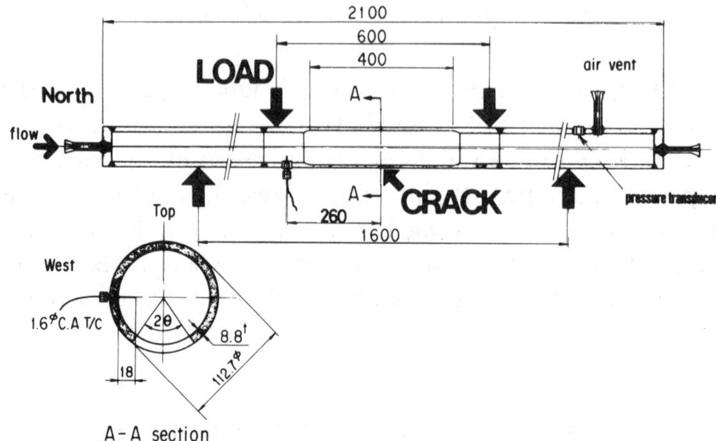

Fig. 2. 4 in. test pipe for JAERI's leak-rate test.

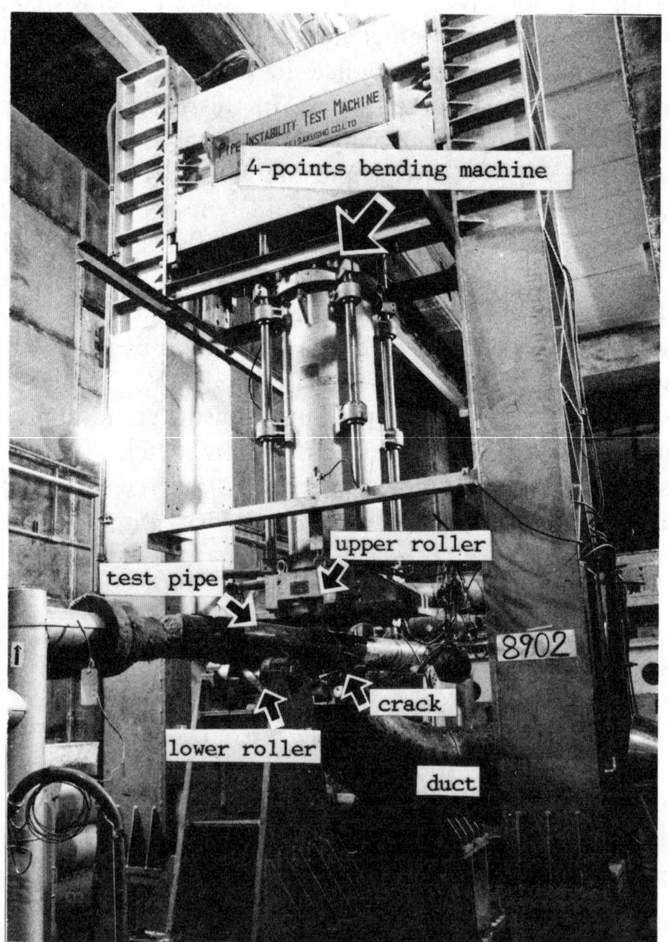

Fig. 3. J6SS68 test pipe and four-point bending machine.

roller so that the leak went downwards. Figure 3 shows the 6 in. (165 mm) test pipe after the heat insulation was removed and the four-point bending machine. The upper loading span was 600 mm, and the lower span was 1600 mm. The pipe was subjected to increasing four-point bending load with a jack of 2 mm/kN compliance. A 2 mm or 5 mm range clip gage was mounted across the crack at the pipe bottom to measure COD during the leak test.

3 TEST RESULTS

Table 1 shows the dimensions of JAERI's and Tanaka *et al.*'s[9,11] test pipe. The present paper refers to their data frequently, to compare and supplement the current test results. The crack length was so determined that a leak rate of 2, 8 or 20 kg/min would be expected, under the loading condition that $P_m + P_b$ be equal to 0·5 S_m, where the S_m value is 115 MPa for SUS304 steel, and 123 MPa for STS42 steel, respectively.[16]

Table 2 shows the leak-rate test results. The pressure in the test pipe was up to 7 MPa; a saturated condition could not be achieved during the leak tests

TABLE 1
Dimension of Test Pipe

Pipe identification[a]	*Material*	*Outer diameter d (mm)*	*Thickness t (mm)*	*Crack length 2a (mm)*	*Crack angle 2θ (degrees)*	*Expected leak rate (kg/min)*
J4SS102	SUS304	112·7	8·7	101·8	104	20
J4SS82	SUS304	112·7	8·7	82·4	84	8
J4CS83	STS42	112·7	8·9	82·5	84	8
J6CS104	STS42	163·4	8·7	104·2	73	8
J6SS68	SUS304	163·4	9·6	67·5	47	2
J6SS105	SUS304	164·4	9·6	105·2	73	8
J6SS139	SUS304	163·3	9·3	138·6	97	20
T4CS60	STS42	114·3	8·6	60	60	—
T4CS97	STS42	114·3	8·6	97	97	—
T4SS60	SUS316	114·3	8·6	60	60	—
T4SS129	SUS316	114·3	8·6	129	129	—
T8CS98	STS42	216·3	12·7	98	52	—
T8CS135	STS42	216·3	12·7	135	72	—

[a] In the '*Pipe identification*' column, the prefix 'J' means JAERI and 'T' means Tanaka *et al.*; the numbers 4, 6 or 8 denote the nominal pipe diameter in inches; CS and SS, respectively, mean that the test pipe is made of carbon steel STS42 and stainless steel SUS304, respectively, in JAERI's experiments, and made of SUS316 in Tanaka *et al.*'s experiments; the last numeral is the outer crack length 2*a* (in mm).

TABLE 2
Test Results

Test number	*p (MPa)*	*T (°C)*	*sat. T (°C)*	*sub. T (°C)*	*L (kN)*	*W (kg/min)*		
						DP	*LT*	*Ve*
J4CS83								
1	7·15	270	287	17	0·0	1·3	—	1·3
2	7·13	272	287	15	12·6	3·8	—	3·8
3	6·86	273	284	11	19·6	7·3	—	7·3
J4SS82								
1	6·76	271	283	12	0·0	3·5	—	
2	6·70	276	283	7	12·5	5·3	—	5·2
3	6·79	275	284	9	12·5	4·6		
J4SS102								
1	7·19	268	287	19	0·0	2·8	—	
2	6·27	257	278	21	12·7	20·3	—	
3	7·06	250	286	36	0·0	7·9	—	
4	6·17	267	277	10	12·7	24·0	—	
J6CS104								
1	6·57	273	281	8	0·0	7·0	—	6·9
2	6·75	270	283	13	24·0	—	9·7	9·8
3	5·78	273	273	0	61·9	—	62·0	—
J6SS68								
1	6·60	270	282	12	0·0	4·1	—	4·0
2	6·66	274	282	8	23·9	—	12·6	13·2
3	5·68	276	272	−4	62·4	—	46·2	—
4	5·64	272	272	0	100·5	—	66·3	—
J6SS105								
1	6·66	263	282	19	0·0	—	11·7	10·9
2	6·96	281	285	4	23·9	—	44·6	44·0
3	6·37	276	279	3	61·9	—	70·4	—
J6SS139								
1	6·22	278	278	0	0·0	—	58·0	—
2	6·27	271	278	8	23·9	—	74·0	—

[a] *T* is the water temperature measured with a 1·6 mm CA thermocouple attached through the pipe wall; *L* is a jack load; *W* is a leak rate; *DP* is the leak rate measured by the differential pressure transducer attached to a 0·2 m^3 vessel; *LT* is the leak rate measured on a 4 m^3 large vessel and *Ve* is the leak rate measured by the Venturi flow meter.

TABLE 3
Material Data at 300°C (JAERI), 286°C (Tanaka *et al.*)[9,11]

Pipe identification	*Yield strength* σ_y *(MPa)*	*Ultimate tensile strength* σ_u *(MPa)*	*Flow stress* σ_f *(MPa)*	*Critical half crack angle* θ_c *(degree)*
J4CS	196	431	314	114
J4SS	127	421	274	111
J6CS	186	431	309	104
J6SS	147	461	304	106
T4CS	193	395	294	111
T4SS	197	479	338	115
T8CS	209	439	324	107

because there was no heat source in the 0·2 m^3 vessel or in the test pipe. The leak rates measured by differential pressure and the Venturi flow meter are in good agreement.

Table 3 shows the authors' tensile test results at 300°C, plus those of Tanaka *et al.* at 286°C.[9]

4 DISCUSSION

4.1 Leak rate and bending stress

Figures 4 and 5 show the relationship between leak rate and bending stress including the data of Tanaka *et al.*[9] These two figures show that the leak rate increases with increasing bending stress and initial crack length.

4.2 Leak rate in net section collapse criteria diagram

Figures 6–9 show the relationship between the nondimensional moment, $M/(4\sigma_f R^2 t)$, and nondimensional half-crack angle, $\Theta = \theta/\theta_c$, where θ_c, given in Table 3, is defined as the critical half-crack angle when the pipe undergoes an unstable fracture by a single pressure load without any bending load. The curve in the diagram is obtained by the net section collapse criterion[15] written as

$$M_c/(4\sigma_f R^2 t) = \sin\beta - 0{\cdot}5\sin\theta \tag{1}$$

where

$$\beta = \frac{\pi - \theta}{2} - \left(\frac{\pi}{4} \times \frac{pR}{t\sigma_f}\right) \tag{2}$$

This curve is obtained from the flow stress and piping geometry; however, it

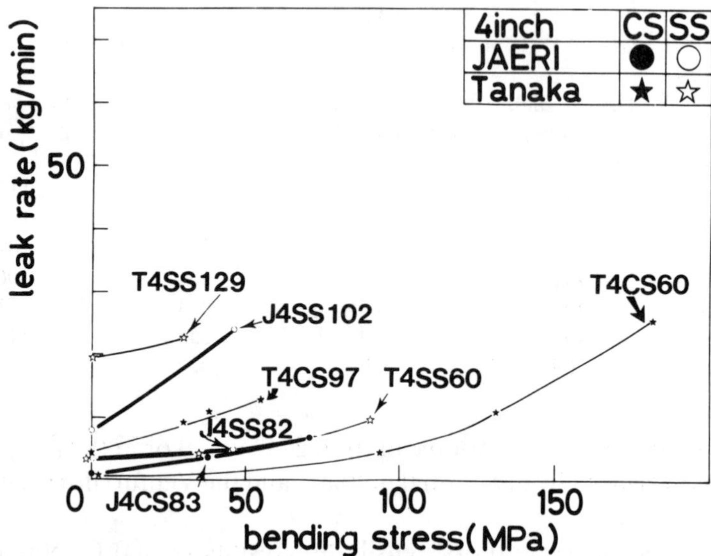

Fig. 4. Leak rate of 4 in. STS and SUS piping with and without bending load.

is similar for each pipe in spite of the difference of diameter, wall thickness and flow stress. The pipe subjected to both pressure and bending loads will undergo an unstable fracture when the applied moment is equal to the critical moment, M_c.

The coordinates of the points ○, ●, ☆, and ★ indicate the crack angle

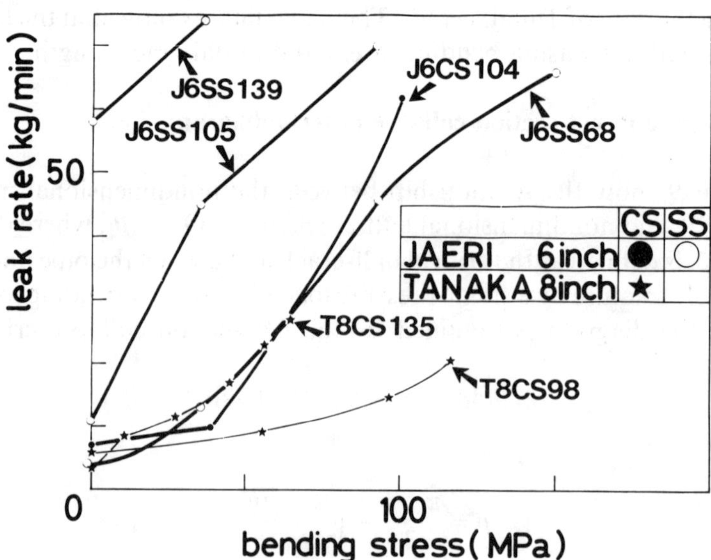

Fig. 5. Leak rate of 6 in., 8 in., STS and SUS piping with and without bending load.

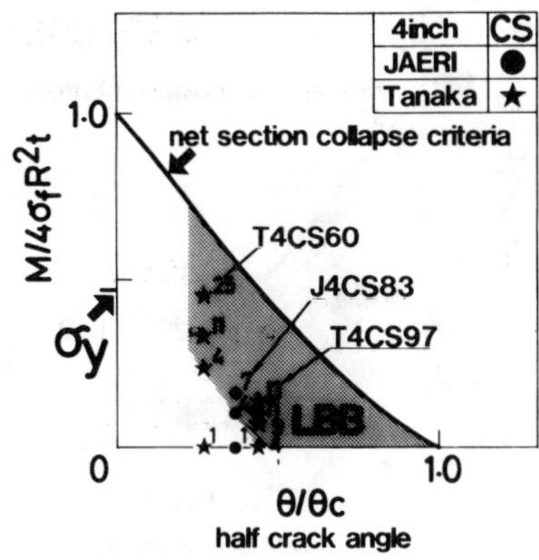

Fig. 6. Test result of 4 in. STS piping on net section collapse diagram. Numbers adjacent to the data points are the measured leak rates.

and the applied moment in the experiments. Numerals near the points are the leak rates. By introducing these two parameters Θ and $M/(4\sigma_f R^2 t)$, we can estimate the distribution of leak rate like a contour map. Unfortunately, with a nondimensional expression of this type, it is difficult to evaluate the exact bending stress. As a guide, the bending stress relative to the value of σ_y is given on the left side of the figures. We calculate that $\sigma_y \times Z/(4\sigma_f R^2 t)$ is 0·47 for STS42 steel and 0·38 for SUS steel in a $M/(4\sigma_f R^2 t)$ scale for all the 4, 6 and 8 in. pipes. Material constant σ_y is obtained as an average value of each yield strength given in Table 3. With this scale, we can immediately know the bending stress. For example, Fig. 6 shows that the maximum load applied to the 4 in. carbon-steel piping imposes a bending stress of about σ_y. No unstable fracture was observed either in the authors' or Tanaka *et al.*'s experiments.

The minimum detectable leak rate specified for Japanese BWR piping is 4 kg/min. LBB is considered acceptable when a leak rate greater than the minimum value specified is attained before the crack becomes unstable. The

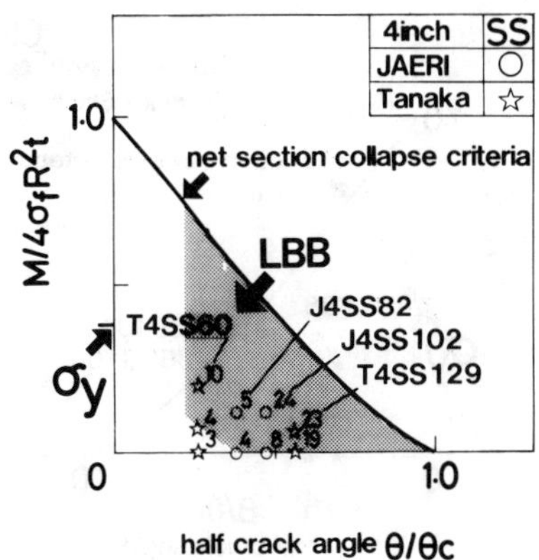

Fig. 7. Test result of 4 in. SUS piping on net section collapse diagram. Numbers adjacent to the data points are the measured leak rates.

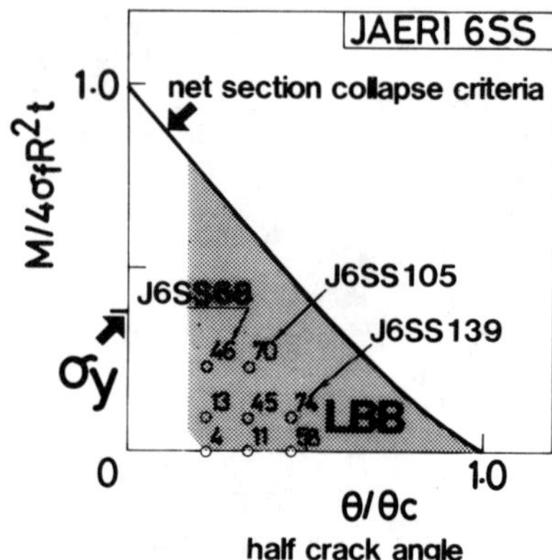

Fig. 8. Test result of 6 in. SUS piping on net section collapse diagram. Numbers adjacent to the data points are the measured leak rates.

region between the $M_c/(4\sigma_f R^2 t)$ curve and a curve corresponding to a leak rate of 4 kg/min is represented by the shaded areas in Figs 6–8. This is called the LBB region. If the 4 kg/min contour lines lie close to the collapse moment line, thenLBB is not accepted, but if the minimum leak rate lines lie near the origin, then LBB is established with wide margins, as shown in Figs 6–9.

The influence of pipe diameter on the LBB region for the carbon steel STS42 is indicated by comparing Fig. 6 with Fig. 9. The 4 kg/min contour line is closest to the M_c with a 4 in. pipe, and furthest in tests on 8 in. pipes. This means that from an LBB viewpoint, an 8 in. pipe is the most advantageous among 4, 6 and 8 in. pipes. A comparison of Figs 7 and 8 shows that a similar trend is observed for stainless-steel piping. When the results on the 4 in. SUS and STS42 steel pipes are compared (Figs 6 and 7), the 4 kg/min line for SUS steel lies closer to the origin than that of STS42 steel, which indicates that 304 or 316 stainless steels have advantages over STS42 carbon steel for LBB.

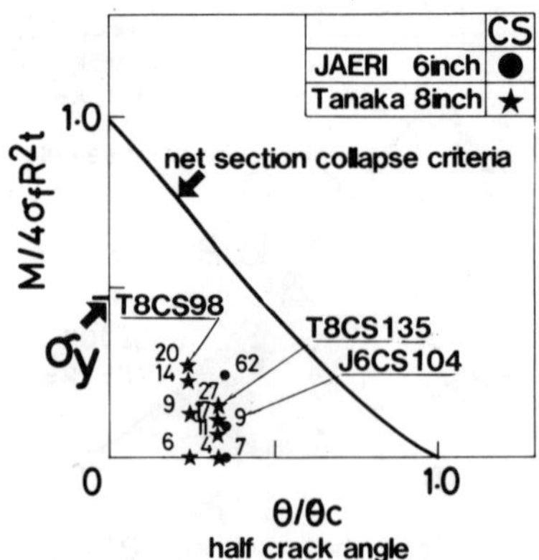

Fig. 9. Test result of 6 and 8 in. STS piping on net section collapse diagram. Numbers adjacent to the data points are the measured leak rates.

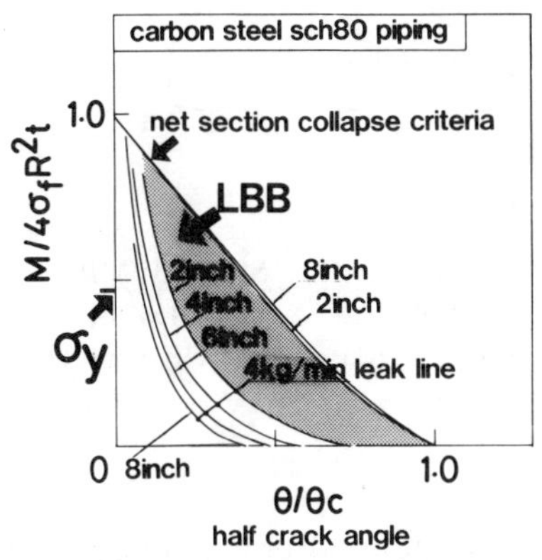

Fig. 10. Calculated 4 kg/min leak line for 2, 4, 6 and 8 in. STS piping on net section collapse diagram.

4.3 Limit-line calculation for 4 kg/min leak rate

In this section, the relationship between the nondimensional moment $M/(4\sigma_f R^2 t)$ and angle Θ in producing a 4 kg/min leak is calculated.[17] The moment M is simply given by $\sigma_b[\mathrm{Pa}] \times Z[\mathrm{m}^3]$. A computer program has been written to calculate

(1) the crack opening area (COA) under bending and pressure load,[18] and
(2) the critical mass flux by Henry's model.[1,19,20]

In calculating CPA, equivalent yield strength defined as $\sigma_Y = (3\sigma_y + \sigma_u)/4$ was used instead of the single σ_y in Tada's equation.[21] Figures 10 and 11 show the 4 kg/min leak contour line for 2, 4, 6 and 8 in. STS and SUS steel piping. The left-side scale of Figs 10 and 11 is the same as that of Figs 6 and 7. In this calculation, nominal schedule 80 piping geometry was used. Constants used in the calculation are summarized in Table 4. Surface

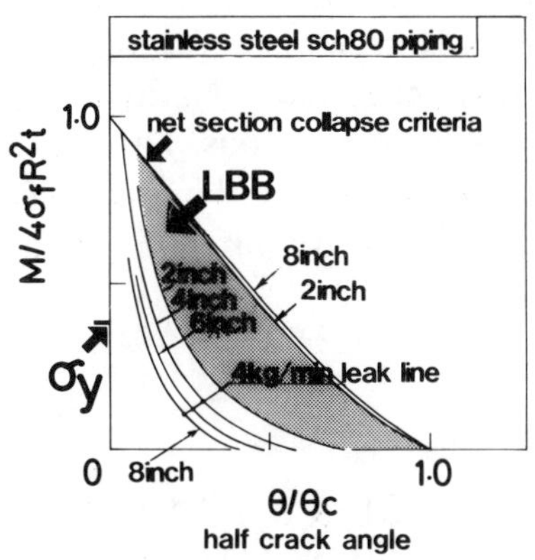

Fig. 11. Calculated 4 kg/min leak line for 2, 4, 6 and 8 in. SUS piping on net section collapse diagram.

TABLE 4
Constants used in Calculating Leak Rate

Pressure	7·0 MPa	
Temperature	286°C	
Surface roughness[a]	30 μm	
Yield strength at 300°C[a]	STS42: 196 MPa,	SUS: 157 MPa
Ultimate tensile strength at 300°C[a]	STS42: 424 MPa,	SUS: 454 MPa
Young's modulus at 300°C[16]	STS42: 180 320 MPa,	SUS: 178 360 MPa

[a] Measured value.

roughness on the crack was measured by a profilometer after a cracked pipe was split open. A DIN R_z value is used here. The ultimate tensile strength was determined by averaging the values of the tensile test results of the authors' and Tanaka *et al.*'s (Table 3). Young's modulus is derived from the MITI's code.[16] There is little difference of the curves for net section collapse between the 2 and 8 in. piping. As described in Section 4.2, the LBB region increases as the diameter of the pipe increases.

5 CONCLUSIONS

The leak rate through the through-wall crack in pipes was obtained for carbon- and stainless-steel piping of 4 and 6 in. diameters under BWR and four-point bending load conditions. Stainless steel is better than carbon steel and a pipe with a large diameter is better than a pipe with a small diameter because of their larger margin for LBB.

6 ACKNOWLEDGEMENTS

The authors would like to thank Dr T. Uga, Messrs T. Ohba, R. Kawamura, H. Ohura and R. Yagioka for their sincere assistance in performing the difficult experiments. Messrs Y. Tanaka, K. Matsumoto and T. Narabayashi are also acknowledged for giving the authors kind suggestions. This work was performed in fiscal year 1988 under a contract between the Science & Technology Agency of Japan (STA) and JAERI, to prove the safety in the primary coolant circuits of nuclear power plants.

REFERENCES

1. Collier, R. P., Stulen, F. B., Mayfield, M. E., Pape, D. B. & Scott, P. M., Two-phase flow through intergranular stress corrosion cracks and resulting acoustic emission. EPRI NP-3540-LD, April 1984.

2. Amos, C. N. & Schrock, V. E., Critical discharge of initially subcooled water through slits. NUREG/CR-3475, LBL-1636, 1983.
3. Yano, T., Matsushima, E. & Okamoto, A., Leak flow rate from through-wall crack in pipe. 2nd ASME-JSME Thermal Engineering Joint Conference, Hawaii, Vol. 111, March 1987, pp. 301–08.
4. Yano, T., Matsushima, E. & Okamoto, A., Leak flow rate from a through-wall crack in pipe. *JSME Int. J.*, Series II, **31**(3) (1988) 494–504.
5. Yano, T., Matsushima, E. & Okamoto, A., Evaluation of leak flow rate and jet impingement related to leak-before-break. *Nuclear Engineering and Design*, **111** (1989) 197–205.
6. Matsushima, E., Yano, T. & Okamoto, A., Experimental study of leak flow through rectangular slits. *Trans. JSME*, **54**(506) (1988) 2785–91 (in Japanese).
7. John, H., Reimann, J. & Eisele, G., Kritische Leckströmung aus rauhen Rissen in Druckbehältern. KfK4192, 1987.
8. John, H., Reimann, J., Westphal, F. & Friedel, L., Critical two-phase flow through rough slits. *Int. J. Multiphase Flow*, **14**(2) (1988) 155–74.
9. Tanaka, Y., Matsumoto, K. & Narabayashi, T., Experimental studies on high temperature water leak rates through pipe cracks. Paper presented at the 4th Japanese–German Joint Seminar on Structural Strength and NDE Problems in Nuclear Engineering, 1988, pp. 383–402.
10. Narabayashi, T., Ishiyama, T., Fujii, M., Matsumoto, M., Horimizu, Y. & Takaka, Y., Study on coolant leak rates through pipe cracks: Part 1—Fundamental Test. *ASME PVP*, **165** (1989).
11. Matsumoto, M., Nakamura, S., Gotoh, N., Narabayashi, T., Tanaka, Y. & Horimizu, Y., Study on coolant leak rates through pipe cracks: Part 2—Pipe Test. *ASME PVP*, **165** (1989).
12. Abdollahian, D. & Chexal, B., Calculation of leak rates through cracks in pipes and tubes. EPRI NP-3395, December 1983.
13. Mayfield, M. E., Forte, T. P., Rodabaugh, E. C., Leis, B. N. & Eiber, R. J., Cold leg integrity evaluation. NUREG/CR-1319 R5, February 1980.
14. Schrock, V. E., Revankar, S. T., Lee, S. Y. & Wang, C. H., A computational model for critical flow through intergranular stress corrosion cracks. NUREG/CR-5133 LBL-21967, 1988, pp. 113–20.
15. Kanninen, M. F., Marschall, C. W., Rybicki, E. F., Sampath, S. G., Simonen, F. A. & Wilkowski, G. M., Mechanical fracture predictions for sensitized stainless steel piping with circumferential crack. EPRI NP-192, 1976, pp. 121–7.
16. Japanese MITI code 501, Denryoku-sinpou-sya, 1986 (in Japanese).
17. Shinokawa, H., Shibata, K. & Isozaki, T., Development of leak analysis from through-wall crack, JAERI-M 90-050, 1990 (in Japanese).
18. Paris, P. C. & Tada, H., The application of fracture-proof design method using tearing-instability theory to nuclear piping postulating circumferential through-wall cracks. NUREG/CR-3464, 1983.
19. Henry, R. E., The two-phase critical discharge of initially saturated or subcooled liquid. *Nucl. Sci. & Eng.*, **41** (1970) 336–42.
20. Henry, R. E. & Fauske, H. K., The two-phase critical flow of one-component mixtures in nozzles, orifices, and short tubes. *J. of Heat Trans., Trans. ASME*, **93** (1971) 179–87.
21. Shibata, K., Chujo, N., Onizawa, K. & Miyazono, S., Progress and evaluation of test results on JAERI's ductile pipe fracture test program. Paper presented at the 4th Japanese–German Joint Seminar on Structural Strength and NDE Problems in Nuclear Engineering, 1988, pp. 347–64.

Int. J. Pres. Ves. & Piping **43** (1990) 413–424

Leak Rate Experiments for Through-Wall Artificial Cracks

J. M. Boag, M. T. Flaman & B. E. Mills

Ontario Hydro Research Division (OHRD), 800 Kipling Avenue, Toronto, Ontario, Canada M8Z 5S4

ABSTRACT

An experimental program of leak rate testing has been carried out for different geometries of artificial cracks in piping under various thermo-hydraulic conditions. The objective of these tests was to obtain a database of measured leak rates from artificial piping cracks and to compare test results with leak rate predictions obtained from the EPRI developed analytical model called PICEP. Experimental results for leak rates associated with rectangular and flare crack geometries are presented. Initial results of comparing analytical leak rate predictions with the experimental data is also presented.

1 INTRODUCTION

The possibility of a sudden and complete rupture of large diameter piping in nuclear power plants represents a serious concern: it can have an impact on public and plant personnel safety and can result in severe and costly damage to reactor buildings and equipment. Large diameter piping is used in many areas of nuclear reactors and considerable cost is involved in attempting to design plants which minimize the consequences of a complete pipe rupture during operation.

At Ontario Hydro, the Leak-Before-Break (LBB) approach has been employed to demonstrate that sudden, catastrophic failure of critical nuclear piping will not occur. This LBB approach is based on the concept that pipe deterioration would be progressive and not a sudden occurrence,

Int. J. Pres. Ves. & Piping 0308-0161/90/$03·50

and would first manifest itself in a through-wall breach of the pressure boundary resulting in a non-catastrophic leak; this leak would be detectable and remedial action could be taken well before the dangerous situation of a large break would arise.

This paper describes an experimental leak rate testing program, which is intended to support analytical efforts regarding use of the LBB concept at Ontario Hydro. Details of the test fixture and measurement techniques will be presented, as well as some experimental leak rate results. In addition, a comparison between these experimental results and leak rates predicted by PICEP, an EPRI-developed analytical model[1] is presented.

2 EXPERIMENTAL TEST SET-UP

2.1 General

The experimental program involves the use of small test fixtures containing precise artificial crack geometries to obtain leak rate information for various thermo-hydraulic conditions.

The leak rate tests have been carried out in the OHRD-Burst Test Facility. Special attention has been paid to the accurate measurement of leak rates, process parameters, through-wall fluid temperatures and pressures, and crack face metal temperatures.

2.2 Leak rate test fixture

A schematic diagram of the leak rate test fixture is shown in Fig. 1. The fixture consists of two precisely-machined stainless steel cylindrical halves.

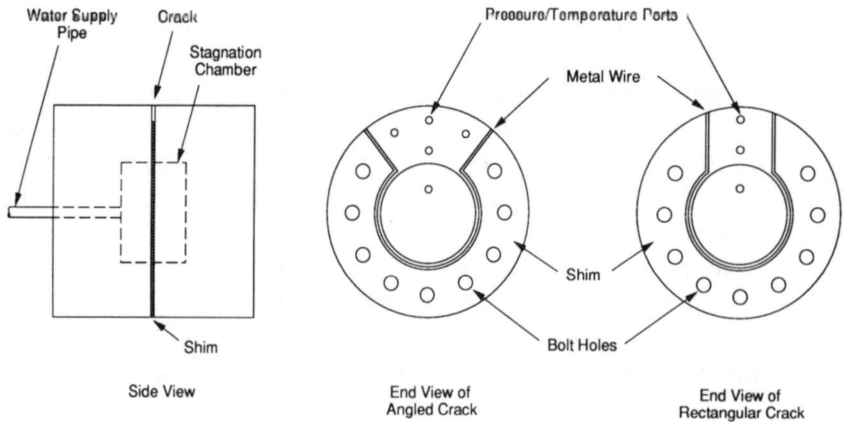

Fig. 1. Schematic diagram of leak rate test fixtures.

This results in a smooth-surfaced, straight-sided crack with a clearly-defined geometry. This is different from typical fatigue cracks, which are rough-surfaced with many turns, and with complex geometries. However, the purpose of this experimental program was not to simulate actual fatigue cracks, but rather to supply points of reference for the leak rate model. Therefore, the simplest, most clearly-defined geometry possible was chosen for the test fixture. A small continuous groove was machined into the face of one half of the fixture to outline the desired through-thickness crack shape. A precisely measured shim (which determines the crack opening displacement or COD) is positioned on the fixture face just outside the continuous groove. An aluminum wire, slightly larger in diameter than the depth of the groove, is then placed into the groove and the two halves of the fixture are bolted together. The aluminum wire crushes to the thickness of the shim, thus providing a leak-tight boundary (similar to an 'O' ring seal) and a well-defined artificial crack geometry.

Ports were machined into the test fixture to obtain precise temperature and pressure data in the stagnation chamber as well as through the depth of the artificial crack. It should be noted that there is virtually no flow in the leak rate fixture. However, the fluid velocity within the crack is much higher than even the velocity of the fluid flowing within real pipes. Therefore, the flow in real pipes is not expected to have a significant impact on the leak rate or the pressure and temperature distributions in the test fixture.

The parameters of the three artificial crack geometries discussed in this paper are given in Table 1. The aspect ratios of these artificial cracks are typical of the geometries expected in large thick-walled piping.

The fixture was designed to be rigid to minimize any changes in COD that might occur during testing. (Any change in COD occurring during the experiments could significantly affect the comparison with predicted results obtained from the analytical leak rate model because the model assumes a constant value for the COD.)

Following the leak rate testing, some initial measurements of COD changes in the test fixture were made. These measurements were performed

TABLE 1

Test Fixture Crack Geometries

Test no.	*Fixture no.*	*Crack shape*	*COD (mm)*	*Crack length at the outer surface (mm)*	*Average surface roughness (mm)*	*Crack depth (mm)*
1	SS-A1	Flare (75·2°)	0·102	98·6	$2{\cdot}03 \times 10^{-4}$	38·1
2	SS-A2	Rectangular	0·109	48·8	$2{\cdot}03 \times 10^{-4}$	38·1
3	SS-A3	Flare (75·2°)	0·508	98·6	$2{\cdot}03 \times 10^{-4}$	38·1

under actual test conditions, including temperature gradients, pressure profiles, and two-phase flow. The preliminary results of these measurements indicate that COD changes of less than 5% occur during testing. Since the flow in the test fixture cracks is generally turbulent, this COD variation will affect the measured leak rates by less than 5%.

2.3 Weigh balance water supply system

An important requirement for the experimental program was the accurate measurement of leak rates for each test fixture crack geometry. For these tests, a pressure vessel with a capacity of approximately 200 kg of water was used as a reservoir to supply water to the leak-rate fixture under different thermo-hydraulic conditions. At a maximum allowable CANDU reactor piping leak rate of 0·5 kg/s, the inventory of water in the pressure vessel allowed for approximately 400 s of continuous testing.

The weight loss of water in the pressure vessel (as recorded by a precision load cell) was measured as a function of time to establish leak rates for each test. Rather than directly weighing the pressure vessel to measure water loss, the OHRD approach was to mount the pressure vessel, nitrogen gas cylinders, and all ancillary equipment on a balancing platform as shown in Fig. 2. Small adjustments to the location of the nitrogen tanks in relationship to the pressure vessel were made during initial pressurization tests with nitrogen only. By suitably locating the nitrogen tanks on the weigh balance system, the effect of nitrogen mass shift (from the gas cylinders to the

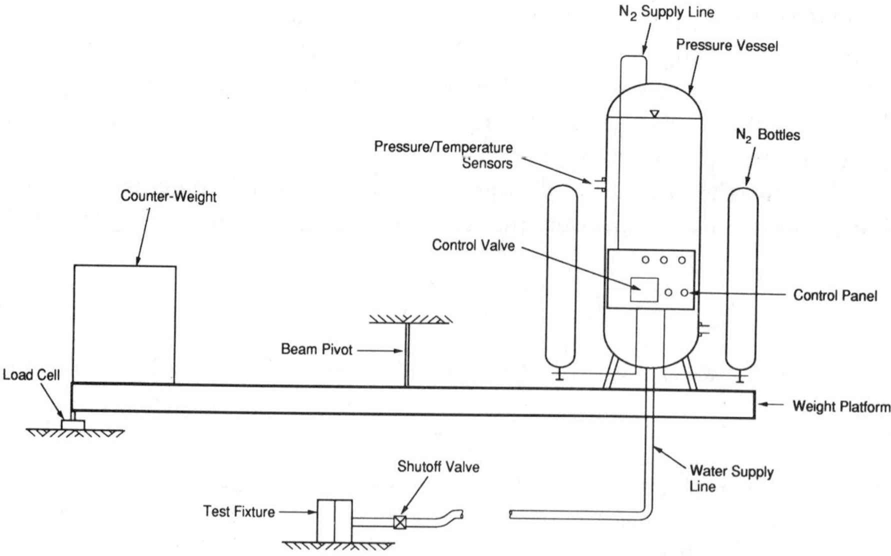

Fig. 2. Schematic diagram of water supply system.

pressure vessel) on the load cell output during the leak rate tests is eliminated. In this manner, the load measuring equipment can be sized to the weight of the water only, rather than to the entire weight of all the equipment, thus yielding results of much greater accuracy.

Calibration of the weigh balance system was performed so that water loss from the pressure vessel (as measured by a precision load cell) could be correlated to fixture leak rates. Water was removed in stages from the pressure vessel and the mass of the water at each stage was determined using a high-accuracy balance and traceable reference weights. The load cell output was recorded against the total mass of water removed from the pressure vessel. A best-fit linear regression was then performed on the calibration data.

Based on this procedure, the calibration was determined to be linear and the accuracy of all reported leak rate measurements is typically within $\pm 0{\cdot}005$ kg/s. Re-calibration of the weighing balance system has been performed periodically during the testing program and has been found to be repeatable to within 0·05%.

3 EXPERIMENTAL RESULTS

3.1 Test procedure

For the tests described in this paper, the pressure vessel was filled with deionized and demineralized water. The water in the pressure vessel was heated to and maintained at a predetermined temperature for each set of tests using a 40 kW immersion heater. The nitrogen cover gas pressure was controlled (via a flow control valve) so that different pressure conditions can be obtained in the test fixture stagnation chamber. Typically, the pressure in the test fixture stagnation chamber could be changed four or five times (at constant temperature) to obtain different amounts of fluid subcooling for each fill-up of the pressure vessel. This approach enabled each set of tests to be performed under relatively constant temperature and water chemistry conditions.

The target thermo-hydraulic water conditions in the test fixture stagnation chamber are given in Table 2.

3.2 Data collection and analysis

A computerized data logger automatically collected data on the water loss, pressure, and temperature at 30 readings/s during the entire test. Typical plots of water loss, pressures and metal temperatures through the depth of the artificial crack versus time are given in Fig. 3. Also shown are typical through-wall pressure and metal temperature profiles.

It can be observed from the metal temperature and pressure graphs (Fig. 3) that equilibrium conditions are achieved within 10 to 20 s of the start of testing.

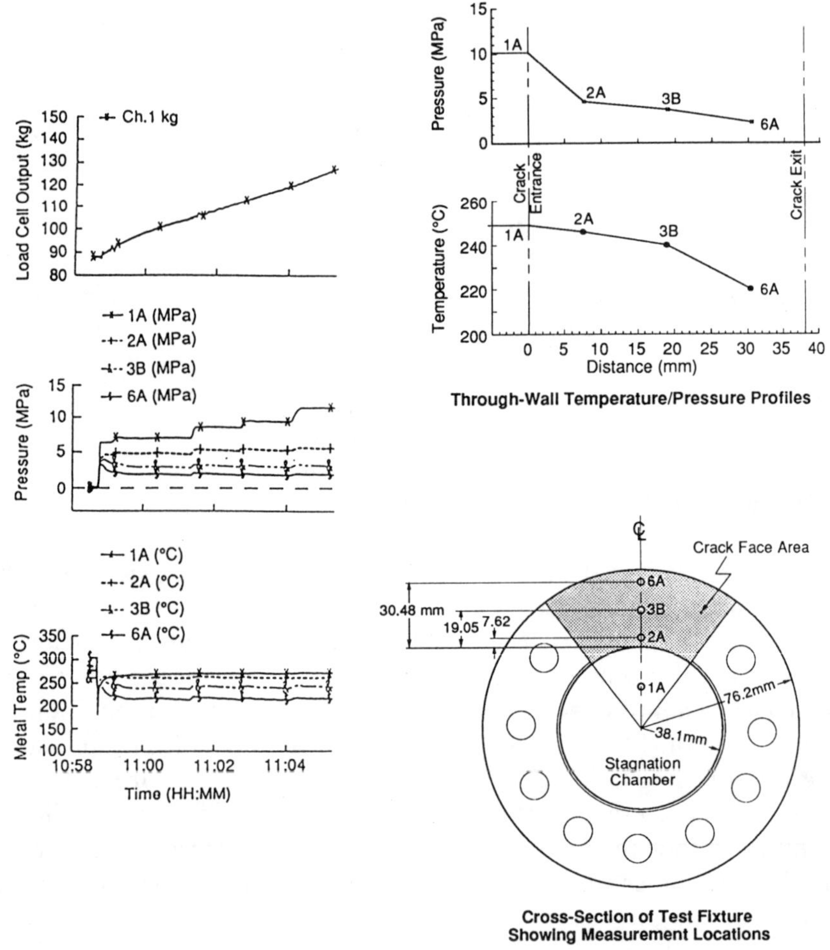

Fig. 3. Typical measurement data.

After removing the data points measured during the transient conditions at the start and end of each test, the 'average' leak rate over the test period for the fixture is determined from the slope of a best fit straight line through the load cell output. Similarly, 'average' pressures and temperatures are determined throughout the fixture using this data analysis technique.

TABLE 2
Test program Target Thermo-hydraulic Conditions

Temperature (°C)	*Pressure (MPa)*
170	2·0
170	4·0
170	8·0
170	10·0
170	12·0
250	6·0
250	8·0
250	10·0
250	12·0
275	7·5
275	9·0
275	10·0
275	12·0
300	10·4
300	11·0
300	12·0

4 LEAK RATE RESULTS

The leak rate results for different crack geometries under various thermo-hydraulic conditions in the fixture stagnation chamber are shown in Figs 4–6.

4.1 Experimental results

The leak rate is a strong function of stagnation pressure, particularly for cases with a large degree of subcooling. This observation is typical of single-phase liquid flow. However, as the amount of subcooling decreases (and the stagnation chamber fluid temperature increases), the leak rate becomes a weaker function of the stagnation pressure. This decrease in dependence on stagnation pressure occurs because as the degree of subcooling is decreased, the inception of choked flow within the crack face moves closer to the crack entrance. This trend is most clearly shown in Figs 5 and 6, i.e. for the smaller cracks, where flashing occurs closer to the crack entrance. It should be noted that in the parallel-sided crack geometry, as opposed to the flared cracks, the fluid flow is friction-restricted. Thus, for the parallel-sided crack geometry, as the degree of subcooling decreases, the friction-induced pressure drop will cause the choking point to move closer to the crack entrance.

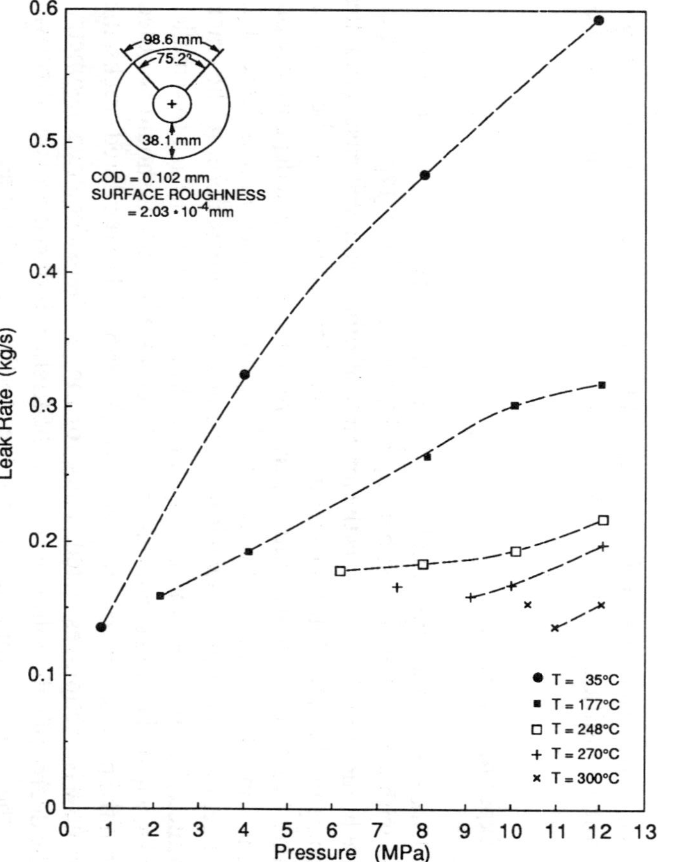

Fig. 4. Leak rate results—Fixture SS-A1.

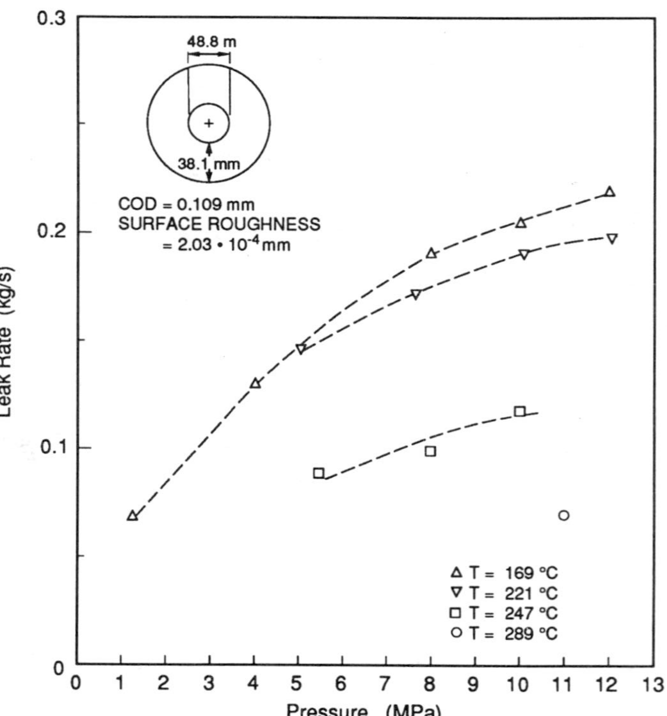

Fig. 5. Leak rate results—Fixture SS-A2.

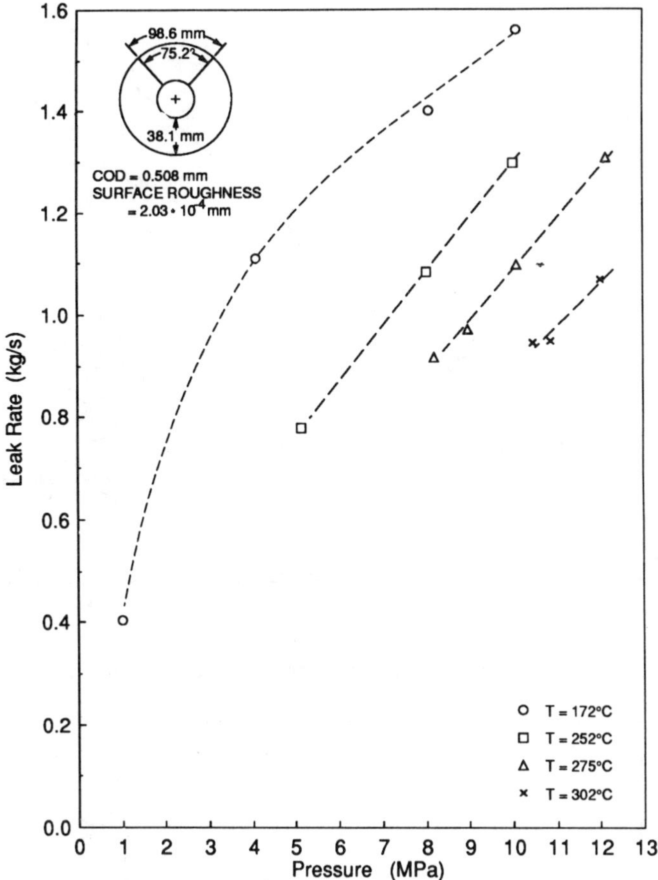

Fig. 6. Leak rate results—Fixture SS-A3.

Fluid choking plays an important role in limiting the leak rate as witnessed by the flattening of the curves in Figs 4–6 for higher stagnation chamber temperatures. This phenomenon has also been observed in other experimental results.[2]

4.2 Comparison with theoretical predictions

Of particular interest in this program was a comparison between the experimentally measured leak rates and those determined by a predictive analytical model. Results of such a comparison are presented in Figs 7–9. These graphs show the measured leak rates versus those predicted by the 'PICEP' analytical model. PICEP, the Pipe Crack Evaluation Program, has been developed by the Electric Power Research Institute (EPRI). Its leak rate calculation is based on a homogeneous nonequilibrium critical flow model

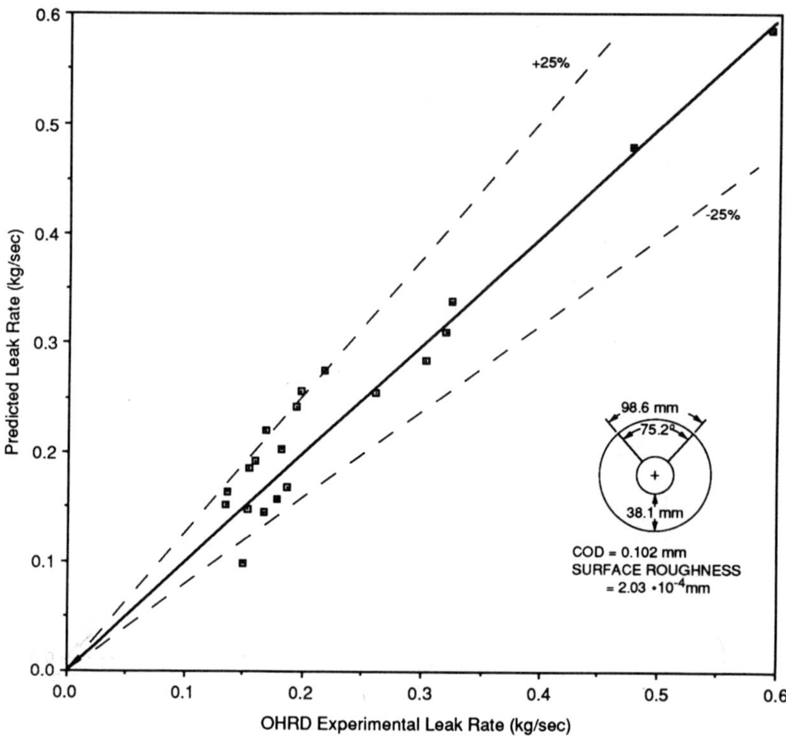

Fig. 7. Comparison of experimental and predicted leak rates—Fixture SS-A1.

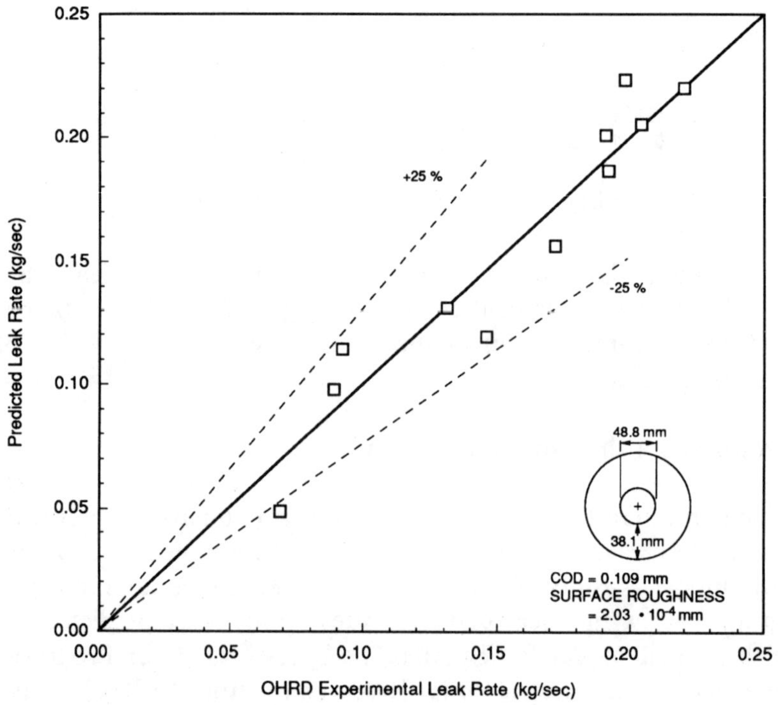

Fig. 8. Comparison of experimental and predicted leak rates—Fixture SS-A2.

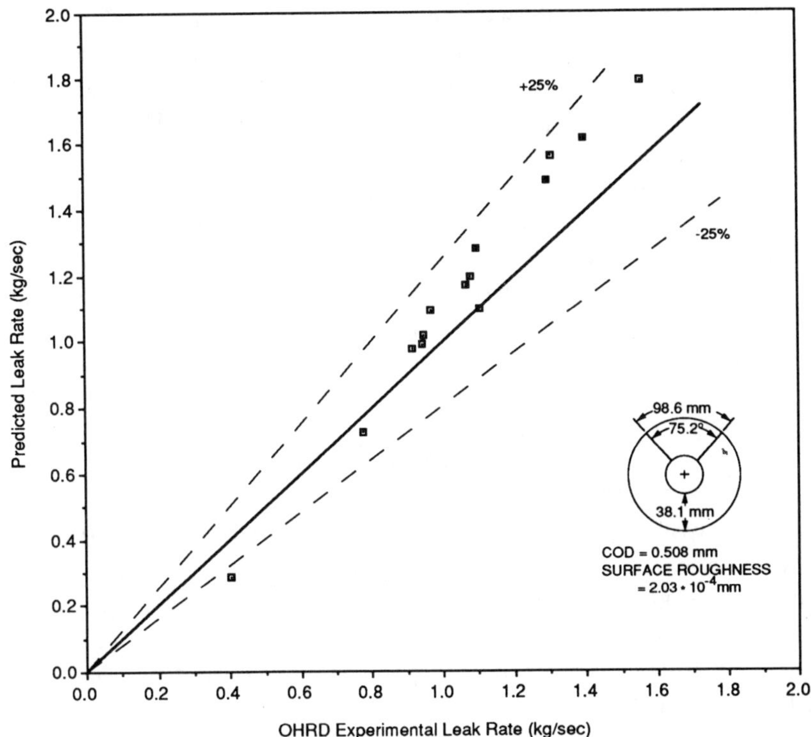

Fig. 9. Comparison of experimental and predicted leak rates—Fixture SS-A3.

with several modifications to account for friction, crack turns, and fluid conditions.

In virtually all cases, the correlation between the predicted and measured leak rates is within $\pm 25\%$. This is considered a reasonable amount of agreement. Comparisons between predicted and measured leak rates by other researchers show similar agreement.[3–5]

5 CONCLUSIONS

(1) A weigh balance water supply system has been designed, constructed, and commissioned to obtain reliable and accurate leak rate data in support of the Ontario Hydro LBB approach. Recent calibration of the weigh balance system has shown that a leak rate data accuracy of $\pm 0{\cdot}005$ kg/s can be readily obtained.

(2) A leak-tight fixture has been designed and tested so that different artificial crack geometries can be easily tested. In addition, through-wall crack pressures and metal temperatures can be accurately measured using this fixture.

(3) Initial leak rate data have been reported for both rectangular (parallel-sided) and flare type artificial crack shapes under various thermo-hydraulic conditions. These cracks are similar in aspect ratio to those that are predicted to be found in CANDU-type nuclear power piping.
(4) Initial comparison of the experimental leak rate data to leak rates predicted by the PICEP computer model show reasonable correlation, i.e. almost all data within $\pm 25\%$. Future work will involve more detailed comparisons, further COD measurements, and the development of Ontario Hydro's own computer model to predict leak rates.

ACKNOWLEDGMENTS

The authors would like to express their gratitude to Messrs S. Ray and T. Caputo who carried out the experiments. The financial and technical support of the Darlington Engineering and Nuclear Safety Departments of Ontario Hydro is also greatly appreciated. Special thanks go to Mr B. Chexal of the Electric Power Research Institute for his permission to use the PICEP computer code.

REFERENCES

1. PICEP (Pipe Crack Evaluation Program) Rev. 1, July 1987. Developed by the Electric Power Research Institute, Palo Alto, CA.
2. Amos, C. N. & Schrock, V. E., Critical discharge of initially subcooled water through slits. NUREG Report CR-3475/LBL-16363, November 1983.
3. Scott, D. A. & Cook, A., Leakage rates through cracks in thick wall piping. AECL Report No. 9003, Toronto, June 1983.
4. Yano, T., Matsushima, E. & Okamoto, A., Leak flow rate from through-wall crack in pipe. Paper presented at the 1987 ASME-JSME Thermal Engineering Joint Conference, Honolulu, HI, March 1987.
5. Chouard, P. & Richard, P., Recent results of leak rate studies at PWR conditions in France. Paper presented at the 1987 Leak-Before-Break Seminar, Tokyo, May 1987.

Int. J. Pres. Ves. & Piping **43** (1990) 425–432

Directed Discussion

Moderated by Prof. E. Smith
Summarized by L. A. Simpson, C. E. Coleman & E. Smith

The directed discussion has proved useful in the context of arriving at conclusions in the annual reviews of work on fracture for the CANDU Owners Group. Prof. E. Smith as a consultant has played an important role as the moderator of these discussions. This approach was adopted for the present Specialist Meeting as a good means to highlight key issues. This summary is based on notes and a tape of the discussions and is believed to be generally accurate. Participants were, however, expressing personal views and have not had the opportunity to confirm the report of their remarks. The remarks should not, therefore, be taken as representing the firm position of individuals or of their organizations.

The summary consists of two parts, an overview by Prof. Smith in which he identifies eight key issues and the discussion by the delegates on those issues.

THE ISSUES

Smith opened his remarks by stating that it was gratifying to attend a meeting which was focussed on an issue, rather than a component or system, as it provides for a transfer of thinking. He referred to his own career, as a consultant in fracture-related problems in several countries, as an example where this has been effective and saving of much time and effort. It was clear to him that Leak-Before-Break (LBB), as a route to guaranteeing structural integrity of reactor components, is with us to stay.

> 'In this meeting we have seen its use discussed for various systems and to varying degrees as part of a defence-in-depth procedure. These systems include zirconium alloy pressure tubes, piping systems and steam generator tubing. One could also have included the vessel in a sodium-cooled fast reactor. Often the application of LBB or proposed application

is underpinned by extensive R & D programs with the following main themes:

—Methodologies: stable and unstable growth
leakage calculations
—Qualification of LBB procedures
—Materials data for input into assessments

We have seen some important new results and perspectives in each of these areas.'

Smith did not propose to highlight the R & D progress but rather to indicate specific outstanding issues that might jeopardize LBB or, put another way, issues whose resolution might allow us to broaden the scope of the LBB route: 'Thus, I shall try to identify areas where I believe that our future efforts should be focussed—in order that we can strengthen the LBB method for guaranteeing structural integrity'.

Issue 1: Sensitivity and reliability of detection devices

When leakage occurs, we want to be able to detect it and furthermore, to identify the leaking component quickly. He posed the question, 'are we satisfied with the current equipment?'

Issue 2: Factors that affect leakage and make detection difficult

Much of the methodology is aimed at calculating crack opening areas and correlations with leak rates for somewhat idealized situations. 'I have a firm view that a disproportionate amount of effort is given to the factors which affect leakage in comparison with efforts which govern unstable crack size.' Factors which were noted to affect leakage include (1) residual stresses, or bending stresses which acted to open or close the crack, (2) crack shape, dependent on growth mechanisms and (3) plugging of the crack by debris such as fractured oxide films.

He asked whether there was further scope for safe stimulation of leakage, as described in the Coleman paper on CANDU tubes, in applications to other systems.

Issue 3: The gradual development of a part-through crack, its shape, and its effect on instability

'The simplest way of viewing Leak-Before-Break is in terms of a through-wall crack size for instability and a through-wall crack size to give a detectable leakage; then think in terms of having a desirable margin between these sizes, that margin being expressed in terms of a size difference with piping or a response time factor in pressure tubes. However, this is too simplistic if, during slow crack growth, lengthwise extension is greater than through-thickness growth. Then, as a start, it is more appropriate to use the sort of LBB diagram as we saw in the Clayton–Sharples paper.'

Professor Smith went on to emphasize the undesirability of situations where one gets 'canoeing' as in the Brocks paper, or tunnelling as described by Moan for CANDU pressure tubes. A related undesirable situation was where one had the possibility of a line initiator.

For the case of the Duane Arnold, safe-end type of crack, which showed very uneven growth, he left that even the Clayton–Sharples diagram was too simplistic. Here, we have the potential for ligament breakage and then unstable circumferential growth assisted by the presence of the part-through crack around the circumference acting like a side-groove on a toughness specimen.

Clearly then, these comments focus on:

(a) the importance of the development of the part-through crack and its shape;
(b) instability criteria for the part-through crack. This is getting much less attention then the through-wall crack.

Issue 4: Correct consideration of weld properties

Smith posed the question as to whether we were satisfied that we were dealing with cracks in welds in an appropriate manner and developing appropriate methodologies. Here, we are dealing with a composite situation, very often the weld material is softer than the alloy matrix. It is, therefore, good enough to take the J_R curve from small specimens of weld metal and apply them to the real situation? Similarly, in estimating leakage sizes do we always allow for the fact that the hardness of the weld region could restrict crack opening?

Issue 5: Application of LBB methodology to non-ideal regions

The key issue here is that much of the LBB methodology has been developed for ideal regions such as straight pipe sections and there has been only limited effort on regions such as T-junctions and elbows. Prof. Smith illustrated this point with an example.

'Unstable crack sizes are often, in cases of piping, determined on the basis of stresses at an assumed crack in a section via an uncracked piping analysis. Because piping is built in at its ends, say into a vessel or steam generator, the critical unstable crack size is underestimated, while the leakage crack size is also underestimated. This situation depends on system flexibility, which depends on position of the crack in the system.'

Issue 6: Probabilistic approach

LBB is not a yes/no situation and it must involve a probabilistic aspect. We have seen several attempts at this meeting to arrive at the probability of LBB and this becomes important when margins are not great. The question is, 'Can we incorporate some of the complicating issues raised in this discussion within the probability umbrella?'

Issue 7: When should we use LBB?

This issue is not unrelated to the probability issue. For instance, we have the elimination of dynamic effects of postulated high energy pipe ruptures from the design basis of power plants unless (quoting Wichman & Lee) 'piping particularly susceptible to failure from potential degradation mechanisms will be excluded from the application of LBB.' In contrast, quoting from the Wong paper (on zirconium alloys), 'For the delayed hydride cracking failure mode, LBB is used as a defence-in-depth against unstable rupture.' In terms of the question, 'Is this seeming difference in approach just a difference in the extent to which LBB is used as a defence-in-depth or, is it a completely different philosophy?', it was Professor Smith's belief that is was the former.

In this context he noted that with respect to piping (Wichman & Lee), susceptibility to failure by fatigue or SCC is excluded from LBB considerations because they may introduce flaws whose geometry may not be bounded by postulated through-wall flaws in LBB analyses; clearly an issue related to the shape development issue mentioned earlier.

Issue 8: Incorporation within codes

One of the stated goals of this meeting was to establish how LBB methodology might be, or is being, incorporated into codes to demonstrate safety of pressure boundary components. Codes are usually simple in concept and application and, therefore, how can we envisage them incorporating complicated issues? Should we therefore, limit codes and treat complicated issues on an ad hoc basis?

HIGHLIGHTS FROM THE DELEGATES' RESPONSES

Professor Smith suggested the discussion commence with the issue 'when to use LBB?' He called on B. A. Cheadle (Canada) to indicate how LBB fits into the overall integrity situation for CANDU. Cheadle stated than many of the papers he had listened to dealt with somewhat hypothetical situations of low probability cracking mechanisms. In CANDU we are dealing with a real problem on components that (because they are located in the center of the core) are continuously degrading with time. The margin for LBB is a decreasing quantity and we need to define what the limiting value of that margin is in order to define a safe limit for the end of life condition of a pressure tube.

S. Lee (USA) was invited to comment. Lee saw a fundamental difference in view between USNRC and COG which, in his view, was related to an interpretation of terminology. LBB is a simple term used in many different situations. The USNRC position was related to licensing and the design basis issue of eliminating pipe whip. They do not intend that LBB be used for evaluating a flaw in service. They call this 'flaw evaluation' and the NRC requires a fracture mechanics analysis to determine the fitness of that component. One exception was the application of LBB to assessment of thermal stratification effects on PWR surge line integrity.

G. A. Barthelome (FRG) indicated that in Germany, LBB can be used as an additional defence-in-depth for reactor coolant lines, main steam lines and feedwater lines. In application, they must first show that basic safety exists (code requirement), that is they must show that the defect will not grow in the reactor operating period (similar to USNRC rule). They then allow the crack to grow to leakage beyond the reactor operating period and the crack must be shown to be detectable with margin. Calculational procedures must be verified. Smith asked if the basic difference between the FRG and USA was that the former allowed for the possibility of the crack growing through the wall. Barthelome answered in the affirmative.

W. Grant (Canada) gave the regulator's view of the application of LBB to CANDU components. The Canadian regulators have some concerns regarding the application of LBB to pressure tubes beyond the first few years of operation and consequently have intensified the mandatory inspection program in the Canadian CANDU reactors. This includes the requirement that the annulus gas system be continuously monitored for leakage and that the gas flow through the annulus be dynamic as opposed to static. They have also instituted a periodic inspection program which includes taking scrape samples from the tubes to determine the pick-up of hydrogen isotopes. Where steel piping is concerned, they tend to follow the USA practice. E. G. Price (Canada) then commented that the application of LBB to CANDU pressure tubes is an economic issue as opposed to one of safety since the simultaneous failure of a pressure tube and its calandria tube is a licensible event.

Returning to the issue of crack shape (Issue 3), Smith asked whether G. Wilkowski (USA) had any comments on the Duane Arnold type of crack.

Wilkowski responded that complex crack geometries can arise from severe SCC or thermal fatigue in feedwater lines. The presence of the Duane Arnold type of surface crack can constrain the plasticity at the tip of an associated through-thickness crack and lower the fracture toughness for circumferential propagation. The Duane Arnold surface crack was 75% through wall and the reduction in tearing modulus of a compact specimen with such a side groove was 14 times that for the smooth-sided specimen.

W. Brocks (FRG) stated that we need to understand crack shape development to show that wall penetration will occur at lengths much less than the critical crack length. To do this we need a criterion for crack growth in three dimensions and material data that are independent of size and geometry. Smith commented that Brocks did not get much lengthwise extension in the cracks described in his paper and asked how sensitive crack shape development was to the directional dependences of the J_R curve. Brocks felt that the sensitivity was significant.

W. Schmidt (FRG) did not believe that J_R curves were critical for LBB because pressures to initiate grwoth (in vessels) are far beyond operating pressures. Information on SCC and fatigue was needed rather than on toughness and there was also a need for tools other than finite element methods for analysis.

The discussion then turned to probabilistic issues (Issue 6) and Smith noted that papers had been presented on both the CANDU reactor and light water reactors. He asked J. Walker (Canada) if he had received useful input to his program from hearing the LWR presentations. Walker said yes and went on to comment that the cost of computer time for the probabilistic studies was trivial when compared with the investment in nuclear power stations. He asked for more and better data from the experimenters, especially on crack shape, to provide more confidence in the assessments.

S. Beliczey (FRG) noted that because of the difficulty in handling complex crack shapes (the probability of any particular one occurring could be very small) it is necessary to use statistics. Variability of the properties of weld materials also demands a probabilistic analysis.

B. Brickstad (Sweden) agreed that probabilistic approaches have merit when dealing with complex crack shapes during sub-critical growth. He then described a flaw indication in a stainless steel pipe in Sweden and raised the issue of whether it would have been appropriate to allow that pipe to continue in service if an LBB analysis showed that it would leak before becoming unstable. Lee commented that in the USA the weld would have been repaired with an overlay and re-inspected in one year.

The discussion then turned to codes (Issue 8). Smith remarked that he was not an expert on codes but felt they should deal with simplified situations. He asked if we should leave codes for simple situations and use ad hoc analyses for the complicated ones.

On the subject of large-scale verification tests (Issue 5), Smith commented that they seemed to concentrate on simple geometries and there did not seem to be much work on components such as T-junctions or nozzles. He saw some scope existing for verification tests on non-ideal regions of primary circuit components. M. Kozluk (Canada) commented that it was a case of walking before one could run. Such tests are very complex involving many in-plane and out-of-plane moment forces. Simply preparing cracks in such specimens can be extremely difficult because there are many choices for their location. The tendency has therefore been to use bounding cracks of simple geometries in the full-scale tests to verify models and predictions from small specimens. Having verified the analytical procedures with the simple tests one can then apply them to more complicated geometries.

Barthelome (FRG) pointed out that, for piping, LBB is generally applied to welds only (Issue 4). He asked about applications to base materials, for instance elbows. W. Stoppler (FRG) replied that in Germany some of these complicated geometries are being addressed using extensive finite element analysis. P. Monette (Belgium) stated that BWR primary loop piping has a lower bound toughness that is much lower than that of the weldments. This is found in elbows which are cast structures and subject to thermal aging. This degraded toughness is used as the bounding case for LBB.

Smith stated that depending on size and geometry, a plastic zone could be confined to the weld metal or extend beyond it into the heat affected zone or the parent material. The CEGB R6 procedure uses the flow properties of the parent material and the J_R properties of the weld in stability assessments. The question therefore arises, 'Are we being unnecessarily conservative in this approach?'

Beliczey raised the issue of thermal fatigue due to thermal stratification and asked how one should deal with this form of thermal fatigue (Issue 3). C. Faidy (France) described a French program to study stratification which involved large-scale loop testing to determine the causes of stratification. They hope to complete the study by the end of 1990.

Smith turned the discussion to the subject of leakage (Issues 1 and 2). He noted three items in the meeting which were cited as impediments to leakage:

(a) residual stresses or through-wall bending stresses,
(b) crack growth laws which govern crack shape, and
(c) plugging of the orifice by debris.

He invited comments on this issue L. P. Harrop (UK) commented on the factor of 10 required on leak rates in NUREG 1061 as not being based on mechanistic growth rates but rather was a large safety factor to cover such uncertainties as bending stresses, anisotropy of toughness and so forth. G. S. Holman (USA) agreed that the factor of 10 has no mechanistic basis but is there to cover a 'multitude of sins'. He said the USNRC was currently rethinking its leak detection guidelines, including the factor of 10.

Int. J. Pres. Ves. & Piping **43** (1990) 433–442

Index